Analysis of Naturally Occurring Food Toxins of Plant Origin

Natural toxins are toxic compounds that are naturally produced by living organisms. These toxins are not harmful to the organisms themselves, but they may be toxic to other creatures, including humans, when eaten. These chemical compounds have diverse structures and differ in biological function and toxicity. Some toxins are produced by plants as a natural defense mechanism against predators, insects, or microorganisms, or as a consequence of infestation with microorganisms, such as mold, in response to climate stress (such as drought or extreme humidity). Well-known groups of natural toxins of plant origin are: *cyanogenic glycosides, pyrrolizidine alkaloids, furocoumarins, lectins,* and *glycoalkaloids*. These plant-origin natural toxins can cause a variety of adverse health effects and pose a serious health threat to both humans and livestock.

Analysis of Naturally Occurring Food Toxins of Plant Origin is divided into three sections that provide a detailed overview of different classes of naturally occurring food toxins of plant origin, including various analytical techniques exploited for their structural characterization, identification, detection, and quantification. It provides in-depth information and comprehensive discussion over different analytical techniques utilized for qualitative and quantitative application purposes of natural toxins of plant origin in food.

Key Features:

- Provides a detailed overview of different classes of plant-origin natural toxins in food.
- Explains how IR, NMR, and mass spectrometry are utilized in characterization and identification.
- Describes applicability of HPLC, LC-MS, GC-MS, and HPTLC techniques for detection and quantification.
- Discusses progress in the field related to capillary electrophoresis, ELISA, and biosensors for quantitative application of these techniques.

Food Analysis and Properties

Series Editor: Leo M. L. Nollet, University College Ghent, Belgium

This CRC series, Food Analysis and Properties, is designed to provide state-of-the-art coverage on topics to the understanding of physical, chemical, and functional properties of foods, including: (1) recent analysis techniques of a choice of food components; (2) developments and evolutions in analysis techniques related to food; (3) recent trends in analysis techniques of specific food components, and/or a group of related food components.

Ambient Mass Spectroscopy Techniques in Food and the Environment
Edited by Leo M. L. Nollet and Basil K. Munjanja

Food Aroma Evolution: During Food Processing, Cooking and Aging
Edited by Matteo Bordiga and Leo M. L. Nollet

Mass Spectrometry Imaging in Food Analysis
Edited by Leo M. L. Nollet

Proteomics for Food Authentication
Edited by Leo M. L. Nollet, Otles, Semih

Analysis of Nanoplastics and Microplastics in Food
Edited by Leo M. L. Nollet and Khwaja Salahuddin Siddiqi

Chiral Organic Pollutants: Monitoring and Characterization in Food and the Environment
Edited by Edmond Sanganyado, Basil Munjanja, and Leo M. L. Nollet

Sequencing Technologies in Microbial Food Safety and Quality
Edited by Devarajan Thangadurai, Leo M. L. Nollet, Saher Islam, and Jeyabalan Sangeetha

Nanoemulsions in Food Technology: Development, Characterization, and Applications
Edited by Javed Ahmad and Leo M. L. Nollet

Mass Spectrometry in Food Analysis
Edited by Leo M. L. Nollet and Robert Winkler

Bioactive Peptides from Food: Sources, Analysis, and Functions
Edited by Leo M. L. Nollet and Semih Ötleş

Nutriomics: Well-being through Nutrition
Edited by *Devarajan Thangadurai, Saher Islam, Leo M. L. Nollet, and Juliana Bunmi Adetunji*

Analysis of Naturally Occurring Food Toxins of Plant Origin
Edited by Leo M. L. Nollet and Javed Ahmad

For more information, please visit the series page: https://www.crcpress.com/Food-Analysis--Properties/book-series/CRCFOODANPRO

Analysis of Naturally Occurring Food Toxins of Plant Origin

Edited by
Leo M. L. Nollet
Javed Ahmad

Boca Raton London New York

CRC Press is an imprint of the
Taylor & Francis Group, an **informa** business

First edition published 2023
by CRC Press
6000 Broken Sound Parkway NW, Suite 300, Boca Raton, FL 33487-2742

and by CRC Press
4 Park Square, Milton Park, Abingdon, Oxon, OX14 4RN

CRC Press is an imprint of Taylor & Francis Group, LLC

© 2023 selection and editorial matter, Leo M. L. Nollet and Javed Ahmad; individual chapters, the contributors

Reasonable efforts have been made to publish reliable data and information, but the author and publisher cannot assume responsibility for the validity of all materials or the consequences of their use. The authors and publishers have attempted to trace the copyright holders of all material reproduced in this publication and apologize to copyright holders if permission to publish in this form has not been obtained. If any copyright material has not been acknowledged please write and let us know so we may rectify in any future reprint.

Except as permitted under U.S. Copyright Law, no part of this book may be reprinted, reproduced, transmitted, or utilized in any form by any electronic, mechanical, or other means, now known or hereafter invented, including photocopying, microfilming, and recording, or in any information storage or retrieval system, without written permission from the publishers.

For permission to photocopy or use material electronically from this work, access www.copyright. com or contact the Copyright Clearance Center, Inc. (CCC), 222 Rosewood Drive, Danvers, MA 01923, 978-750-8400. For works that are not available on CCC please contact mpkbookspermissions@tandf.co.uk

Trademark notice: Product or corporate names may be trademarks or registered trademarks and are used only for identification and explanation without intent to infringe.

ISBN: 978-1-032-08030-7 (hbk)
ISBN: 978-1-032-11930-4 (pbk)
ISBN: 978-1-003-22219-4 (ebk)

DOI: 10.1201/9781003222194

Typeset in Sabon
by MPS Limited, Dehradun

Contents

Preface		vii
About the Editors		ix
Contributors		xi

SECTION I AN INTRODUCTION TO NATURALLY OCCURRING FOOD TOXINS

Chapter 1	Naturally Occurring Food Toxins – An Overview *Leo M. L. Nollet and Javed Ahmad*	3
Chapter 2	Cyanogenic Glycosides as Food Toxins *Michael Amoafo Mensah*	25
Chapter 3	Pyrrolizidine Alkaloids as Food Toxins *Javed Ahamad, Mohd. Javed Naim, Subasini Uthirapathy, and Showkat R. Mir*	53
Chapter 4	Furanocoumarins and Lectins as Food Toxins *Kavita Munjal, Saima Amin, Showkat R. Mir, Vinod Kumar Gauttam, and Sumeet Gupta*	67
Chapter 5	Glycoalkaloids as Food Toxins *Javed Ahamad, Subasini Uthirapathy, Esra T. Anwer, Mohd. Javed Naim, and Showkat R. Mir*	81

SECTION II ANALYTICAL TECHNIQUES EXPLOITED IN STRUCTURAL CHARACTERIZATION AND IDENTIFICATION: *QUALITATIVE APPLICATION*

Chapter 6	Infrared Spectroscopy *Mohd. Javed Naim, Javed Ahamad, and Esra T. Anwer*	97
Chapter 7	Mass Spectrometry in Analysis of Food Toxins *Mohd Javed Naim and Javed Ahamad*	115

v

vi   Contents

Chapter 8	NMR in Analysis of Food Toxins	131

Naila H. Alkefai, Showkat R. Mir, Saima Amin, Javed Ahamad, Kavita Munjal, and Vinod Kumar Gauttam

SECTION III ANALYTICAL TECHNIQUES EXPLOITED IN DETECTION AND QUANTIFICATION: *QUANTITATIVE APPLICATION*

Chapter 9   HPLC for Detection and Quantification of Food Toxins of Plant Origin   143

Devarajan Thangadurai, D. Divya, Suraj S. Dabire, Jeyabalan Sangeetha, Mahantesh Dodamani, Saher Islam, Ravichandra Hospet, Simran Panigatti, Muniswamy David, Zaira Zaman Chowdhury, and Vishal Ahuja

Chapter 10   Analytical Determination of Food Toxins of Plant Origin Using LC-MS   161

Jeyabalan Sangeetha, D. Divya, Pavitra Chippalakatti, Devarajan Thangadurai, Jarnain Naik, Saher Islam, Ravichandra Hospet, Muniswamy David, and Zaira Zaman Chowdhury

Chapter 11   Quantitative Determination of Food Toxins of Plant Origin by GC-MS   179

Devarajan Thangadurai, D. Divya, Poojashree Nagappa Kunnur, Saher Islam, P. Lokeshkumar, Jeyabalan Sangeetha, Pavitra Chippalakatti, Ravichandra Hospet, Muniswamy David, Zaira Zaman Chowdhury, Vishal Ahuja, and Vedavyas Shivanand Chavan

Chapter 12   High Performance Thin Layer Chromatography   195

Semih Ötles and Vasfiye Hazal Özyurt

Chapter 13   Capillary Electrophoresis   201

Suraj Singh S. Rathod, Gagandeep Kaur, Neha Sharma, Navneet Khurana, and Awanish Mishra

Chapter 14   Enzyme-Linked Immunosorbent Assay   215

Ritam Bandopadhyay, Suraj Singh S. Rathod, and Awanish Mishra

Chapter 15   Detection and Quantification of Food Toxins of Plant Origin Using Biosensors   245

Saher Islam, Devarajan Thangadurai, and Ravichandra Hospet

Index   259

Preface

Natural toxins are toxic compounds that are naturally produced by living organisms. These toxins are not harmful to the organisms themselves, but they may be toxic to other creatures, including humans, when eaten. These chemical compounds have diverse structures and differ in biological functions and toxicity.

Some toxins are produced by plants as a natural defense mechanism against predators, insects, or microorganisms, or as consequence of infestation with microorganisms, such as mold, or in response to climate stress (such as drought or extreme humidity).

Natural toxins can cause a variety of adverse health effects and pose a serious health threat to both humans and livestock. Some of these toxins are extremely potent.

Adverse health effects can be acute poisoning ranging from allergic reactions to severe stomachache and diarrhea, and even death.

Long-term health consequences include effects on the immune, reproductive, or nervous systems, and also cancer.

A scientific expert committee jointly arranged by WHO and the Food and Agriculture Organization of the United Nations (FAO) (JECFA) is the international body responsible for evaluating the health risk from natural toxins in food.

International standards and codes of practice to limit exposure to natural toxins from certain foods are established by the Codex Alimentarius Commission, based on JECFA assessments.

Well-known groups of natural toxins are:

- cyanogenic glycosides
- furocoumarins
- lectins
- glycoalkaloids
- pyrrolizidine alkaloids

Although many foods contain toxins as naturally occurring constituents or, are formed as the result of handling or processing, the incidence of adverse reactions to food is relatively low. The low incidence of adverse effects is the result of some pragmatic solutions by the U.S. Food and Drug Administration (FDA) and other regulatory agencies through their advice. Manufacturers have also played a role by setting limits on certain substances and developing mitigation procedures for process-induced toxins. Regardless of measures taken by regulators and food producers to protect consumers from natural food toxins, consumption of small levels of these materials is unavoidable. Although the risk for toxicity due to consumption of food toxins is quite low, there is always the possibility of toxicity due to contamination, overconsumption, allergy, or an unpredictable idiosyncratic response.

The purpose of this book is to provide an overview of analysis and detection methods of a number of the toxins present in some commonly consumed foods and, where possible, detail the pros and cons of different analysis methods. These analysis techniques include, HPLC, GC, MS, and IR. This list is by no means exhaustive.

This book is structured in three sections.

Section I is an introduction to five groups of naturally occurring food toxins: cyanogenic glycosides, furocoumarins, lectins, glycoalkaloids, and pyrrolizidine alkaloids.

Section II details qualitative application aspects of analytical techniques exploited in the structural characterization and identification of naturally occurring food toxins.

In Section III, the quantitative analytical techniques of detection and quantification of these five groups of toxins are discussed.

Some topics are discussed only now and then and not in detail. These are natural toxins present in foods of animal origin, toxins involving algae or from seafood sources, toxins from poisonous mushrooms, and mycotoxins.

The editors wish to thank all contributors for their excellent work and appreciated commitment.

"Decide. Commit. Succeed."

Anonymous

Leo M. L. Nollet and Javed Ahmad

About the Editors

Leo M. L. Nollet earned an MS (1973) and PhD (1978) in biology from the Katholieke Universiteit Leuven, Belgium. He is an editor and associate editor of numerous books. He edited for M. Dekker, New York – now CRC Press of Taylor & Francis Publishing Group – the first, second, and third editions of *Food Analysis by HPLC* and the *Handbook of Food Analysis*. The last edition is a two-volume book. Dr. Nollet also edited the *Handbook of Water Analysis* (first, second, and third editions) and *Chromatographic Analysis of the Environment*, third and fourth editions (CRC Press). With F. Toldrá, he co-edited two books published in 2006, 2007, and 2017: *Advanced Technologies for Meat Processing* (CRC Press) and *Advances in Food Diagnostics* (Blackwell Publishing – now Wiley). With M. Poschl, he co-edited the book *Radionuclide Concentrations in Foods and the Environment*, also published in 2006 (CRC Press). Dr. Nollet has also co-edited with Y. H. Hui and other colleagues on several books: *Handbook of Food Product Manufacturing* (Wiley, 2007); *Handbook of Food Science, Technology, and Engineering* (CRC Press, 2005); *Food Biochemistry and Food Processing* (first and second editions; Blackwell Publishing – now Wiley – 2006 and 2012); and the *Handbook of Fruits and Vegetable Flavors* (Wiley, 2010). In addition, he edited the *Handbook of Meat, Poultry, and Seafood Quality*, first and second editions (Blackwell Publishing – now Wiley – 2007 and 2012). From 2008 to 2011, he published five animal product–related volumes with F. Toldrá: *Handbook of Muscle Foods Analysis*, *Handbook of Processed Meats and Poultry Analysis*, *Handbook of Seafood and Seafood Products Analysis*, *Handbook of Dairy Foods Analysis* (second edition in 2021), and *Handbook of Analysis of Edible Animal By-Products*. Also, in 2011, with F. Toldrá, he co-edited two volumes for CRC Press: *Safety Analysis of Foods of Animal Origin* and *Sensory Analysis of Foods of Animal Origin*. In 2012, they published the *Handbook of Analysis of Active Compounds in Functional Foods*. In a co-edition with Hamir Rathore, *Handbook of Pesticides: Methods of Pesticides Residues Analysis* was marketed in 2009; *Pesticides: Evaluation of Environmental Pollution* in 2012; *Biopesticides Handbook* in 2015; and *Green Pesticides Handbook: Essential Oils for Pest Control* in 2017. Other finished book projects include *Food Allergens: Analysis, Instrumentation, and Methods* (with A. van Hengel; CRC Press, 2011) and *Analysis of Endocrine Compounds in Food* (Wiley-Blackwell, 2011). Dr. Nollet's recent projects include *Proteomics in Foods* with F. Toldrá (Springer, 2013) and *Transformation Products of Emerging Contaminants in the Environment: Analysis, Processes, Occurrence, Effects, and Risks* with D. Lambropoulou (Wiley, 2014). In the series Food Analysis and Properties, he edited (with C. Ruiz-Capillas) *Flow Injection Analysis of Food Additives* (CRC Press, 2015) and *Marine Microorganisms: Extraction and*

Analysis of Bioactive Compounds (CRC Press, 2016). With A. S. Franca, he co-edited *Spectroscopic Methods in Food Analysis* (CRC Press, 2017), and with Horacio Heinzen and Amadeo R. Fernandez-Alba he co-edited *Multiresidue Methods for the Analysis of Pesticide Residues in Food* (CRC Press, 2017). Further volumes in the series Food Analysis and Properties are *Phenolic Compounds in Food: Characterization and Analysis* (with Janet Alejandra Gutierrez-Uribe, 2018); *Testing and Analysis of GMO-containing Foods and Feed* (with Salah E. O. Mahgoub, 2018); *Fingerprinting Techniques in Food Authentication and Traceability* (with K. S. Siddiqi, 2018); *Hyperspectral Imaging Analysis and Applications for Food Quality* (with N. C. Basantia, Leo M. L. Nollet, Mohammed Kamruzzaman, 2018); *Ambient Mass Spectroscopy Techniques in Food and the Environment* (with Basil K. Munjanja, 2019); *Food Aroma Evolution: During Food Processing, Cooking, and Aging* (with M. Bordiga, 2019); *Mass Spectrometry Imaging in Food Analysis* (2020); *Proteomics in Food Authentication* (with S. Ötleş, 2020); *Analysis of Nanoplastics and Microplastics in Food* (with K. S. Siddiqi, 2020); *Chiral Organic Pollutants, Monitoring and Characterization in Food and the Environment* (with Edmond Sanganyado and Basil K. Munjanja, 2020); *Sequencing Technologies in Microbial Food Safety and Quality* (with Devarajan Thangardurai, Saher Islam, Jeyabalan Sangeetha, 2021); *Nanoemulsions in Food Technology: Development, Characterization, and Applications* (with Javed Ahmad, 2021); *Mass Spectrometry in Food Analysis* (with Robert Winkler, 2022); *Bioactive Peptides from Food: Sources, Analysis, and Functions* (with Semih Ötles, 2022); and *Nutriomics: Well-being through Nutrition* (with Devarajan Thangadurai, Saher IslamJuliana Bunmi Adetunji, 2022).

Javed Ahmad is an assistant professor at the Department of Pharmaceutics, College of Pharmacy, Najran University, Saudi Arabia. He received his doctorate degree in pharmaceutical sciences (pharmaceutics) from the School of Pharmaceutical Education and Research, Jamia Hamdard, New Delhi, India. He was the recipient of the CSIR-Senior Research Fellowship and the International Travel Award from the Department of Science and Technology, Government of India, during his PhD. After receiving his PhD, he joined NIPER, Raebareli (an autonomous institute of Department of Pharmaceuticals, Ministry of Chemicals and Fertilizers, Government of India) as a faculty member in the Department of Pharmaceutics. He has published >100 high-quality research and review articles in peer-reviewed journals of international repute. He has also published various book chapters (>35) for edited books and book series. He has been the editor/coeditor of different books including recently published books such as *Bioactive Phytochemicals: Drug Discovery to Product Development* (Bentham Science Publishers, 2020) and *Nanoemulsions in Food Technology: Development, Characterization, and Applications* (CRC Press, 2021). His current h-index is 28 with 2,000 citations of his publications. He has been a reviewer of many peer-reviewed journals of international repute. He has also received many awards for his scientific findings and reviewing tasks. Recently, he has been listed among the World's Top 2% Scientists for the year 2021 in the field of pharmacology and pharmacy, a list created by Stanford University. His current research interest lies in formulation, design, and characterization of pharmaceuticals/nutraceuticals/cosmeceuticals to improve efficacy and performance.

Contributors

Javed Ahamad
Department of Pharmacognosy, Faculty of
Pharmacy
Tishk International University
Erbil, Kurdistan Region, Iraq

Javed Ahmad
Department of Pharmaceutics
Najran University
Najran, Kingdom of Saudi Arabia

Vishal Ahuja
Department of Biotechnology
Himachal Pradesh University
Shimla, Himachal Pradesh, India

Naila H. Alkefai
Department of Pharmaceutical Chemistry,
Faculty of Pharmacy
University of Hafer Albatin
Hafer Albatin, Saudi Arabia

Saima Amin
Department of Pharmacognosy and
Phytochemistry, School of
Pharmaceutical Education
and Research
Jamia Hamdard, New Delhi, India

Esra T. Anwer
Department of Pharmaceutics, Faculty of
Pharmacy
Tishk International University
Kurdistan Region, Iraq

Ritam Bandopadhyay
Department of Pharmacology, School of
Pharmaceutical Sciences
Lovely Professional University
Phagwara, India

Vedavyas Shivanand Chavan
Department of Botany
Karnatak University
Dharwad, Karnataka, India

Pavitra Chippalakatti
Department of Botany
Karnatak University
Dharwad, Karnataka, India

Zaira Zaman Chowdhury
Nanotechnology and Catalysis Research
Center (NANOCAT)
Institute of Advanced Studies (IAS)
University of Malaya
Kuala Lumpur, Malaysia

Suraj S. Dabire
Department of Zoology
Karnatak University
Dharwad, Karnataka, India

Muniswamy David
Department of Zoology
Karnatak University
Dharwad, Karnataka, India

D. Divya
Pinnacle Biosciences
Kanyakumari, Tamil Nadu, India

Mahantesh Dodamani
Department of Zoology
Karnatak University
Dharwad, Karnataka, India

Vinod Kumar Gauttam
IES Institute of Pharmacy
Bhopal, India

Sumeet Gupta
Department of Pharmacology
M. M. College of Pharmacy
M. M. (Deemed to be University)
Haryana, India

Ravichandra Hospet
Department of Food Protectants and
    Infestation Control, CSIR-Central Food
    Technological Research Institute
Mysore, Karnataka, India

Saher Islam
Department of Biotechnology
Lahore College for Women University
Lahore, Pakistan

Gagandeep Kaur
Department of Pharmacology, School of
    Pharmaceutical Sciences
Lovely Professional University
Phagwara, India

Navneet Khurana
Department of Pharmacology, School of
    Pharmaceutical Sciences
Lovely Professional University
Phagwara, India

Poojashree Nagappa Kunnur
Department of Botany
Karnatak University
Dharwad, Karnataka, India

P. Lokeshkumar
Department of Zoology
Karnatak University
Dharwad, Karnataka, India

Michael Amoafo Mensah
Freelance
Gent, Belgium

Showkat R. Mir
Department of Pharmacognosy and
    Phytochemistry, School of
    Pharmaceutical Education
    and Research
Jamia Hamdard, New Delhi, India

Awanish Mishra
Department of Pharmacology and
    Toxicology, National Institute of
    Pharmaceutical, Education and
    Research (NIPER)
Guwahati, India

Kavita Munjal
Department of Pharmacognosy
M. M. College of Pharmacy
M. M. (Deemed to be University)
Haryana, India

Jarnain Naik
Department of Zoology
Karnatak University
Dharwad, Karnataka, India

Mohd. Javed Naim
Department of Pharmaceutical Chemistry,
    Faculty of Pharmacy
Tishk International University
Erbil, Kurdistan Region, Iraq

Leo M. L. Nollet
University College Ghent
Gent, Belgium

Semih Ötles
Food Engineering Department
Ege University
Izmir, Turkey

Vasfiye Hazal Özyurt
Gastronomy and Culinary Arts
Mugla Sitki Kocman University
Mugla, Turkey

Simran Panigatti
Department of Botany
Karnatak University
Dharwad, Karnataka, India

Suraj Singh S. Rathod
Department of Pharmacology, School of
Pharmaceutical Sciences
Lovely Professional University
Phagwara, India

Jeyabalan Sangeetha
Department of Environmental Science
Central University of Kerala
Kasaragod, Kerala, India

Neha Sharma
Department of Pharmacology, School of
Pharmaceutical Sciences
Lovely Professional University
Phagwara, India

Devarajan Thangadurai
Department of Botany
Karnatak University
Dharwad, Karnataka, India

Subasini Uthirapathy
Department of Pharmacology, Faculty of
Pharmacy
Tishk International University
Erbil, Kurdistan Region, Iraq

SECTION I

An Introduction to Naturally Occurring Food Toxins

CHAPTER 1

Naturally Occurring Food Toxins – An Overview

Leo M. L. Nollet[1] *and Javed Ahmad*[2]

[1]University College Ghent, Gent, Belgium
[2]Department of Pharmaceutics, Najran University, Najran, KSA

CONTENTS

1.1	Naturally Occurring Food Toxins	4
1.2	Naturally Occurring Food Toxins of Plant Origin	5
	1.2.1 β-Thujone	5
	1.2.2 Prussic Acid	6
	1.2.3 Hypericin	6
	1.2.4 Goitrogens	7
	1.2.5 Erucic Acid	8
	1.2.6 Furanocoumarins	8
	1.2.7 α-Amylase Inhibitors	9
	1.2.8 Lectins	9
	1.2.9 Anti-Thiamine Compounds	9
	1.2.10 Pyrrolizidine Alkaloids	10
	1.2.11 Oxalates	10
	1.2.12 Cucurbitacins	11
	1.2.13 Phytates and Phytic Acid	12
	1.2.14 Hypoglycin – Ackee Fruit	14
	1.2.15 Safrole	14
	1.2.16 Myristicin	15
	1.2.17 Japanese Star Anise	15
	1.2.18 Cyanogenic Glycosides	16
	1.2.19 Glycoalkaloids – Solanines and Chaconine	16
	1.2.20 Cycasin	17
	1.2.21 Sanguinarine	17
1.3	Environmental Contaminants	17
1.4	Contaminants Formed During Processing	17
1.5	Toxins in Seafood	17
References		18

DOI: 10.1201/9781003222194-2

1.1 NATURALLY OCCURRING FOOD TOXINS

Natural toxins are compounds naturally produced by living organisms. These toxins are not harmful to the producing organisms but they may be toxic to other creatures, including humans, when consumed. These chemical compounds have diverse structures and differ in biological function and toxicity.

Some natural toxins can be found in food as a result of defense mechanisms of plants, through their infestation with toxin-producing mold, or through ingestion by animals of toxin-producing microorganisms.

Natural toxins can cause a variety of adverse health effects and pose a serious health threats to both humans and livestock. Some of these toxins are extremely potent.

Adverse health effects can be acute poisoning ranging from allergic reactions to severe stomachache and diarrhea, and even death.

Long-term health consequences include effects on the immune, reproductive, or nervous systems, and also cancer.

A scientific expert committee jointly convened by WHO and the Food and Agriculture Organization of the United Nations (FAO) – called JECFA – is the international body responsible for evaluating the health risk from natural toxins in food.

The Codex Committee on Contaminants in Foods (CCCF) (1) is set up

a. to establish or endorse permitted maximum levels or guidelines levels for contaminants and naturally occurring toxicants in food and feed;
b. to prepare priority lists of contaminants and naturally occurring toxicants for risk assessment by the Joint FAO/WHO Expert Committee on Food Additives;
c. to consider methods of analysis and sampling for the determination of contaminants and naturally occurring toxicants in food and feed;
d. to consider and elaborate standards or codes of practice for related subjects; and
e. to consider other matters assigned to it by the Commission in relation to contaminants and naturally occurring toxicants in food and feed.

International standards and codes of practice to limit exposure to natural toxins from certain foods are established by the Codex Alimentarius Commission based on JECFA assessments.

The reader is directed to specialized literature (2–4)

In references 5 and 6 the reader finds information on chemical, metals, natural toxins, and pesticides guidance documents and regulations of the Food and Drug Administration (U.S.).

EFSA is a European agency funded by the European Union that operates independently of the European legislative and executive institutions (Commission, Council, Parliament) and EU Member States (7).

It was set up in 2002 following a series of food crises in the late 1990s to be a source of scientific advice and communication on risks associated with the food chain. The agency was legally established by the EU under the General Food Law – Regulation 178/2002.

Some of the most commonly found natural toxins that can pose a risk to our health are described below.

How to minimize the health risk from natural toxins?

When it comes to natural toxins it is important to note that they can be present in a variety of different crops and foodstuff. In a usual balanced, healthy diet, the levels of natural toxins are well below the threshold for acute and chronic toxicity.

To minimize the health risk from natural toxins in food, people are advised to:

- not assume that if something is 'natural' it is automatically safe;
- throw away bruised, damaged, or discolored food, and in particular moldy foods;
- throw away any food that does not smell or taste fresh, or has an unusual taste; and
- only eat mushrooms or other wild plants that have definitively been identified as nonpoisonous.

WHO, in collaboration with FAO, is responsible for assessing the risks to humans of natural toxins – through contamination in food – and for recommending adequate protections.

Risk assessments of natural toxins in food done by the Joint FAO/WHO Expert Committee on Food Additives (JECFA) are used by governments and by the Codex Alimentarius Commission (the intergovernmental standards-setting body for food) to establish maximum levels in food or provide other risk management advice to control or prevent contamination. Codex standards are the international reference for national food supplies and for trade in food, so that people everywhere can be confident that the food they buy meets the agreed standards for safety and quality, no matter where it was produced.

JECFA sets the tolerable intake level for natural toxins. JECFA or ad hoc FAO/WHO scientific expert groups consist of independent, international experts who conduct scientific reviews of all available studies and other relevant data on specific natural toxins. The outcome of such health risk assessments can either be a maximum tolerable intake (exposure) level, or other guidance to indicate the level of health concern (such as the Margin of Exposure), including advice on risk management measures to prevent and control contamination, and on the analytical methods and monitoring and control activities.

Exposure to natural toxins needs to be kept as low as possible to protect people. Natural toxins not only pose a risk to both human and animal health, but also impact food security and nutrition by reducing people's access to healthy food. The WHO encourages national authorities to monitor and ensure that levels of the most relevant natural toxins in their food supply are as low as possible and comply with both national and international maximum levels, conditions, and legislation (4).

Other sources of natural toxins may be microscopic algae and plankton in oceans or sometimes in lakes that produce chemical compounds that are toxic to humans but not to fish or shellfish that eat these toxin-producing organisms. When people eat fish or shellfish that contain these toxins, illness can rapidly follow. These compounds are not treated in the following chapters. See references 8 and 9.

In the next paragraphs, naturally occurring food toxins of plant origin are discussed.

1.2 NATURALLY OCCURRING FOOD TOXINS OF PLANT ORIGIN

1.2.1 β-Thujone

Thujone is found in a number of plants, such as arborvitae (genus *Thuja*), Nootka cypress, some junipers, mugwort, oregano, common sage, tansy, and wormwood, most notably grand wormwood (*Artemisia absinthium*), usually as a mix of isomers in a 1:2 ratio. It is also found in various species of *Mentha* (mint) (10).

Essential oils from these plants are used in herbal medicines, as flavorings in alcoholic drinks and frangrances.

In addition to the naturally occurring (–)-α-thujone and (+)-β-thujone, two other forms are possible: (+)-α-thujone and (–)-β-thujone.

Oil of thuja contains the terpene thujone which has been studied for its GABA receptor antagonizing effects, with potentially lethal properties. Cedarwood oil and cedar leaf oil, which are derived from *Thuja occidentalis*, have different properties and uses.

Joint FAO/WHO Expert Committee on Food Additives considers that the amounts of thuyone isomers in foods and beverages resulting from the addition of thuyone-containing agents (e.g. sage) should be reduced to the lowest practicable level. No ADI is allocated (11).

1.2.2 Prussic Acid

HCN is obtainable from fruits that have a pit, such as cherries, apricots, apples, and bitter almonds, from which almond oil and flavoring are made. Many of these pits contain small amounts of cyanohydrins, such as mandelonitrile and amygdalin, which slowly release hydrogen cyanide.

So-called "bitter" roots of the cassava plant may contain up to 1 g of HCN per kilogram.

Prussic acid is formed when cyanogenic glycosides come into contact with *beta*-glycosidase or emulsion enzymes (12).

As the first frost date approaches, producers often have concerns about the risk of prussic acid poisoning in livestock. Certain forage plants, especially sorghums and related species, are associated with an increased risk of death loss because of prussic acid poisoning.

Prussic acid, also known as hydrocyanic acid or cyanide, is a rapidly acting, lethal toxin. Prussic acid inhibits oxygen utilization by the animal at the cellular level resulting in suffocation. Ruminants are more susceptible because the rumen microbes have enzymes that release the prussic acid in the digestive tract. Death often occurs within minutes of exposure.

Some plants, particularly sorghums and sudangrass, accumulate cyanogenic glucosides in the outer tissue layers of the plant. The enzymes that would trigger the prussic acid production and located in other plant tissues, specifically the leaf (13).

Under normal conditions, these is no contact between these compounds and therefore no risk of poisoning. However, any factor that causes the plant cells to rupture and these compounds to combine can lead to prussic acid release. The damage could be caused by frost and freezing, or anything else that leads to cell rupture such as crushing, trampling, chewing, or chopping.

1.2.3 Hypericin

Hypericin is a naturally occurring substance found in the common St. John's wort (*Hypericum* species). As the main component of *Hypericum perforatum,* it has traditionally been used throughout the history of folk medicine. In the last three decades, hypericin has also become the subject of intensive biochemical research and is proving to be a multifunctional agent in drug and medicinal applications. Recent studies report antidepressive, antineoplastic, antitumor, and antiviral (human immunodeficiency and hepatitis C virus) activities of hypericin.

It shows a photo-dynamic effect, which can cause skin irritation.

The fresh plant of *H. perforatum* contains hypericin, pseudohypericin, and isohypericin; protohypericin, protopseudohypericin (biosynthetic precursors of hypericin and pseudohypericin, respectively), and cyclopseudohypericin. In general, hypericin and pseudohypericin are referred to as total hypericins (14,15).

St. John's wort might cause serious interactions with some medications (16).

St. John's wort is most commonly used for depression and mood disorders. There is some strong scientific evidence that it is effective for mild to moderate depression. St. John's wort had comparable efficacy as SSRIs for mild-moderate depression (17).

1.2.4 Goitrogens

Goitrogens are compounds of various chemical structures, the administration of which to humans or animals for a certain time results in an increase of thyroid size, which is called goiter (18). Several groups of organic compounds are introduced into food chains and other natural cycles by human activities.

Foods that have been demonstrated to have goitrogenic effects include soy, cassava (when crushed and not detoxified by soaking), vegetables in the genus *Brassica* (such as broccoli and cabbage), and other cruciferous vegetables.

The Rosaceae family of fruits, which includes almonds, apricots, cherries, peaches, pears, plums, raspberries, and strawberries, is also goitrogenic.

In places where iodine deficiency exists in tandem with millet being a major component of the diet, millet consumption can contribute to thyroid enlargement which is the start of endemic goiter (19–21).

There are three main types of goitrogens found in foods (19):

1. Goitrins
2. Thiocyanates
3. Flavonoids

Goitrins and thiocyanates are produced when plants are damaged, such as when they are sliced or chewed. Goitrin is a sulfur-containing oxazolidine, a cyclic thiocarbamate. It is found in cruciferous vegetables such as cabbage, brussels sprouts, and rapeseed oil (22).

Flavonoids are naturally present in a wide variety of foods. Some examples include the resveratrol in red wine and the catechins in green tea.

Flavonoids are generally considered to be healthy antioxidants, but some of them can be converted into goitrogenic compounds by our gut bacteria (23).

Brassica vegetables are common components of the diet and have beneficial as well as potentially adverse health effects. Following enzymatic breakdown, some glucosinolates in brassica vegetables produce sulforaphane, phenethyl, and indolylic isothiocyanates that possess anticarcinogenic activity. In contrast, progoitrin and indolylic glucosinolates degrade to goitrin and thiocyanate, respectively, and may decrease thyroid hormone production (24).

Cooking lowers the goitrogenic content of foods. Steaming crucifers until fully cooked reduces goitrogens by two-thirds. Boiling crucifers for 30 minutes destroys 90% of the goitrogens by stimulating the production of myrosinase, an enzyme that helps deactivate goitrogenic glucosinolates (25).

In contrast to cooking, fermenting increases the goitrogen content of cabbage, but it simultaneously decreases the level of nitriles (26). Because nitriles are more harmful than goitrogens, the overall effect of fermentation is probably positive.

1.2.5 Erucic Acid

Erucic acid is the trivial name of the fatty acid cis-13-docosenoic acid and occurs at high concentrations mainly in the seeds of species of the Brassicaceae (e.g., rape seed or mustard seed). The European Commission requested EFSA to deliver a scientific opinion on the risks for animal and human health related to the presence of erucic acid in feed and food (27,28). For most humans, the main contributor to dietary exposure to erucic acid is fine bakery products. The heart is the principal target organ for toxic effects after exposure.

The dietary exposure assessment has concluded that the majority of exposure to erucic acid by the general population in Australia and New Zealand would come from the consumption of canola oil (29). The dietary intake of erucic acid by an individual consuming at the average level is well below the PTDI, therefore, there is no cause for concern in terms of public health and safety. However, the individual consuming at a high level has the potential to approach the PTDI. This would be particularly so if the level of erucic acid in canola oil was to exceed 2% of the total fatty acids.

In 2016, the European Food Safety Authority (EFSA) proposed a lower maximum content of erucic acid in edible oils of 2% (instead of 5%) and also suggested a tolerable daily intake of 7 mg erucic acid per kg body weight. Salmon fillet (erucic acid content: wild catch ~ farmed salmon < organic salmon) which, together with other fish, is verified as a relevant but unregulated source of erucic acid intake. Salmon also contained an isomer of erucic acid whose content was even higher than erucic acid (30).

Based on the low consumption rate of mustard oil in Germany and the EU, table mustard, rapeseed oil, and fish/seafood still remain the major sources of erucic acid intake in Germany and the EU.

1.2.6 Furanocoumarins

The furanocoumarins have only limited distribution in the plant kingdom and are found primarily within the families of Apiaceae, Moraceae, Rosaceae, Rutaceae, and Fabaceae. These toxins are present in many plants, such as celery roots, citrus plants (lemon, lime, grapefruit, bergamot) and some medicinal plants. Furanocoumarins are stress toxins and are released in response to stress, such as physical damage to the plant. Furanocoumarins may be produced by plants as a defense mechanism against predators such as insects and mammals. Furanocoumarins are also related to a plant's natural defense against fungal attack. In particular, the linear furanocoumarins, psoralen, bergapten, and methoxsalen are known to be toxic to fungi.

Some of these toxins can cause gastrointestinal problems in susceptible people. Furanoocoumarins are phototoxic, they can cause severe skin reactions under sunlight (UVA exposure). The phototoxic action of the furanocoumarins is due to their ability to react with nucleobases in DNA under the influence of UV-A radiation. While mainly occurring after dermal exposure, such reactions have also been reported after consumption of large quantities of certain vegetables containing high levels of furanocoumarins (31,32).

Chapter 4 details furanocoumarins.

1.2.7 α-Amylase Inhibitors

α-Amylase inhibitors have been purified and partially characterized from different varieties of common beans, including white kidney beans, red kidney beans, and black kidney beans. The content of α-amylase inhibitors differs greatly among legumes, with the highest amounts found in dry beans. α-amylase inhibitors were found in common beans and runner beans (*Phaseolus coccineus*) at levels of 2–4 kg^{-1} of seed meal. Field beans, black-eyed peas, and chickpeas contain low levels of 0.1–0.2 kg^{-1} of seed meal (33).

α-Amylase inhibitors inhibit the activity of salivary and pancreatic amylase *in vitro* and *in vivo*. They can impair the growth and metabolism of animals when given at high levels in the diet but may have beneficial uses in treatment of obesity or diabetes (34).

Amylase inhibitors are found in the seeds of plants such as cereal grains (wheat, maize, rice, barley) and legumes (kidney beans, cowpea, adzuki beans). Amylase inhibitors inhibit amylases of insects in general and inhibit the growth of insects, and thus serve as defense proteins in both cereal grains and bean seeds (35).

There are the different forms of amylase inhibitors, namely, Alpha-amylase inhibitor isoform 1 (Alpha-AI1), Alpha-AI2, and Alpha-AIL, which can be found in in the embryonic axes and cotyledons in the seed of common beans (*Phaseolus* spp.). These so-called glycoproteins bind to alpha-amylase non-covalently, mainly through hydrophobic interaction, by completely blocking access to the active site of the alpha-amylase. The Alpha-AI1 isoform is the one with anti-amylase bioactivity in humans, and, therefore, inhibits the starch digestion (36).

1.2.8 Lectins

Lectins are ubiquitous and versatile proteins of nonimmune origin that bind reversibly and specifically to carbohydrates. Their multivalent structure enables their well-known ability to agglutinate cells (37).

Many types of beans contain lectins, and kidney beans have the highest concentrations, especially red kidney beans. As few as four or five raw beans can cause severe stomachache, vomiting, and diarrhea. Lectins are destroyed when the dried beans are soaked for at least 12 hours and then boiled vigorously for at least 10 minutes in water. Tinned kidney beans have already had this process applied and so can be used without further treatment.

Some lectins are beneficial while others may be powerful toxins, such as ricin. Ricin is an highly potent toxin produced in the seeds of the castor oil plant, *Ricinus communis* (38).

1.2.9 Anti-Thiamine Compounds

Two thiamin-cleaving enzymes have been identified, called thiaminase I and thiaminase II.

Thiaminase I is found in shellfish, clams (but not oysters), some freshwater fish viscera, crustacea, and certain ferns, but very few higher plants. Also, certain species of *Bacillus* and *Clostridium*, which are components of the human and animal intestinal flora, have been found to produce this enzyme. The enzyme catalyzes an exchange reaction, in which the thiazol moiety of the molecule is displaced by another N-containing base or a SH-compound.

Thiaminase I may also be produced by the rumen microflora of ruminants or by plants and, in the presence of suitable cosubstrates, e.g., niacin, or pyridoxine, and certain anti-helmintics, seems to be responsible for the ruminant CNS disorder, polioencephalomalacia.

It has been found in certain species of *Bacillus* and *Clostridium* bacteria isolated from the intestinal flora of Japanese people and has been implicated in 'thiaminase disease', a beriberi-like condition with an incidence as high as 70% of the population in certain Japanese cities.

Thiaminase II is of bacterial origin (predominantly *Bacillus, Candida*, and *Oospora*) and breaks down the free vitamin, but not the thiamin pyrophosphate, into pyrimidin and thiazol components. More prevalent are thermostable thiamin-inactivating factors of plant origin, e.g., polyphenolic substances such as flavonoids and catechol derivatives in fermented tea, ferns, sweet potatoes, and betel nuts, and, in small quantities, in other leaves, fruits, and roots. These can decompose the thiazol component of the vitamin, accelerate the oxidation to the disulfide form, or form unabsorbable adducts with thiamin (39).

The physiological significance of thiaminase is not known. Thiaminase activity does not seem to be present in tissue until the tissue is ground. Tissue which is ground and then found to be free of thiamine may be shown to be originally high in thiamine if the thiaminase is first inactivated by subjecting the tissue to boiling buffer solution before grinding. Thiaminase has limited importance in human nutrition, being of possible consequence only in individuals whose intestinal tract harbors thiaminase-containing bacteria, or in those people who eat considerable amounts of raw clams, raw fish containing thiaminase, or processed foods in which ground raw whole fish or shell fish was incorporated prior to heat treatment (40).

1.2.10 Pyrrolizidine Alkaloids

Pyrrolizidine alkaloids (PAs) are toxins produced by an estimated 600 plant species. The main plant sources are the families *Boraginaceae, Asteraceae,* and *Fabaceae*. Many of these are weeds that can grow in fields and contaminate food crops. PAs can cause a variety of adverse health effects. Certain representatives of this class and the plants in which they occur are hepatotoxic as well as mutagenic and hepatocarcinogenic. They can be acutely toxic and of main concern is the DNA-damaging potential of certain PAs, potentially leading to cancer.

PAs are stable during processing, and have been detected in herbal teas, honey, herbs and spices, and other food products, such as cereals and cereal products. Human exposure is estimated to be low, however (41). Due to the complexity of the subject and the large number of related compounds, the overall health risk has not been fully evaluated yet. Guidance is set up by the FAO/WHO Codex Committee on Contaminants in Food on management strategies to prevent PA-containing plants from entering the food chain.

CXC 74–2014 is the Code of Practice for Weed Control to Prevent and Reduce Pyrrolizidine Alkaloid Contamination in Food and Feed (42).

See also Chapter 3 of this book.

1.2.11 Oxalates

Oxalates protect plants from being eaten by critters and are found in most plants, nuts, and seeds. Spinach, okra, sweet potato, elderberry, figs, leeks, buckwheat, celery, other leafy greens, and dandelions are some sources. For frame of reference, ingesting 250 mg of oxalates is considered high, and 1 cup of raw spinach contains a whopping 656 mg.

Oxalic acid binds minerals like calcium and potassium, making them insoluble and less bioavailable. Oxalate salts increase the risk of kidney stones, especially in patients with gut dysbiosis. Gut bacteria are responsible for breaking down oxalates, but when the microbiota are compromised, oxalates can enter the blood, turn into crystals, and get stored in tissues like the kidney. Unfortunately, cooking does not decrease the oxalate content much.

Oxalate salts are poorly soluble at intestinal pH, and oxalate is known to decrease calcium absorption in monogastric animals. For instance, oxalate binds to calcium to form complexes (calcium oxalate crystals). These oxalate crystals prevent the absorption and utilization of calcium by the body, causing diseases such as rickets and osteomalacia. The calcium crystal may also precipitate around the renal tubules, thereby causing renal stones. The formation of oxalate crystal is said to take place in the digestive tract. Oxalate is a concern in legumes because high-oxalate diets can increase the risk of renal calcium absorption, as calcium is made unavailable to the body due to the presence of oxalate. Legumes such as lentils, red kidney beans, and white beans have been analyzed for oxalate. The highest and lowest contents are present in Anasazi beans (80 mg, 100 g, 100 g^{-1} wet weight) and black-eyed peas (4 mg, 100 g, 100 g^{-1} wet weight), respectively.

The consumption of high-oxalate foods is more likely to pose health problems in those who have an unbalanced diet or those with intestinal malfunction. A diet high in oxalate and low in essential minerals, such as calcium and iron, is not recommended. Vegans and lactose-intolerant persons may have a high-oxalate and low-calcium diet unless their diet is supplemented. Vegetarians who consume greater amounts of vegetables will have a higher intake of oxalates, which may reduce calcium availability (43,44).

1.2.12 Cucurbitacins

Cucurbitacins are tetracyclic terpenes with steroidal structures. Cucurbitacins are derived from cucurbitane, a triterpene hydrocarbon – specifically, from the unsaturated variant cucurbit-5-ene, or 19(10→9β)-abeo-10α-lanost-5-ene. They often occur as glycosides. Cucurbitacins and their derivatives are found in many plant families, including Brassicaceae, Cucurbitaceae (pumpkins, gourds, and cucumbers), Scrophulariaceae, Begoniaceae, Elaeocarpaceae, Datiscaceae, Desfontainiaceae, Polemoniaceae, Primulaceae, Rubiaceae, Sterculiaceae, Rosaceae, and Thymelaeaceae; in some mushrooms (including *Russula* and *Hebeloma*); and even in some marine mollusks.

Twenty cucurbitacins have been isolated from the ≈965 known species (in ≈95 genera) of Cucurbitaceae (45,46). See Table 1.1 (45).

Nor-cucurbitacins are substances that can be seen as deriving from cucurbita-5-ene skeleton by loss of one of the methyl groups. They are found in the leaves of *Mareya micrantha* (47).

The plants of genera *Momordica* contain a special group of cucurbitacins called momordicosides (48).

Medicinal uses include emetics, narcotics, and antimalarials. Cucurbitacin E (also called α-elaterin) has antitumor properties. Cucurbitacins are under basic research for their biological properties, including toxicity and potential pharmacological uses in development of drugs for inflammation, cancer, cardiovascular diseases, and diabetes, among others.

TABLE 1.1 (45) Molecular Formulae and Physical Properties of Cucurbitacins

Cucurbitacin	Nature	Formula	UV max. (ethanol) nm	Mass	m.p.
A	Crystals	$C_{32}H_{46}O_9$	229, 290	574.314	207–208°
B	Crystals	$C_{32}H_{46}O_8$	–	558.3192	184–186°
C	Needles	$C_{32}H_{48}O_8$	231–298	560.3348	207–207.5°
D	Needles	$C_{30}H_{44}O_7$	230	516.3087	151–153°
E	Crystals	$C_{32}H_{44}O_8$	234, 268	556.3035	234.5°
F	Needles	$C_{30}H_{46}O_7$	–	518.3243	244–245°
G	Crystals	$C_{30}H_{52}O_9$	–	534.3192	150–152°
H	Amorphous solid	$C_{30}H_{46}O_8$	–	534.3192	150–152°
I	Needles	$C_{30}H_{42}O_7$	234, 266	514.293	148–148.5°
J	Crystals	$C_{30}H_{44}O_8$	270	532.3036	200–202°
K	Needles	$C_{30}H_{44}O_8$	270	532.3036	200–202°
L	Needles	$C_{30}H_{46}O_7$	270	516.3087	137–142°
O	–	$C_{30}H_{46}O_7$	–	518.3243	122–127°
P	–	$C_{30}H_{46}O_7$	–	520.3399	–
Q	–	$C_{32}H_{46}O_8$	–	560.3348	–
S	–	$C_{30}H_{42}O_6$	–	498.298	–

The two most common cucurbitacins are B and E. Cucurbitacin B is extremely toxic when ingested; E is not as toxic but is still harmful if swallowed. In *in vitro* research, cucurbitacins have cytotoxic properties.

Cucurbitacins may be a taste deterrent in plants foraged by some animals and in some edible plants preferred by humans, like cucumbers, zucchinis, melons, and pumpkins.

The toxicity associated with consumption of foods high in cucurbitacins is sometimes referred to as 'toxic squash syndrome'. The high concentration of toxin in the plants could result from cross-pollination with wild cucurbitaceae species, or from plant growth stress due to high temperature and drought (Table 1.2).

1.2.13 Phytates and Phytic Acid

Phytic acid is a sixfold dihydrogenphosphate ester of inositol, inositol hexakisphosphate (IP6), or inositol polyphosphate. At physiological pH, the phosphates are partially ionized, resulting in the phytate anion.

The phytate anion is a colorless species that has significant nutritional role as the principal storage form of phosphorus in many plant tissues, especially bran and seeds. It is also present in many legumes, cereals, and grains. Phytic acid and phytate have a strong binding affinity to the dietary minerals, calcium, iron, and zinc, inhibiting their absorption in the small intestine (66).

The article of Schemmer et al. (66) gives an overview of phytic acid in food and of its significance for human nutrition. It summarises phytate sources in foods and discusses problems of phytic acid/phytate contents of food tables. Data on phytic acid intake are evaluated and daily phytic acid intake depending on food habits is assessed. Degradation of phytate during gastro-intestinal passage is summarized, the mechanism of phytate interacting with minerals and trace elements in the gastro-intestinal chyme described, and

TABLE 1.2 (45) Reported Biological Activities of Cucurbitacins with Probable Mechanism of Action

Activity	Mechanism	Reference
Antitumor activity	Inhibition of Janus kinase/Signal Tranducer Activator of Transcription " (JAK/STAT3) signaling pathway Disruption of F-actin cytoskeleton Down-Regulation of the c-Myc/hTERT/Telomerase Pathway and Obstruction of the Cell Cycle	49–54
Anti-inflammatory	Inhibit the expression of TNF and proinfammatory mediators such as nitric-oxides synthase-2 and cytoaxygenase-2	55–58
Artherosclerosis	Inhibition of lipid-oxidation products Malonaldehyde (MAD) and 4-hydoxynonenal (4-HNE)	57–61
Blood circulation promoter	Inhibition of Na^+/K^+-ATPase	62
Immunosuppressant	By inhibition expression of surface markers CD69 and CD25 requiered for activiation of lymphocytes	63
Antidiabetic	Activitation of AMPK pathway (a major regulatory pathway for GLUT4 translocation	64,65

TNF = tumor necrosis factor, AMPK = 5'-adenosine monophosphate activated protein kinase, GLUT4 = glucose transporter type 4.

the pathway of inositol phosphate hydrolysis in the gut presented. The present knowledge of phytate absorption is summarised and discussed. Effects of phytate on mineral and trace element bioavailability are reported and phytate degradation during processing and storage is described. Beneficial activities of dietary phytate, such as its effects on calcification and kidney stone formation and on lowering blood glucose and lipids, are reported. The antioxidative property of phytic acid and its potentional anticancerogenic activities are briefly surveyed. Development of the analysis of phytic acid and other inositol phosphates is described, problems of inositol phosphate determination and detection discussed, and the need for standardisation of phytic acid analysis in foods argued.

The main sources of phytate in the daily diet are cereals and legumes, including oil seeds and nuts. They are important for human nutrition and represent 140 and 160% of total caloric intake for humans in developed and in developing countries, respectively (67).

In cereals, phytate is located in the aleurone layer and the germ while the endosperm is almost free of phytate.

As most plant food such as whole grain products, cereals and legumes – the main sources of dietary phytate intake – are processed or heat treated either during food production or preparation in one way or the other, phytases in prepared food should probably be inactivated to a large extent. This means that in humans, consuming Western-style diets with low phytase activity, phytate degradation in the stomach and the small intestine by food phytases is very limited.

Phytic acid can have health benefits due to its antioxidant properties. Laboratory and animal studies show that it can protect against DNA damage and cancer cell growth (68).

Phytic acid, the hexaphosphate of myo-inositol, is an abundant component of bran, where it occurs as a magnesium–calcium salt. The phytic acid content of different types of bran can vary widely. Rice bran contains approximately 10% phytic acid, whereas wheat bran and rye bran contain only 5%. The phytic acid content of different cereals depends on the botanic subfamily to which the grain belongs. Wheat and rye are members of the Hordeae subfamily, whereas rice is a member of the Oryzeae. The phytic acid content of oat bran (subfamily: Aveneae) is approximately 6% (69).

1.2.14 Hypoglycin – Ackee Fruit

Ackee fruits (*Blighia sapida*) derive from large green leafy trees of West African origin and are consumed either raw or after boiling in milk or water and served on their own or in meat or fish dishes, such as ackee and salt fish (70). Ackee fruit is a substantial part of the diet in poor, agricultural areas and its taste resembles that of a hazelnut or avocado. Toxicity is related to hypoglycin A and hypoglycin B, the former molecule being more toxic than the latter. The ripe fruit flesh of ackee contains only low quantities of hypoglycins, but concentrations in unripe fruits are 10–100 times greater, depending on the season and exposure to sunlight, which significantly reduces hypoglycin concentrations. In view of several larger series of intoxications in Jamaica in the past, the disease related to ackee fruit poisoning is also termed *Jamaica vomiting sickness*, with a clinical pattern similar to Reye syndrome that includes gastrointestinal symptoms, marked hypoglycemia, and central nervous system abnormalities that typically develop within 6–48 hours of ingestion. Lethality is high, particularly in infants and children. Toxicity is believed to be related to methylenecyclopropylacetic acid, a toxic hypoglycin metabolite.

Relatives of ackee, including lychee, longan, and rambutan, can contain enough α-(methylenecyclopropyl)glycine, a homologue of hypoglycin A, in their fruit to cause hypoglycemic encephalopathy in undernourished children, when consumed in large quantities (71).

Hypoglycin B is a naturally occurring organic compound in the species *Blighia sapida*. It is particularly concentrated in the fruit of the plant especially in the seeds. Hypoglycin B is toxic if ingested and is one of the causative agents of Jamaican vomiting sickness (72).

1.2.15 Safrole

Safrole is the principal component of brown camphor oil, and sassafras oil (made from *Sassafras albidum*).

Safrole is a member of the methylenedioxybenzene group, of which many compounds are used as insecticide synergists; for example, safrole is used as a precursor in the synthesis of the insecticide piperonyl butoxide. Safrole is also used as a precursor in the synthesis of the drug ecstasy (MDMA, N-methyl-3,4-methylenedioxymethamphetamine). Accelerating demand for this party drug during the last few decades has encouraged the unscientific harvesting, Safrole exhibits antibiotic and anti-angiogenic functions.

Safrole, a phenylpropene with a 'candy shop' aroma, is abundant in nature among diverse plant genera such as *Sassafras*, *Ocotea*, *Cinnamomum*, *Myristica*, and *Piper*. *Sassafras* oil has been used extensively for a long time, first by Native Americans and later by European settlers in traditional medicine and as a flavouring agent. Until 1960 the consumption of safrole by the western population, as a flavouring agent in beer, meat, and soft drinks, was unregulated. Later, the recognition of this phytochemical as a weak hepatocarcinogen with demonstrated genotoxicity in rodents led to strict restrictions on its use in food by various regulatory bodies globally. Moreover, in Asian countries, oral carcinogenesis has been linked to safrole through the habit of chewing betel quid. As a separate issue, the identity of safrole has altered with time from a pleasant flavoring agent to a hepatocarcinogen, and more recently as a driver of the destruction of biodiversity (73,74).

Safrole and sassafras oil were banned as food and flavoring additives by the FDA on December 3, 1960. The ban now includes isosafrole and dihydrosafrole (the latter not being known in nature), and sassafras root bark, but in practice both sassafras oil and bark are still widely available in the United States, from health food stores and internet suppliers. European Council's Directive on food flavourings 88/388/EEC, amended by Coumarin and Safrole Content in Cinnamon-Flavored Food Products Journal of Pharmacy and Nutrition Sciences, 2017, Vol. 7, No. 3 125 91/71/EEC and implemented into UK national law in the Flavourings in Food Regulations 1992, limits safrole in foodstuffs to 1 ppm, except for foodstuffs containing nutmeg (15 ppm) or alcoholic drinks >25% volume alcohol (5 ppm) and other alcoholic drinks (2 ppm).

1.2.16 Myristicin

Myristicin, or methoxysafrole, is the principal aromatic constituent of the volatile oil of nutmeg, the dried ripe seed of *Myristica fragrans*. Myristicin can be found in nutmeg, black pepper, and many members of the *Umbelliferae* family including anise, carrots, parsley, celery, dill, and parsnip.

The nutmeg fruit provides a tropical spice with pleasant aromatic fragrance and strong taste, while its special ability is to enhance the taste of food hence widely used as a flavoring agent in cakes, puddings, beverages, meat and sausages. Besides, nutmeg possesses many distinct advantages, some of which are anti-diarrheal activity, antidiabetic, stimulant, antifungal, carminative and anti-inflammatory properties.

Several intoxications have been reported after an ingestion of approximately 5 g of nutmeg, corresponding to 1–2 mg myristicin/kg body weight (b.w.). Although these intoxications may be ascribed to the actions of myristicin, it is likely that other components of nutmeg may also be involved. The metabolism of myristicin resembles that of safrole (75).

Isolated myristicin has proven an effective insecticide against many agricultural pests, including A*edes aegypti* mosquito larvae, *Spilosoma obliqua* (hairy caterpillars), *Epilachna varivestis* (Mexican bean beetles), *Acyrthosiphon pisum* (pea aphids), mites, and *Drosophila melanogaster* (fruit flies). Myristicin was shown to be an effective repellant, and to cause mortality via direct and systemic exposure. It also displayed a synergistic effect when administered to insects in combination with existing insecticides (76).

The characteristics of myristicin acid is that it act as a weak monoamine oxidase inhibitor and some portion of myristicin have structures similar to serotonin agonist. Myristicin may be metabolized to exhibit compounds similar to amphetamine with hallucinogenic effects similar to lysergic acid diethylamide (77,78).

It interacts with many enzymes and signaling pathways in the body, is cytotoxic to living cells, and may also have chemoprotective properties.

1.2.17 Japanese Star Anise

Chinese star anise (*Illicium verum*) is a well-known spice used in many cultures. Many populations use it as a treatment for infant colic. Japanese star anise (*Illicium anisatum*), however, has been documented to have both neurologic and gastrointestinal toxicities.

Recently, concern has been raised regarding the adulteration of Chinese star anise with Japanese star anise. Japanese star anise contains the neurotoxin anisatin (79,80).

1.2.18 Cyanogenic Glycosides

Cyanogenic glycosides are phytotoxins which occur in at least 2,000 plant species (81). A number of these species are used as food in some areas of the world. Cassava, sorghum, stone fruits, bamboo roots, and almonds are especially important foods containing cyanogenic glycosides. Common cyanogenic glycosides include amygdalin found in bitter almonds and peach kernels, a prunasin in wild cherry bark (*Prunus serotina*). Linamarin and lotaustralin are the two most commonly found cyanide glycosides and are reported from many plant families, whereas other cyanide glycosides such as amygdalin common in Rosaceae plants and dhurrin often found in Poaceae plants are more limited in distribution.

The potential toxicity of a cyanogenic plant depends primarily on the potential that its consumption will produce a concentration of cyanide that is toxic to exposed humans. In humans, the clinical signs of acute cyanide intoxication can include: rapid respiration, drop in blood pressure, dizziness, headache, stomach pains, vomiting, diarrhea, mental confusion, cyanosis with twitching, and convulsions followed by terminal coma. Death due to cyanide poisoning can occur when the cyanide level exceeds the limit an individual is able to detoxify.

Although hydrocyanic acid is a violent poison, oral intake of cyanogenic glycosides (for example via food, especially in primitive diets) is not necessarily toxic, particularly in the short term. Hydrolysis of the glycosides in the digestive tract or by the liver leads to a slow release of hydrocyanic acid that is readily detoxified by the body (82–85).

In Chapter 2 of this book, the reader is given further information.

1.2.19 Glycoalkaloids – Solanines and Chaconine

Glycoalkaloids (GA), generally occur as plant steroidal glycosides, are secondary metabolites produced in the leaves, flowers, roots, and edible parts including sprouts and skin of the plants of Solanaceae family. Glycoalkaloids are found in many species of the genera *Solanum* and *Veratrum*. All Solanacea plants, which include tomatoes, potatoes, and eggplant, contain glycoalkaloids, e.g. solanines and chaconine.

There are two main glycoalkaloids found in commercial potatoes. These include α-chaconine and α-solanine, which are glycosylated derivatives of the aglycone solonidine. Wild potatoes (*Solanum chacoense*) and eggplant contain the glycoalkaloid solasonine. The major glycoalkaloid reported in tomatoes is α-tomatine, which is a glycosylated derivative of aglycone tomatidine.

While levels are generally low, higher concentrations are found in potato sprouts and bitter-tasting peel and green parts, as well as in green tomatoes. The plants produce the toxins in response to stresses like bruising, UV light, microorganisms, and attacks from insect pests and herbivores. To reduce the production of solanines and chaconine it is important to store potatoes in a dark, cool, and dry place, and not to eat green or sprouting parts (86).

Steroidal alkaloids and their glycosides present in several species of *Solanum* are known to possess a variety of biological activities, such as antifungal, teratogenic, antiviral, antitumor, and antiestrogenic activities.

GAs are produced by plants as a resistance to challenges such as insects and pests (87). See also Chapter 5.

1.2.20 Cycasin

Cycasin is found in all known cycad genera and is distributed throughout the body of the plant, but with the highest concentration in the seeds. It is one of several toxins found in cycad plants, along with the neurotoxic amino acid BMAA (β-Methylamino-L-alanine).

Cycasin is a carcinogenic and neurotoxic glucoside found in cycads (88).

1.2.21 Sanguinarine

Sanguinarine is an alkaloid found in many medicinal plants. It has diverse biological activities, including modulation of nuclear factor-κB and of several enzymes. It is also known to induce apoptosis, perturb microtubules, and to have antimicrobial effects. This article reviews its cardiovascular properties, including hypotensive, antiplatelet, and positive inotropic effects. Its pharmacokinetics and toxicology, including its carcinogenic potential, are also discussed. Further pharmacological and toxicological studies with sanguinarine are needed before its therapeutic use can be considered (89).

1.3 ENVIRONMENTAL CONTAMINANTS

During growth, storage, and processing, some compounds may enter the food chain.

Selenium (Se) may enter foods via plan and microorganism conversion of inorganic selenium (90,91).

1.4 CONTAMINANTS FORMED DURING PROCESSING

Heterocyclic aromatic amines (HAAs) are a class of compounds in the diet that are receiving widening attention as a risk factor for human cancer. There are two major classes of HAAs. Pyrolytic HAAs are formed during high-termperature pyrolysis (>250°C) of individual amino acids. The second class of HAAs, aminoimidazoarenes (AIAs), is formed in meats cooked at temperatures (150°C–250°C) commonly used in the household kitchen (92).

Other contaminants formed during or after processing are polyclic aromatic hydrocarbons, acrylamide, chloropropanols, furan, trans fatty acids, nitrosamines, and biogenic amines.

1.5 TOXINS IN SEAFOOD

Seafood contaminated with algal toxins may result in paralytic, neurotoxic, amnesic, or diarrhetic shellfish poisoning and ciguatera poisoning (93).

Seafood can contain toxins of non-algae origin: gempylotoxin (94), tetramine (95), and trimethylamine oxide (96).

Finally, we have toxins from animals of non-seafood sources: grayatoxins in honey (97) and tremetol from white snakeroot in milk (98).

REFERENCES

1. http://www.fao.org/fao-who-codexalimentarius/committees/committee/en/?committee=CCCF
2. Witczak, A. and Sikoroski, Z. (2017). *Toxins and Other Harmful Compounds in Foods*, CRC Press, p. 508.
3. Wong, Y.-C. and Lewis, R. J. (2017). *Analysis of Food Toxins and Toxicants*, Wiley, p. 816.
4. Dolan, L. C., Matulka, R. A. and Burdock, G. A. (2010). Naturally occuring food toxins. *Toxins*, 2, 2289–2332. 10.3390/toxins2092289
5. https://www.fda.gov/food/chemicals-metals-pesticides-food/natural-toxins-and-mycotoxins
6. https://www.fda.gov/food/guidance-documents-regulatory-information-topic-food-and-dietary-supplements/chemical-metals-natural-toxins-pesticides-guidance-documents-regulations
7. https://www.efsa.europa.eu/en/aboutefsa
8. Rossini, G. P. (2014). *Toxins and Biologically Active Compounds from Microalgae*, CRC Press, Volume 1, p. 542.
9. Rossini, G. P. (2014). *Toxins and Biologically Active Compounds from Microalgae*, CRC Press, Volume 2, p. 700.
10. Zámboriné Németh, É. and Thi Nguyen, H. (2020). Thujone, a widely debated volatile compound: What do we know about it? *Phytochem Rev*, 19, 405–2423. 10.1007/s11101-020-09671-y
11. https://apps.who.int/food-additives-contaminants-jecfa-database/chemical.aspx?chemID=192
12. Vetter, J. (2000). Plant cyanogenic glycosides. *Toxicon*, 38(1), 11–236. 10.1016/S0041-0101(99)00128-2
13. Whittier, J. C.(2011). mountainscholar.org. https://mountainscholar.org/bitstream/handle/10217/183206/AEXT_ucsu2062216122011.pdf?sequence=1
14. Kubin, A., Wierrani, F., Burner, U., Alth, G. and Grunberger, W. (2005). Hypericinj – The facts about a controversial agent. *Current Pharmaceutical Design?*, 11(2), 233–253. 10.2174/1381612053382287
15. Mukherjee, P. K. (2019). *Quality Control and Evaluation of Herbal Drugs*, Elsevier, p. 762. 10.1016/C2016-0-04232-8
16. Nicolussi, S., Drewe, J., Butterweck, V. and Meyer zu Zwabedisschen, H. E. (2020). Clinical relevance of St. John's Wort drug interactions revisited. *British J Pharmacol*, 177, 1212–1226. 10.1111/bph.14936
17. Xiang Ng, Q., Venkatanarayanan, N. and Yih Xian Ho, C. (2017). Clinical use of hypericum perforatum (st john's wort) in depression: A meta-analysis. *Journal of Affective Disorders*, 210, 211–221. 10.1016/j.jad.2016.12.048
18. Langer, P. and Michajlovskii (1992). Naturally occurring goitrogens. In Tu, A. (ed.). *Food Poisoning*, 1st Edition, Routledge, pp. 101–129. 10.1201/9780203752708
19. Vanderpas, J. (2006). Nutritional epidemiology and thyroid hormone metabolism. *Annu. Rev. Nutr*, 26, 293–322. 10.1146/annurev.nutr.26.010506.10381
20. Gaitan, E. (1990). Goitrogens in food and water. *Annual Review of Nutrition*, 10(1), 21–37. 10.1146/annurev.nu.10.070190.000321
21. Gaitan, E., Lindsay, R. H., Reichert, R. D., Ingbar, S. H., Cooksey, R. C., Legan, J., Meydrech, E. F., Hill, J. and Kubota, K. (1989). Antithyroid and goitrogenic effects

of millet: Role of C-Glycosylflavones. *The Journal of Clinical Endocrinology & Metabolism*, 68(4), 707–714. 10.1210/jcem-68-4-707

22. Lüthy, J., Carden, B., Friederich, U. and Bachmann, M. (1984). Goitrin — A nitrosatable constituent of plant foodstuffs. *Experientia*, 40(5), 452–453. 10.1007/BF01952381

23. de Souza Dos Santos, M. C., Gonçalves, C. F., Vaisman, M., Ferreira, A. C. and de Carvalho, D. P. (2011). Impact of flavonoids on thyroid function. *Food Chem Toxicol*, 49(10), 2495–2502. 10.1016/j.fct.2011.06.074

24. Felker, P., Bunch, R., Leung, A. M. (2016). Concentrations of thiocyanate and goitrin in human plasma, their precursor concentrations in brassica vegetables, and associated potential risk for hypothyroidism. *Nutrition Reviews*, 74(4), 248–258. 10.1093/nutrit/nuv110

25. McMillan, M., Spinks, E. A. and Fenwick, G. R. (1986). Preliminary observations on the effect of dietary brussels sprouts on thyroid function. *Hum Toxicol*, 5(1), 15–19. 10.1177/096032718600500104

26. Tolonen, M., Taipale, M., Viander, B., Pihlava, J.-M., Korhonen, H. and Ryhänen, E.-L. (2002). Plant-derived biomolecules in fermented cabbage. *J Agric Food Chem*, 50(23), 6798–6803. 10.1021/jf0109017

27. https://www.efsa.europa.eu/en/efsajournal/pub/4593

28. https://efsa.onlinelibrary.wiley.com/doi/10.2903/j.efsa.2016.4593

29. Food Standards Australia New Zealand. (2003, June). Erucic Acid in Food: A Toxicological Review and Risk Assessment Technical Report, Series No. 21.

30. Vetter, W., Darwisch, V. and Lehnert, K. (2020). Erucic acid in *Brassicaceae* and salmon – An evaluation of the new proposed limits of erucic acid in food. *NFS Journal*, 19, 9–15. 10.1016/j.nfs.2020.03.002

31. Buckle, J. (2015). *Clinical Aromatherapy, Essential Oils in Healthcare*, Third Edition, Elsevier, p. 432.

32. Christensen, L. P. (2018). Polyphenols and polyphenol-derived compounds from plants and contact dermatitis. In Watsen, R. R., Preedy, V. R. and Zibadi, S. (eds.). *Polyphenols: Prevention and Treatment of Human Disease*, Second Edition, Elsevier, Volume 2, pp. 349–379.

33. Mohan, V. R., Tresina, P. S. and Daffodil, E. D. (2016). Antinutritional factors in legume seeds: Characteristics and determination. In Caballero, B., Finglas, P. M. and Toldrá, F. (eds.). *Encyclopedia of Food and Health*, Academic Press, pp. 211–220.

34. Grant, G., Duncan, M., Alonso, R. and Marzo, M. (2003). Peas and lentils. In Caballero, B., Trugo, L. and Finglas, P. M. (eds.). *Encyclopedia of Food Sciences and Nutrition*, Second Edition, Elsevier, pp. 4433–4440. 10.1016/B0-12-227055-X/00899-3

35. Yamane, H., Konno, K., Sabelis, M., Takabayashi, J., Sassa, T. and Oikawa, H. (2010). Chemical defence and toxins of plants. In Ben, H.-W. and Lew, M. (eds.). *Comprehensive Natural Products II*, Elsevier, Volume 4, pp. 339–385. 10.1016/B978-008045382-8.00099-X

36. Udani, J., Tan, O. and Molina, J. (2018). Systematic review and Meta-analysis of a proprietary alpha-amylase inhibitor from white bean (*Phaseolus vulgaris* L.) on weight and fat loss in humans. *Foods*, 7, 63–73. 10.3390/foods7040063

37. Kalač, P. (2016). *Edible Mushrooms, Chemical Composition and Nutrional Value*, Elsevier, p. 143.

38. Chan, C. K. F., Ransom, R. C. and Longaker, M. T. (2016). Skeletal Stam cells: Lectins bring benefits to bones. *eLife*, 5, e18782. 10.7554/eLife.22926

39. Bitsch, R. (2003). Thiamin | physiology. In Caballero, B. (ed.). *Encyclopedia of Food Sciences and Nutrition*, Second Edition, Elsevier, pp. 5772–5780.

40. Sebrel, W. H. and Harris, R. S. (1972). *Thiamine in The Vitamins*, Sebrel, W.H., & Harris, R.S., Second Edition, Volume 5.

41. Aronson, J. K. (2016). *Meyler's Side Effects of Drugs: The International Encyclopedia of Adverse Drug Reactions and Interactions*, Sixteenth Edition, Elsevier, p. 5, 1069.

42. http://www.fao.org/fao-who-codexalimentarius/committees/committee/related-standards/en/?committee=CCCF

43. Mohan, V. R., Tresina, P. S. and Daffodil, E. D. (2016). Antinutritional factors in legume seeds: Characteristics and determination In Caballero, B., Finglas, P. M. and Toldrá, F. (eds.). *Encyclopedia of Food and Health*, Elsevier, pp. 211–220.

44. Morrison, S. C. and Savage, G. P. (2003). Oxalates. In Caballero, B. (ed.). *Encyclopedia of Food Sciences and Nutrition*, Second Edition, Elsevier, pp. 4282–4287.

45. Kaushik, U. and Mir, S. R. (2015). Cucurbitacins – An insight into medicinal leads from nature. *Pharmacogn Rev*, 9(17), 12–18. 10.4103/0973-7847.156314

46. Gry, J., Søborg, H. and Andersson, C. (2006). Cucurbitacins in plant food. *TemaNord*, Nordic Council of Ministers, Copenhagen, 2006, 556.

47. Yoh, N., Douhoré, T., Attioua, K. B., Soro, Y., Kabran, F. A., Kablan, L. C. A., Vedrenne, M., Mathieu, C. and Vaca-Garcia, C. (2020). Nor-cucurbitacins from the leaves of mareya micrantha (Benth.) Müll. *Arg. (Euphorbiaceae), Fitoterapia*, 143, 104538. 10.1016/j.fitote.2020.104538

48. Karale, P. A., Dhawale, S. and Karale, M. (2021, May 4). Phytochemical profile and antiobesity potential of *Momordica charantia* linn. *IntechOpen*, 2021. 10.5772/intechopen.96808.

49. Bowman, T., Yu, H., Sebti, S., Dalton, W. and Jove, R. (1999). Signal transducers and activators of transcription: Novel targets for anticancer therapeutics. *Cancer Control*, 6, 427–435. 10.1177/107327489900600501

50. Turkson, J. and Jove, R. (2000). Stat proteins: Novel molecular targets for cancer drug discovery. *Oncogene*, 19, 6613–6626. 10.1038/sj.onc.1204086

51. Bowman, T., Garcia, R., Turkson, J. and Jove, R. (2000). Stats in oncogenesis. *Oncogene*, 19, 2474–2488. 10.1038/sj.onc.1203527

52. Dong, Y., Lu, B., Zhang, X., Zhang, J., Lai, L., Li, D., Wu, Y., Song, Y., Luo, J., Pang, X., Yi, Z. and Liu, M. (2010). Cucurbitacin E, a tetracyclic triterpenes compound from chinese medicine, inhibits tumor angiogenesis through VEGFR2 mediated JAK2/STAT3 signaling pathway. *Carcinogenesis*, 31, 2097–2104. 10.1093/carcin/bgq167

53. Duncan, K. L., Duncan, M. D., Alley, M. C., Sausville, E. A. (1996). Cucurbitacin E-induced disruption of the actin and vimentin cytoskeleton in prostate carcinoma cells. *Biochem Pharmacol*, 52, 1553–1560. 10.1016/S0006-2952(96)00557-6

54. Duangmano, S., Dakeng, S., Jiratchariyakul, W., Suksamra, A., Smith, D. R. and Patmasiriwat, P. (2010). Antiproliferative effects of cucurbitacin B in breast cancer cells: Down-regulation of the C-myc/Htert/Telomerase pathway and obstruction of the cell cycle. *Int J Mol Sci*, 11, 5323–5338. 10.3390/ijms11125323

55. Jayaprakasam, B., Seeram, N. P. and Nair, M. G. (2003). Anticancer and anti-inflammatory activities of cucurbitacins from *Cucurbita andreana*. *Cancer Lett*, 189, 11–16. 10.1016/S0304-3835(02)00497-4

56. Escandell, J. M., Recio, M. C., Manez, S., Giner, R. M., Cerda-Nicolas, M., Gil-Benso, R. and Rios, J. L. (2007). Dihydrocucurbitacin B inhibits delayed type hypersensitivity reactions by suppressing lymphocyte proliferation. *J Pharmacol Exp Ther*, 322, 1261–1268. 10.1124/jpet.107.122671

57. Park, C. S., Lim, H., Han, K. J., Baek, S. H., Sohn, H. O., Lee, D. W., Kim, Y. G., Yun, H. Y., Baek, K. J. and Kwon, N. S. (2004). Nhibition of nitric oxide generation by 23,24-dihydrocucurbitacin D in mouse peritoneal macrophages. *J Pharmacol Exp Ther*, 309, 705–710. 10.1124/jpet.103.063693

58. Yuan, G., Mark, L. W., Guoqing, H., Min, Y. and Li, D. (2006). Natural products and anti-inflammatory activity. *Asia Pac J Clin Nutr*, 15, 143–152.

59. Esterbauer, H. (1993). Cytotoxicity and genotoxicity of lipid oxidation products. *Am J Clin Nutr*, 57(5), 779S–785S. 10.1093/ajcn/57.5.779S

60. Tannin-Spitz, T., Bergman, M. and Grossman, S. (2007). Cucurbitacin glucosides: Antioxidant and free-radical scavenging activities. *Biochem Biophys Res Commun*, 364, 181–186. 10.1016/j.bbrc.2007.09.075

61. Saba, A. B. and Oridupa, A. O. (2010). Search for a novel antioxidant, anti-inflammatory/analgesic or anti-proliferative drug: Cucurbitacins hold the ace. *J Med Plants Res*, 4, 2821–2826. 10.5897/JMPR.9001120

62. Chen, R. J., Jin, T. R., Chen, Y. C., Chung, T. Y., Yang, W. H. and Tzen, J. T. (2010). Active ingredients in many chinese medicines promoting blood circulation are na+/K+-ATPase inhibitors. *Acta Pharmacol Sin*, 32, 141–151. 10.1038/aps.2010.197

63. Yaowalak, U., Usaneeporn, L., Weena, J. and Tanawan, K. (2010). Immunosuppressive effects of cucurbitacin B on human peripheral blood lymphocytes. *J Med Plants Res*, 4, 2340–2347. 10.5897/JMPR10.459

64. Ye, J. M., Ruderman, N. B. and Kraegen, E. W. (2005). AMP-activated protein kinase and malonyl-CoA: Targets for treating insulin resistance? *Drug Disc Today Ther Strateg*, 2, 157–163. 10.1016/j.ddstr.2005.05.019

65. Iglesias, M. A., Ye, J. M., Frangioudakis, G., Saha, A. K., Tomas, E., Ruderman, N. B., Cooney, G. J. and Kraeger, E. W. (2002). AICAR administration causes an apparent enhancement of muscle and liver insulin action in insulin-resistant high-fat-fed rats. *Diabetes*, 51, 2886–2894. 10.2337/diabetes.51.10.2886

66. Schlemmer, U., Wenche, F., Prieto, R. M. and Grases, F. (2009). Phytate in foods and significance for humans: Food sources,intake, processing, bioavailability, protective role and analysis. *Mol. Nutr. Food Res.*, 53, S330–S375. 10.1002/mnfr.200900099

67. Faostat, FAO, Rome (2007). http://faostat.fao.org

68. Shamsuddin, A. (2002). Anti-cancer function of phytic acid. *Food Science and Technology*, 37(7), 769–782. 10.1046/j.1365-2621.2002.00620.x

69. Zitterman A. (2003). Dietary fiber – bran. In Caballero, B. (ed.). *Encyclopedia of Food Sciences and Nutrition*, Second Edition, Elsevier, pp. 1844–1850.

70. Seeff L., Stickel F. and Navarro V. J. (2013). Hepatotoxicity of herbals and dietary supplements. In Kaplowitz, N. and DeLeve, L. D. (eds.). *Drug-Induced Liver Disease*, Third Edition, Elsevier, pp. 631–657. 10.1016/B978-0-12-387817-5.00035-2

71. Spencer, P. S., Palmer, V. S. and Mazumder, R. (2015). Probable toxic cause for suspected lychee-linked viral encephalitis. *Emerg Infect Dis*, 21(5), 904–905. 10.3201/eid2105.141650

72. Surmaitis R. and Hamilton R. J. (2021). Ackee fruit toxicity. *StatPearls [Internet]*, StatPearls Publishing.

73. Kemprai, P., Mahanta, B., Sut, D., Barman, R., Banik, D., Lal, M., Saikia, S. P., Haldar, S. (2020). Review on safrole: Identity shift of the 'candy shop' aroma to a carcinogen and deforester. *Flavour and Franfrance Journal*, 35(1), 5–23. 10.1002/ffj.3521

74. Solaiman, R. and Al-Zehour, J. (2017). Coumarin and safrole content in cinnamon-flavored food products on the syrian market. *Journal of Pharmacy and Nutrition Sciences*, 7, 124–129. 10.6000/1927-5951.2017.07.03.8

75. Hallström, H. and Thuvander, A. (1997). Toxicological evaluation of myristicin. *Natural Toxins*, 5(5), 186–192. 10.1002/19970505NT3

76. Lichtenstein, E. P. and Casida, E. J. (1963). Naturally occurring insecticides, myristicin, an insecticide and synergist occurring naturally in the edible parts of parsnips. *Agric. Food Chem.*, 11(5), 410–415. 10.1021/jf60129a017

77. Stein, U., Greyer, H. and Hentschel, H. (2001). Nutmeg (myristicin) poisoning — Report on a fatal case and a series of cases recorded by a poison information centre. *Forensic Science International*, 118(1), 87–90. 10.1016/S0379-0738(00)00369-8

78. Rahman, N. A. A., Fazilah, A. and Effarzah, M. E. (2015). Toxicity of nutmeg (myristicin): A review. *International Journal on Advanced Science Engineering Information Technology*, 5(3), 61–64.

79. Madden, G. R., Dchmitz, K. H. and Fullerton, K. A. (2012). Case of infantile star anise toxicity. *Pediatric Emergency Care*, 28(3), 284–285. 10.1097/PEC.0b013e3182495ba7

80. Ize-Ludlow, D., Ragone, S., Bruck, I. S., Bernstein, J. N., Duchowny, M. and Garcia Pena, B. M. (2004). Neurotoxicities in infants seen with the consumption of star anise tea. *Pediatrics*, 114, e653–e656. 10.1542/peds.2004-005882

81. Bone K. and Mills S. (2013). *Principles and Practice of Phytotherapy*, Second Edition, Churchill Livingstone, p. 1056.

82. Nyirenda K. N. 2020). Toxicity potential of cyanogenic glycosides in edible plants. In Erkekoglu Pınar and Ogawa Tomohisa (eds.). *Edible Plants, Medical Toxicology*, IntechOpen. 10.5772/intechopen.91408

83. Yamane H., Konno K., Sabelis M., Takabayashi J., Sassa T. and Oikawa H. (2010). Chemical defence and toxins of plants. In Hung-Wen, B. and Mander, L. (eds.). *Comprehensive Natural Products II*, Elsevier, volume 4, pp. 339–385.

84. Stegelmeier B. L., Field R., Panter K. E., Hall J. O., et al. (2013). Selected poisonous plants affecting animal and human health. In Haschek, W. M., Rousseau, C. G., Wallig, M. A., Bolon, B., Ochoa, R. and Mahler, B. W. (eds.). *Haschek and Rousseaux's Handbook of Toxicologic Pathology*, Third Edition, Elsevier, pp. 1260–1311.

85. https://inchem.org/documents/jecfa/jecmono/v30je18.htm

86. Navarre D. A., Shakya R. and Hellmann H. (2016). Vitamins, phytonutrients, and minerals in potato. In Singh, J. and Kaur, L. (eds.). *Advances in Potato Chemistry and Technology*, Second Edition, Elsevier, pp. 128–134.

87. Siddique M. A. B. and Brunton N. (2019). ood glycoalkaloids: Distribution, structure, Cytotoxicity, extraction, and biological activity. In Kurek, J. (ed.). *Alkaloids – Their Importance in Nature and Human Life*. Intechopen, pp. 47–72. 10.5772/intechopen.82780

88. Matsumoto H. 1983). Cycasin. In Rechcigl, M. (ed.). *Handbook of Naturally Occurring Food Toxicants*. CRC Press, pp. 43–63.

89. Mackraj, I., Govender, T. and Gathiram, P. (2008). Sanguinarine. *Cardiovascular Drug Reviews*, 26(1), 75–83. 10.1111/j.1527-3466.2007.00037.x

90. Kieliszek, M. (2019). Selenium–fascinating microelement, properties and sources in food. *Molecules*, 24, 1298. 10.3390/molecules24071298 Methyl mercury may be formed by bacterial action in an aquatic environment (89).

91. Zhang, Y., Soerensen, A. L., Schartup, A. T. and Sunderland, E. M. (2020). A global model for methylmercury formation and uptake at the base of marine food webs. *Global Biogeochemical Cycles*, 34(2), e2019GB006348. 10.1029/201 9GB006348

92. Turesky R. J. (2010). Heterocyclic aromatic amines: Potential human carcinogens. In Fishbein, J. C. *Advances in Molecular Toxicology*, Elsevier, Volume 4, pp. 37–83. ISSN 1872-0854, ISBN 9780444535849, https://doi.org/10.1016/S1872-0854(10) 04002-6

93. Bari M. L. and Yamazaki K. (2019). *Seafood Safety and Quality*, CRC Press, p. 332.

94. Aldsworth T. 2017). Fish: Escolar and oilfish. In Dodd, C., Aldsworth, T. and Riemann, H. P. (eds.). *Foodborne Diseases*, Third edition, Elsevier, pp. 527–533.

95. Chau, C. M., Leung, A. K. H. and Tan, I. K. S. (2005). Tetramine poisoning. *Hong Kong Med J*, 11, 511–514.

96. Velasquez, M. T., Ramezani, A., Manal, A. and Raj, D. S. (2016). Trimethylamine N-oxide: The good, the bad and the unknown. *Toxins*, 8, 326–337. 10.3390/ toxins8110326

97. Jansen, S. A., Kleerekooper, I., Hofman, Z. L. M., Kappen, I. F. P. M., Stary-Weinzinger, A. and van der Heyden, M. A. G. (2012). Grayanotoxin poisoning: 'Mad honey disease' and beyond. *Cardiovasc Toxicol*, 12, 208–215. 10.1007/s12 012-012-9162-2

98. Hartmann, A. F., Hartmann, A. F. and Purkerson, M. L. (1963). Wesley. Tremetol poisoning—not yet extinct. *JAMA*, 185(9), 706–709. 10.1001/jama.1963. 03060090038014

CHAPTER 2

Cyanogenic Glycosides as Food Toxins

Michael Amoafo Mensah

Freelance, Gent, Belgium

CONTENTS

2.1	Background	26
	2.1.1 Cyanogenic Glycosides as Natural Toxins	26
	2.1.2 Distribution of Cyanogenic Glycosides	26
	2.1.3 Cyanogenic Glycosides as Secondary Metabolites	27
	2.1.4 Synthesis/Production of Cyanogenic Glycosides	27
	2.1.5 Cyanogenesis	28
2.2	Examples of Cyanogenic Glycosides in Food Plants	28
	2.2.1 Cyanogenic Glycoside Content of Some Important Plant Foods	28
	2.2.1.1 Cassava	30
	2.2.1.2 Bamboo Shoots	31
	2.2.1.3 Sorghum	31
	2.2.1.4 Lima Beans	32
	2.2.1.5 Linseed	32
	2.2.1.6 Apples	32
	2.2.1.7 Apricot Fruits	32
2.3	Biosynthesis of Cyanogenic Glycosides	33
	2.3.1 Enzymatic Hydrolysis of Cyanogenic Glycosides in Food Plants	34
	2.3.2 Toxicity of Cyanogenic Glycosides (Cyanide Toxicity)	35
	2.3.3 Mechanism of How Cyanogenic Glycosides Affects the Body	37
	2.3.4 Mechanism of Cyanide Detoxification in Human	38
	2.3.5 Human Health Effect of Cyanogenic Glycosides	39
	2.3.6 Treatment of Cyanide Poisoning	40
2.4	Preventing the Effects of Cyanogenic Glycosides (Cyanide-Induced Diseases)	40
	2.4.1 Effect of Processing on Cyanogenic Glycosides	40
	2.4.1.1 Effect of Soaking and Fermentation on Cyanogenic Glycosides	41
	2.4.1.2 Effect of Drying on Cyanogenic Glycosides	42
	2.4.1.3 Effect of Cooking on Cyanogenic Glycosides	42
2.5	Diseases Caused by Cyanide	42
	2.5.1 Iodine Deficiency Diseases (Goitre and Cretinism)	42
	2.5.2 Konzo	43
	2.5.3 Tropical Ataxic Neuropathy (TAN)	43

DOI: 10.1201/9781003222194-3

2.5.4	Growth Retardation	43
2.5.5	Cyanide Poisoning	43
2.6	Quantification of Cyanogenic Glycosides	44
2.7	Conclusion and Recommendations	44
References		45

2.1 BACKGROUND

2.1.1 Cyanogenic Glycosides as Natural Toxins

Food plants contain many substances culpable of posing potential risks to consumers, of which cyanogenic glycosides are no exception. Cyanogenic glycosides (CNGs) are natural plant toxicants (Bolarinwa and Orfila 2017), and are chemical compounds contained in foods that release hydrogen cyanide (HCN) when chewed or digested (Vetter 2000; Bolarinwa Oke et al. 2016). The act of chewing or digestion leads to hydrolysis of the substances, causing cyanide (CN) to be released (Bolarinwa *et al.* 2016) which in an aqueous environment, always exists as a mixture of non-dissociated acid (hydrogen cyanide, HCN) and its dissociated form (cyanide ion, CN^-) (Schrenk *et al.* 2019).

Cyanogenic glycosides serve as important chemical weapons in the defense of the plant against herbivores because of their potential to generate toxic hydrogen cyanide (Bolarinwa and Orfila 2017). They are the most important defense-related secondary metabolites and a class of natural plant toxins in plant foods (Folashade 2013). Because CNGs protect plants from herbivore attacks, they are referred to as 'phytoanticipins' (Gleadow and Møller 2014).

In addition to their defense function in plants, cyanogenic glycosides have been reported to assist important functions in primary metabolism. Møller (2010) reported that a high concentration of cyanogenic glycosides in seeds during germination can serve to provide aspartate or asparagine for transamination reactions which are needed to balance amino acid supply to the developing seedling. More so, Poulton (1990) reported that cyanogenic glycosides serve as storage forms for reduced nitrogen. Cyanogenic glycosides are believed to represent a pool of nitrogen to be used by the plant if needed (Gleadow and Møller 2014).

Due to the pivotal roles CNGs play in the organization of chemical defense systems in plants and plant-insect interactions (Ganjewala *et al.* 2010), despite their toxicity, farmers continue to grow crops with high levels of cyanogenic glycosides because of their ability to act as natural pesticides to protect crops against animal pests (Folashade 2013).

2.1.2 Distribution of Cyanogenic Glycosides

Cyanogenic glycosides are said to be widely distributed in the plant kingdom including many that are commonly consumed by humans (Bolarinwa and Orfila 2017). They are present in more than 2,500 species (Vetter 2000; Ganjewala *et al.* 2010) and a lot of such taxa belong to families such as Fabaceae, Rosaceae, Leguminosae, Linaceae, Compositae, and others. Identification of their constituents is, therefore, a useful tool for informative taxonomic markers (Vetter 2000).

2.1.3 Cyanogenic Glycosides as Secondary Metabolites

More than 300,000 plants produce different secondary metabolites, including cyanogenic glycosides (Zagrobelny *et al.* 2008). Cyanogenic glycosides, therefore, belong to the products of secondary metabolism, to the natural products of plants, and are compounds composed of an α-hydroxynitrile type aglycone and a sugar moiety (Vetter 2000).

More so, cyanogenic glycosides are amino-acid-derived constituents of plants produced as secondary metabolites (FAO/WHO 1993). Products of these secondary metabolites are species-specific and give the plant its particular characteristics such as plant pigments, flavors, and compounds that serve to protect the plants. Some of these secondary metabolic products also impart toxicity to the individual when taken orally. These substances may be growth inhibitors, neurotoxins, carcinogens, and teratogens (Omaye 2004).

Despite a great deal of structural diversity in cyanogenic glycosides, almost all of them are believed to be derived from only six different amino acids, which are L-valine, L-isoleucine, L-leucine, L-phenylalanine, L-tyrosine, and cyclopentenyl-glycine (a non-protein amino acid) (Vetter 2000; Ganjewala *et al.* 2010). The use of these L-amino acids in producing cyanogenic glycosides is explained under section 3.0 (Biosynthesis of Cyanogenic Glycosides). All known cyanogenic glycosides are β-linked, mostly with D-glucose (Vetter 2000).

2.1.4 Synthesis/Production of Cyanogenic Glycosides

Cyanogenic glycosides are synthesized in some economically important food plants such as linamarin in cassava and butter bean; dhurrin in sorghum and macadamia nut; and amygdalin in almond, peach, sweet cherry, and sour cherry (Jones 1998; Vetter 2000). It is important to note that great variabilities exist between the levels of these cyanogenic glycosides found in these plants, even among the same cyanogenic plant species. These differences can be attributed to a combination of endogenous factors such as the age and variety of the plant, and exogenous factors such as the environment (Vetter 2000).

For instance, cassava crops grown in low altitude areas have been reported to contain high levels of cyanogenic glycosides while those grown in high altitude areas contain low levels of cyanogenic glycosides (Oluwole *et al.* 2007). More so, the levels of cyanogenic glycosides in different parts of the plants vary and also vary between the same parts of different individual plants of the same species. While high concentrations of cyanogenic glycosides are usually found in plant leaves, cyanogenic glycosides may also be concentrated in plant roots, seeds, or other plant tissues (Seigler 1975).

The major cyanogenic glycosides such as amygdalin (almonds); dhurrin (sorghum); linamarin (cassava, lima beans); lotaustralin (cassava, lima beans); prunasin (stone fruit); and taxiphyllin (bamboo shoots) are found in the edible parts of plants (FSANZ 2005). These foods accumulate significant amounts of cyanogenic glycosides and therefore, represent potential sources of hydrogen cyanide (HCN) (FAO/WHO 1993).

Cyanogenic glycosides are water-soluble compounds that are chemically quite stable and are not toxic on their own because, the CNGs and their catabolic enzymes (β-glycosidases and hydroxynitrile lyases) are stored in separate compartments in intact plant cells (Schrenk *et al.* 2019). However, when cell structures of plants are disrupted (e.g., when cyanogenic plants are chewed by herbivores or when the plants are disintegrated during processes, such as grinding, pounding, soaking or fermentation, or physical processes such as maceration or freezing during food processing (Gleadow and Woodrow 2002; CFS 2007)), hydrolysis of these cyanogenic glycosides occur. This hydrolysis of cyanogenic glycosides yields cyanide (which is the toxic substance) in the process known as cyanogenesis.

2.1.5 Cyanogenesis

Cyanogenesis can, therefore, be defined as the ability of plants and other living organisms to produce hydrogen cyanide (Poulton 1990; Zagrobelny *et al.* 2008). As already stated in the aforementioned paragraph, cyanogenic glycosides are not toxic when intact, but it is the enzymatic action on the cyanogenic plants that causes the release of hydrogen cyanide which in turn causes potential toxicity issues for animals and humans (Folashade 2013). While cyanogenic glycosides are stored in vacuoles within the cell, the enzymes are stored in the cytoplasm but, upon disruption of the plant tissues by herbivores or during processing, hydrolysis of cyanogenic glycosides occur by the plant endogenous enzymes; β-glucosidase and α-hydroxynitrile lyases (Zagrobelny *et al.* 2004) with subsequent breakdown of the cyanogenic glycoside to sugar and a cyanohydrin, and this cyanohydrin rapidly decomposes to hydrogen cyanide and an aldehyde or a ketone (Moller and Seigler 1999; Haque and Bradbury 2002). The glycosides, cyanohydrins, and hydrogen cyanide are collectively known as cyanogens and this interaction of cyanogenic glycosides and hydrolytic enzymes is how cyanogenic plants are protected against predators (Moller and Seigler 1999). The process of cyanogenesis is sometimes also called the 'cyanide bomb' as reported by Morant *et al.* (2008).

2.2 EXAMPLES OF CYANOGENIC GLYCOSIDES IN FOOD PLANTS

Different kinds of cyanogenic glycosides may be found in different cyanogenic food plants (FAO/WHO 1993). Seigler (1991) reports that at least 60 different kinds of cyanogenic glycosides have been identified in plants, and approximately 25 of these cyanogenic glycosides that are known are generally found in the edible parts of plants (FSANZ 2005; Bolarinwa *et al.* 2016).

Amygdalin, which is said to be the most common of the cyanogenic glycosides, can be found in the kernels and seeds of fruits such as almonds, apples, apricots, cherries, plums, peaches, and nectarines (Folashade 2013; Bolarinwa *et al.* 2015; Tamer *et al.* 2019). Linamarin is found in cassava, taxiphyllin is found in bamboo shoots (FAO/WHO 1993). Commercial crop plants, such as sorghum and barley, contain cyanogenic glycosides (Bolarinwa *et al.* 2016). Cyanogenic glycosides are also found in stone fruit, pome fruit, linseed/flaxseed, lima beans, cocoyam, chickpeas, cashews, and kirsch (Haque and Bradbury 2002; Codex 2008).

FAO/WHO (1993) reports that some food ingredients with flavoring properties such as ground almonds powder or paste, marzipan, stone fruit, and alcoholic drinks made from stone fruits may contain cyanogenic glycosides. These foods, therefore, represent potential sources of hydrogen cyanide (Bolarinwa *et al.* 2016). Figure 2.1 is showing the chemical structure of some cyanogenic glycoside and their sugar moiety.

2.2.1 Cyanogenic Glycoside Content of Some Important Plant Foods

Cyanogenic glycosides have been reported to be in high amounts in many nutritious dense plants, and this limits the use of some of these plants as food (Bolarinwa *et al.* 2016), but

Configuration	Glycoside	Sugar
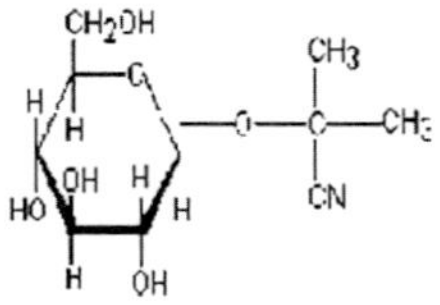	Linamarin	D-glucose
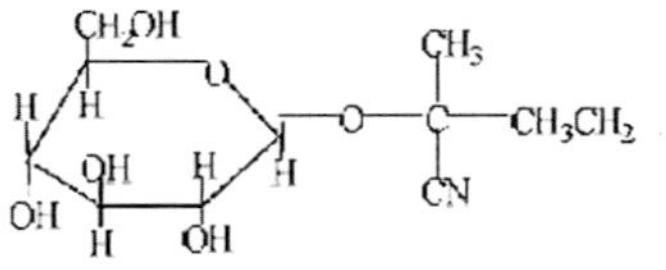	Lotaustralin	D-glucose
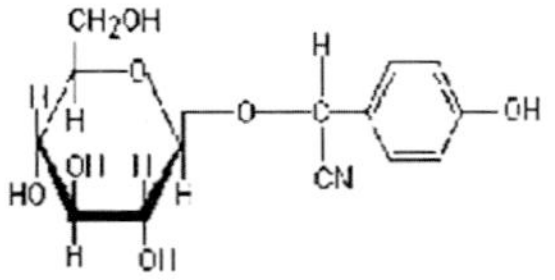	Dhurrin	D-glucose
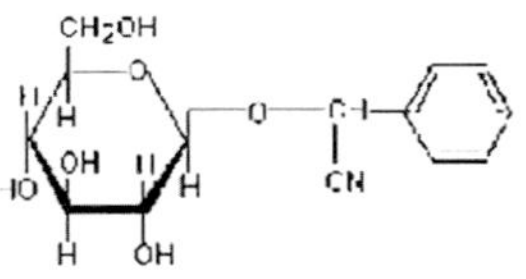	Prunasin	D-glucose
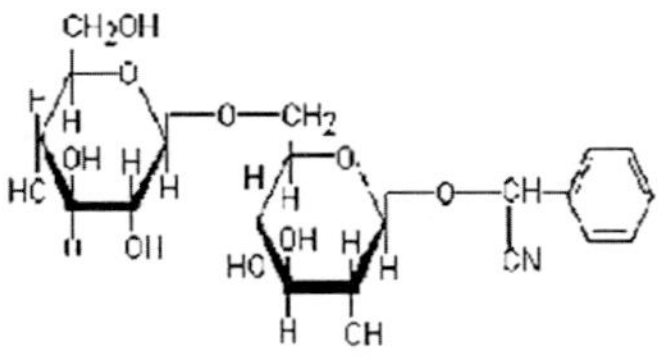	Amygdalin	Gentiobiose
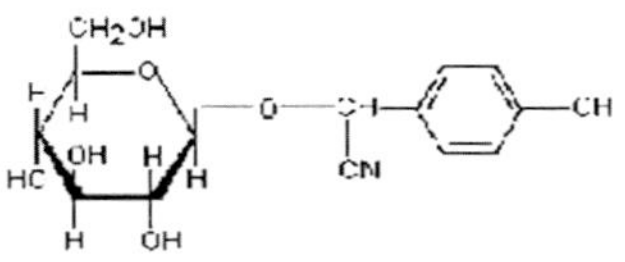	Taxiphyllin	D-glucose
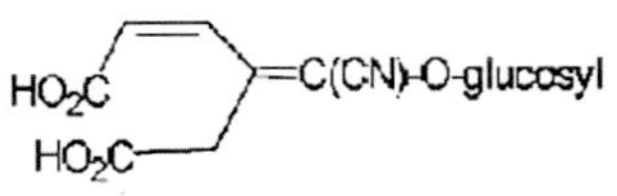	Triglochinin	D-glucose

FIGURE 2.1 The chemical structure of some cyanogenic glycosides and their sugar moiety.

the high levels of these cyanogenic glycosides do not deter the use of some of these plants because of their economic importance and nutritional benefits. Some examples are shown below.

2.2.1.1 Cassava

Cassava (*Manihot esculenta*) is one of the plants with high amounts of cyanogenic glycosides. Cassava grows well in a tropical climate (FSANZ 2005). It is a crop of economic importance in Africa and a staple food in most African communities (Nhassico *et al.* 2008). Cassava is also a valued crop in the Pacific Island countries (FSANZ 2005), South (Latin) America, and Southeast Asia (Bolarinwa *et al.* 2016). Although cassava consumption in Europe is low, it is on the rise due to the high numbers of immigrants from Africa (Kolind-Hansen and Brimer 2010).

According to the Food and Agriculture Organization, cassava is the third most important source of calories in the tropics, after rice and corn (FAO 1991; Kwok 2008). The root tubers of cassava are a rich source of starch but contain only a little protein (less than 5% of the dry weight), whereas the leaves contain valuable proteins, minerals, vitamin A and vitamin C and provide some dietary fiber (Montagnac *et al.* 2009). The EFSA Panel on Contaminants in the Food Chain (CONTAM) reports that both the roots and leaves of bitter cultivars of cassava contain high levels of CNGs (linamarin and lotaustralin in a 20:1 ratio) with the concentration of CNGs in leaves being about tenfold higher than in the root tubers (Schrenk *et al.* 2019).

Cassava root can be processed into various forms before consumption. The root can be boiled and eaten as whole root, or processed into flour for the production of "thick gruel" commonly consumed in Africa, and can also be converted into chips for the production of flour, tapioca, and other foods (Bolarinwa *et al.* 2016). In some African countries, the leaves of cassava are used to produce soup or sauce, following processes such as scrubbing, fermentation, pounding, and boiling (Bolarinwa *et al.* 2016). Apart from human consumption, cassava roots and peels are also consumed by animals (Kwok 2008).

In cassava, the major cyanogenic glycoside is linamarin, while a small amount of lotaustralin (methyl linamarin) is also present, as well as an enzyme linamarase which catalyzes linamarin and then rapidly hydrolyzes it to glucose and acetone cyanohydrin, and the lotaustralin is hydrolyzed to a related cyanohydrin and glucose (Bolarinwa *et al.* 2016). Under neutral conditions, acetone cyanohydrin decomposes to acetone and hydrogen cyanide (Bolarinwa *et al.* 2016). The reaction for linamarin is shown in Figure 2.2.

There are several varieties of cassava, each of which has a different cyanide level. Cassava roots have been reported in the literature to contain values ranging from 15–400 mg/kg fresh weight of hydrogen cyanide (FSANZ 2005). FSANZ (2004) reports

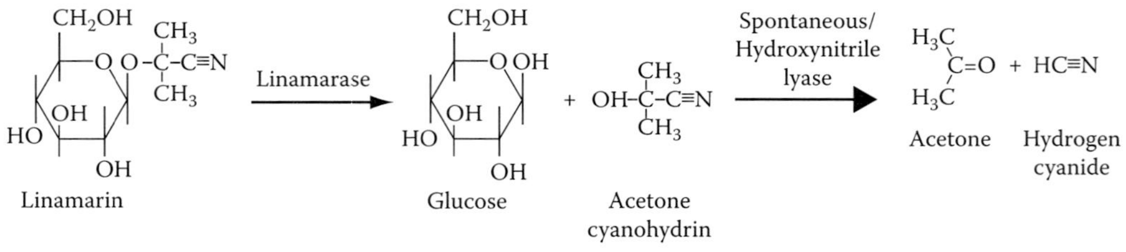

FIGURE 2.2 The reaction of linamarin with linamarase in cassava (Bolarinwa *et al.* 2016).

that the sweet varieties of cassava that contain low cyanide content will typically contain approximately 15–50 mg/kg hydrogen cyanide on a fresh-weight basis. These sweet varieties of cassava can be processed adequately by peeling and cooking (e.g., boiling, roasting, or baking), whereas bitter varieties of cassava (which contain high cyanide content) require more extensive processing, involving techniques such as heap fermentation which take several days (FSANZ 2004). Even though processing methods can reduce linamarin and cyanide levels in food, improperly processed cassava products will contain some amount of residual linamarin and hydrogen cyanide which can result in the potential toxicity of the cassava products, thus, it is recommended to increase the production of cassava breeds with low cyanogen content which will improve the healthy livelihoods of cassava producers, processors, and consumers (Ganjewala *et al.* 2010; Bolarinwa *et al.* 2016).

2.2.1.2 Bamboo Shoots

Bamboo shoots grow in a variety of climates and approximately 1,200 species of bamboo are available, even though only a small number are used as food (FSANZ 2005). Fresh, immature bamboo shoots are consumed as vegetables in some Asian countries as these shoots contain appreciable quantities of vitamin C, carbohydrates, and protein (Bhargava 1996). Besides their nutritive value, bamboo shoots also contain lethal concentrations of cyanogenic glycosides.

In bamboo, the cyanogenic glycoside present in the shoot is taxiphyllin, which is decomposed quickly in boiling water (Bolarinwa *et al.* 2016). This taxiphyllin, which is a p-hydroxylated mandelonitrile tiglochinin, is hydrolyzed to glucose and hydroxybenzaldehyde cyanohydrin. Benzaldehyde cyanohydrin then decomposes to hydroxybenzaldehyde and hydrogen cyanide (FSANZ 2005).

The cyanide content of bamboo shoots has been reported to range from 1,000 to 8,000 mg/kg hydrogen cyanide (Ferreira 1995). Bamboo shoots may contain significantly higher levels of hydrogen cyanide (as much as 1,000 mg/kg hydrogen cyanide) than cassava tubers, however, the cyanide content is reported to decrease substantially following harvesting and processing (FSANZ 2004; FSANZ 2005). The bamboo shoots sold commercially as food can be processed adequately (for instance by boiling and canning) before consumption and this process of canning bamboo shoots liberates and adequately removes hydrogen cyanide (FSANZ 2005).

2.2.1.3 Sorghum

Sorghum (*Sorghum vulgare and Sorghum bicolor*) is a major cereal in the semi-arid regions of the world grown for food and feed crop (Kulamarva *et al.* 2009). The United States of America is a major producer of sorghum mainly as animal fodder with only a small fraction being consumed as human food, but, in the semi-arid tropics of Africa and India, almost all the sorghum produced is used directly as human food as it forms the staple diet for large populations (Kulamarva *et al.* 2009). Sorghum is a gluten-free cereal that is reported to be of significance to people who are gluten intolerant as the grains of sorghum contain phenolic compounds like flavonoids that have been found to inhibit tumour development (Huang and Ferraro 1992; Shahidi and Naczk 1995). The sugars and starches in sorghum are reported to be released more slowly than in other cereals and can therefore be beneficial to diabetics (Toomey 1988; Klopfenstein and Hoseney 1995; Ratnavathi and Tonapi 2020). Sorghum sprouts are used to produce beer by fermentation and this beer is widely consumed in various African countries such as Benin, Cameroon, Ghana, Nigeria, and Togo (Tokpohozin *et al.* 2016).

In sorghum, the major cyanogenic glucoside is dhurrin which represents 30% of the dry weight of shoot tips of seedlings, in addition to amygdalin (Bolarinwa *et al.* 2016). Dhurrin in the young sorghum leaves is localized in the vacuoles of the plant and the corresponding enzymes for its hydrolysis to hydrogen cyanide are located in the cytoplasm while in intact leaves, the tissues are free from cyanide due to compartmental separation of the enzyme and the substrate (Bolarinwa *et al.* 2016). Immature sorghum leaves, therefore, contain high concentrations of dhurrin and the concentration decreases as the plant ages (Poulton 1990). This is also the case of the cyanide content of sorghum as it is also high during the early growth stage and declines as the plant ages (Busk and Møller 2002). Sorghum grains have been reported to contain 122.31 mg/g of amygdalin (Bolarinwa *et al.* 2016).

2.2.1.4 Lima Beans

Lima beans (*Phaseolus lunatus* L.) constitute one of the most widely cultivated pulse crops in temperate and subtropical regions (Adeparusi 2001). It contains antinutrients such as inhibitors of trypsin and amylase but some cultivars have high levels of CNGs, in particular linamarin (Brimer 2010). Methods of reducing CNGs in lima beans include soaking which causes only a moderate decline of cyanogens (30% reduction after 9 hours) and by more effective methods such as autoclaving and by toasting, which leads to a non-detectable level after 20 minutes with both methods (Schrenk *et al.* 2019).

2.2.1.5 Linseed

Linseed (*Linum usitatissimum*), also called flaxseed, has been cultivated for more than 8,000 years in Europe and Asia for its fiber and oil, and more lately for its beneficial micronutrients, in particular highly unsaturated fatty acids and hormonally active lignans (Schrenk *et al.* 2019). Linseed also contains considerable amounts of cyanogenic glycosides, primarily linustatin and neolinustatin together with small amounts of linamarin (Brimer 2010). Conventional methods, such as roasting, boiling, autoclaving, microwave, extrusion, and solvent extraction have been used in reducing the cyanogens but these methods have been reported to have the disadvantage of incomplete degradation of CNGs and partial removal of beneficial constituents (Feng *et al.* 2003; Barthlet and Bacala 2010; Schrenk *et al.* 2019). Yamashita *et al.* (2007) developed a more effective method that lowered the residual cyanide content below the detection limit without affecting the protein, fat, fiber, and lignan content of the linseed.

2.2.1.6 Apples

Apples (*Malus domestica*) are the most cherished fruit among the Rosaceae family. Apples can be consumed raw or processed into an alcoholic (cider) or nonalcoholic beverages (apple juice) or applesauce. Apples are rich sources of vitamins and other nutrients; however, the seeds contain high levels of cyanogenic glycosides (Bolarinwa *et al.* 2016). Even though humans do not consume apple seeds, apple juice is generally produced from whole apples including the seeds thus, the apple seeds disintegrate during juice production and contaminate the juice (Bolarinwa *et al.* 2015). Amygdalin content of apple seeds ranged from 1 to 4 mg/g while that of apple juice was reported to be between 0.001 and 0.08 mg/ml (Bolarinwa *et al.* 2014; Bolarinwa *et al.* 2015).

2.2.1.7 Apricot Fruits

Apricot fruits (*Prunus armeniaca*) are cultivated in middle Asia, Africa, America, and Europe. The fruits are eaten raw or used in dry form for confectioneries while the kernels are mostly processed before use (Bolarinwa *et al.* 2016). There are two varieties of apricot kernels; bitter and sweet. The bitter apricot kernels contain high amounts of the cyanogenic glycosides, amygdalin, which has the potential to cause cyanide toxicity problems at high doses and thus, unsafe for consumption (Bolarinwa *et al.* 2016). The sweet apricot kernels, as well as apricot flesh, are safe for human consumption because of their low level of cyanogens (Canadian Food Inspection Agency 2009; Bolarinwa *et al.* 2015).

The concentration of hydrogen cyanide in apricot kernels varies widely (49–4,000 mg/kg), depending on whether the skin on or off varieties is surveyed (Bolarinwa *et al.* 2016). The cyanide levels of raw or inadequately processed apricot kernels can cause serious acute problems that could lead to death (Haque and Bradbury 2002; Codex 2008). Studies have shown that apricot kernels contain a cyanide (CN) content of 1,450 mg/kg, approximately 0.5 mg CN/kernel (Holzbecher 1984). This value is similar to the toxic dose of cyanide (0.5 mg/kg body weight) reported by the World Health Organization in 1993 (FAO/WHO 1993). The Panel on Contaminants in the Food Chain (CONTAM) has established an acute reference dose (ARfD) for cyanide of 0.02 mg/kg bw (20 µg/kg bw) for use in assessing the risks associated with the presence of cyanogenic glycosides in apricot kernels (Schrenk *et al.* 2019).

2.3 BIOSYNTHESIS OF CYANOGENIC GLYCOSIDES

The production of hydrogen cyanide depends on both the biosynthesis of cyanogenic glycosides and on the existence or absence of its degrading enzymes (Folashade 2013). The biosynthetic precursors of the cyanogenic glycosides (as already stated in Section 1.3). Cyanogenic glycosides as secondary metabolites are different L-amino acids (five amino acids; valine, isoleucine, leucine, phenylalanine, and tyrosine, and from the non-proteinogenic amino acid, cyclopentenyl glycine (Vetter 2000; Ganjewala *et al.* 2010).

For example, amygdalin and prunasin are derived from phenylalanine; linamarin and lotaustralin are derived from valine, isoleucine, and leucine; and dhurrin is derived from tyrosine (Seigler 1975; Conn 1979; Bolarinwa *et al.* 2013). These different L-amino acids of the cyanogenic glycosides used as biosynthetic precursors are hydroxylated, then, the N-hydroxyl amino acids are converted to aldoximes and these are converted into nitriles, which are hydroxylated to α-hydroxynitriles and then glycosylated to cyanogenic glycosides (Vetter 2000).

The generation of HCN from CNGs is a two-step process involving deglycosylation, and cleavage of the molecule (regulated by β-glucosidase and α-hydroxynitrile lyase) (Vetter 2000). CNGs are degraded to cyanide by these β-glucosidase and α-hydroxynitrile lyase, two families of enzymes stored separately from the CNGs in plant cells (Schrenk *et al.* 2019). Gleadow and Møller (2014) report cyanogenic glycosides to be typically confined to the vacuoles whereas β-glucosidases may be localized in the apoplastic space, bound to the cell wall, in the cytoplasm, in small vesicles, or the chloroplast, depending on the plant species. The location of the α-hydroxynitrile lyase is less well known but appears to be cytoplasmic (Schrenk *et al.* 2019). The tissue level compartmentalization of CNGs and their hydrolyzing enzymes prevents large-scale hydrolysis in intact plant tissue (Vetter 2000).

The biosynthetic pathway of cyanogenic glycosides is illustrated in Figure 2.3. The first and the second steps of cyanogenic glycoside biosynthesis are catalyzed by a cytochrome P450 (CYP) through two successive N-hydroxylations of the amino group of the parent amino acid, followed by decarboxylation and dehydration; α-hydroxynitrile is

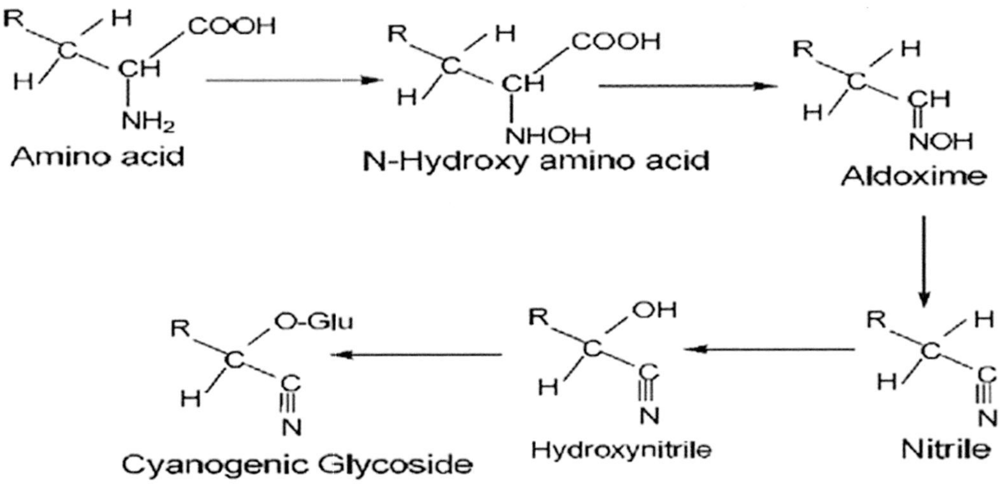

FIGURE 2.3 The general biosynthetic pathway for cyanogenic glycosides from its precursor amino acid.

produced from the action of a second cytochrome P450 on the aldoxime (Kahn *et al.* 1997; Bak *et al.* 1998). Hydroxynitrile, which is also known as cyanohydrin, is generated following an initial dehydration reaction to form a nitrile through the hydroxylation of the alpha carbon and glycosylation of the cyanohydrin moiety (Bolarinwa *et al.* 2013). The final step in cyanogenic glycoside synthesis is catalyzed by UDPG-glycosyltransferase (Jones *et al.* 1999).

2.3.1 Enzymatic Hydrolysis of Cyanogenic Glycosides in Food Plants

Cyanogenesis mostly occurs in plants via the hydrolysis of cyanogenic glycosides that are enzymatically catalyzed (Castada *et al.* 2020). The enzyme activity results in cleavage of the carbohydrate moiety of the cyanogenic glycoside to yield corresponding cyanohydrins which further decompose to release hydrogen cyanide and an aldehyde or ketone (Poulton 1990).

For example, amygdalin (D-mandelonitrile-β-D-gentiobioside) is one of the most common cyanogenic glycosides mostly present in kernels and seeds of fruit (Donald 2009). Enzymatic degradation of amygdalin is divided into three parts: the first part involves the splitting of amygdalin to prunasin and glucose by the enzyme amygdalin lyase; the second part involves the hydrolysis of prunasin to mandelonitrile and glucose by the enzyme prunasin lyase; and the final stage of the hydrolysis is the breaking down of mandelonitrile to benzaldehyde and hydrogen cyanide (HCN) by hydroxynitrile lyase (Haisman and Knight 1967; Bolarinwa *et al.* 2014). Amygdalin and prunasin (Dmandelonitrile-β-D-glucopyranoside) are degraded by endogenous enzymes (β-glucosidases) to form non-glycosidic cyanogen molecules; mandelonitrile and free (CN^-) cyanide ion (Tuncel et al. 1998). Enzymatic hydrolysis of amygdalin is illustrated in Figure 2.4.

Linamarin (α-hydroxybutyronitrile-β-Dglucopyranoside) and lotaustralin (ethyl linamarin) are cyanogenic glycosides present in cassava and these are located in the plant vacuoles while the enzyme is located in the cell wall (Gruhnert *et al.* 1994). Hydrolysis of this linamarin starts from the disintegration of the root tissue during processing or

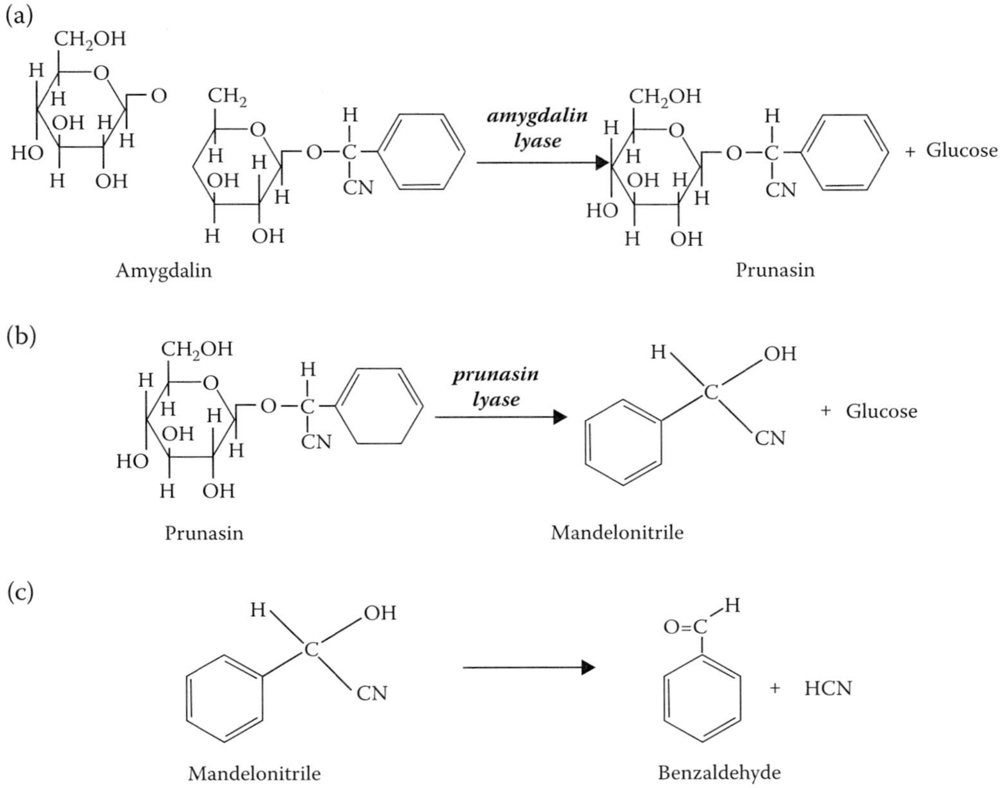

Cyanogenic Glycosides as Food Toxins

(a)

(b)

(c)

FIGURE 2.4 Enzymatic hydrolysis of amygdalin and prunasin to form hydrogen cyanide: (a) splitting of amygdalin to prunasin and glucose by the enzyme amygdalin lyase, (b) the breaking down of prunasin to mandelonitrile and glucose by the enzyme prunasin lyase, and (c) the hydrolysis of mandelonitrile to benzaldehyde and hydrogen cyanide.

chewing to release the endogenous enzyme (*linamarase*) that hydrolyses linamarin to glucose and acetone cyanohydrins. During processing, reduced moisture and increased temperature facilitates the spontaneous breakdown of cyanohydrins to toxic hydrogen cyanide (McMahon 1995).

Dhurrin (4-hydroxymandelonitrile-β-Dglucopyranoside) is the cyanogenic glycoside present in sorghum. It is stored in the plant vacuoles separated from the enzymes responsible for their hydrolysis. When the plant tissue is crushed, the enzymes and dhurrin are brought in contact, and hydrolysis is initiated by β-glucosidase (*dhurrinase*), which hydrolyzes the cyanogenic glycoside to form hydroxymandelonitrile and glucose. Hydroxymandelonitrile further decomposes spontaneously or enzymatically by the action of hydroxynitrile lyase to form hydrogen cyanide and hydroxybenzaldehyde (Wajant et al. 1994). Enzymatic hydrolysis of linamarin and dhurrin is illustrated in Figure 2.5.

2.3.2 Toxicity of Cyanogenic Glycosides (Cyanide Toxicity)

Cyanide produced from cyanogenic glycosides is a highly fatal and rapidly acting poison (Folashade 2013) and as such, much attention must be given to it as consumption of

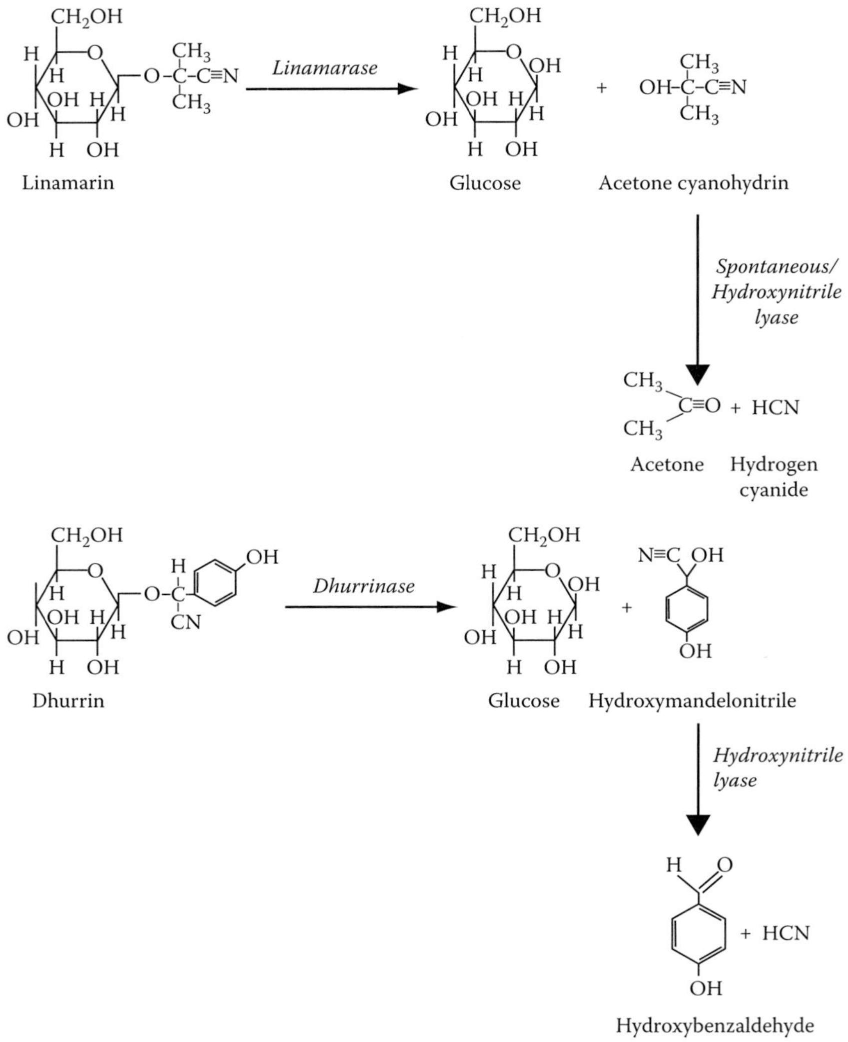

FIGURE 2.5 Enzymatic hydrolysis of linamarin and dhurrin to form hydrogen cyanide.

cyanogenic plants can cause serious health problems for humans. However, the ability of a cyanogenic plant to become toxic predominantly depends on the capability that its consumption will produce a concentration of hydrogen cyanide (HCN) that is toxic to the exposed humans or animals (Speijers 1993; Taylor 2012; Folashade 2013). The factors that are important in this toxicity are said to be: (a) The plant may not be adequately detoxified during preparation or processing and, therefore, hydrogen cyanide may remain in the food; (b) If the plant is consumed raw or inadequately processed, hydrogen cyanide may be released in the body, until the low pH of the stomach deactivates the β-glucosidase enzyme (FSANZ 2005). The enzymatic reactions involved in hydrogen cyanide production are explained in section 3.0, Biosynthesis of cyanogenic glycosides.

Taylor (2012) has reported the potential for toxicity of cyanogenic plants as mg of releasable cyanide per kg of the food commodity. That notwithstanding, factors such as bioavailability, susceptibility, and nutritional status of the individual, and preparations

Cyanogenic Glycosides as Food Toxins 37

and processing of the food plant will determine the final amount of the cyanide toxin presented to the consumer (Taylor 2012).

Cyanide toxicity can occur in humans after the consumption of cyanogenic glycosides in plants at doses between 0.5–3.5 mg kg-1 body weight (0.5–3.5 mg/kg bw) (Speijers 1993; Bolarinwa and Orfila 2017). The European Food Safety Authority (EFSA) Panel on Contaminants in the Food Chain (CONTAM) reported that there are no data indicating that the acute reference dose (ARfD) for cyanide of 0.02 mg/kg bw (20 µg/kg bw) established in 2016 (EFSA Panel CONTAM 2016) should be revised, and that this dose is applicable for acute effects of cyanide regardless of the dietary source (Schrenk *et al.* 2019). The lethal dose of orally ingested hydrogen cyanide for a 60 kg adult man ranges from 30–210 mg equivalent HCN (Nhassico *et al.* 2008). The toxic threshold value for cyanide in the blood is examined to be between 0.5 mg/L (ca. 20 µM) and 1.0 mg/L (ca. 40 µM), and the lethal threshold value spans between 2.5 mg/L (ca. 100 µM) and 3.0 mg/L (ca. 120 µM) (Schrenk *et al.* 2019).

In animals, the lethal doses of hydrogen cyanide are generally reported to be between 0.66 and 15 mg/kg body weight (bw) for various species (Schrenk *et al.* 2019). Chronic sub-lethal dietary cyanide is reported to cause some reproductive effects such as lower birth rates, increased number of neonatal deaths, impaired thyroid function, and behavioral effects that include increasing ambivalence and slower response time (Chandra 2004; FSANZ 2004).

2.3.3 Mechanism of How Cyanogenic Glycosides Affects the Body

When cyanogenic plants are ingested by humans, enzymes produced by the intestinal microflora can hydrolyze intact cyanogenic glycoside to produce hydrogen cyanide in vivo (Taylor 2012) and the mechanism by which the hydrogen cyanide produced affects humans or animals is as explained below.

The toxicity of cyanide is largely attributed to the cessation of aerobic cell metabolism (Hall 2007; Folashade 2013). Hydrogen cyanide exposure to humans or animals either through the skin, eye, or oral ingestion or administration is readily absorbed and rapidly distributed in the body through the blood (FSANZ 2005; Folashade 2013).

In the blood, cyanide combines with iron (Fe^{3+}/Fe^{2+}) present in methaemoglobin and haemoglobin in the erythrocytes causing intracellular hypoxia (deficiency in the amount of oxygen reaching the tissues) by reversibly binding to the cytochrome oxidase a3 (important in the reduction of oxygen to water in oxidative phosphorylation) within the mitochondria (Hall 2007; Folashade 2013). The binding of cyanide to ferric ion in cytochrome oxidase a3 inhibits the terminal enzyme in the respiratory chain and this stalls electron transport and oxidative phosphorylation that is important in the synthesizing of adenosine triphosphate (ATP) and the continuation of cellular respiration (Beasley and Glass 1998; Hamel 2011; Folashade 2013). This results in cytotoxic hypoxia, a shift from aerobic metabolism to anaerobic metabolism (because of a decrease in the ATP/ADP ratio), and an increase in the levels of glucose and lactic acid in the blood (Folashade 2013).

Cyanide also inhibits the tricarboxylic acid cycle by decreasing the rate of glycolysis and activating glycogenolysis, thus, causing a reduction in the energy available for the respiratory system, the heart, and other cells (Speijers 1993). Although, the toxicity of cyanide can develop over minutes or hours after exposure, being exposed to high concentrations of cyanide can cause instant death (Folashade 2013). The cyanide level in different human tissues in a fatal case of hydrogen cyanide poisoning has been reported as gastric content, 0.03; blood, 0.5; liver, 0.03; kidney, 0.11; brain, 0.07; and urine, 0.2 (mg/100 g) (Lang 1933; FSANZ 2004).

Cyanide can also inhibit other several metalloenzymes most of which contain iron, copper, or molybdenum as well as enzymes containing Schiff base intermediates (e.g. 2-keto-4-hydroxyglutarate aldolase) (Speijers 1993; Preedy and Watson 2020). Hydrogen cyanide can reduce the energy availability in all cells with its effects being immediate on the respiratory system and the heart (FSANZ 2004; FSANZ 2005).

2.3.4 Mechanism of Cyanide Detoxification in Human

To counter the toxic effects of cyanide in the body, the major defense of the body is to convert the cyanide to thiocyanate mediated by the enzyme rhodanese (Lang 1933; Speijers 1993; FSANZ 2004) and with the help of sulphur donors, such as sulphur-containing amino acids (cysteine and methionine) or their products of metabolism (Bolarinwa *et al.* 2013; Folashade 2013). In the presence of a sulphur donor (e.g., thiosulfate) and a sulphur transferase (e.g., rhodanese), about 70% of a dose of cyanide is metabolized to thiocyanate, and in contrast to cyanide, thiocyanate does not block the electron transport in the mitochondrial respiratory chain (Schrenk *et al.* 2019). Thiocyanate is excreted in the urine (Banea 2012).

The enzyme rhodanese, a sulphur transferase present in the liver and localized in the mitochondria, is relatively abundant, but in sites, which are not readily accessible to thiosulphate, thus, thiosulphate is the limiting factor for the conversion of cyanide (Bradbury and Holloway 1988; EPA 1990; Rosling 1994; FSANZ 2004; EFSA 2004). The enzyme rhodanese contains an active disulfide group that reacts with the thiosulphate and cyanide to form thiocyanic acid (Schrenk *et al.* 2019).

The rate of detoxification of cyanide in healthy humans as reported by Schulz *et al.* (1982) and Schrenk *et al.* (2019) is only about 1 µg/kg bw per min, which corresponds to about 4.2 mg CN per h in a 70 kg person. Note that when hydrogen cyanide is converted to thiocyanic acid, there is a 200-fold reduction in toxicity. However, the detoxification mechanism in the body will only cope with minimal amounts of cyanide generated from the consumption of small amounts of cyanogenic plants, but not with a toxic amount or a large dose of cyanide introduced artificially to the body (Folashade 2013). High consumption of cyanogenic plants and their products has been reported to result in thiamine deficiency and the depletion of essential amino acids especially when the diet is short of protein and thiamine (Rosling 1994; Padmaja 1996; Ngudi *et al.* 2002; Adamolekun 2010b; Bolarinwa *et al.* 2013).

Another important independent pathway for cyanide detoxification is cyanocobalamin (vitamin B12). Hydroxocobalamin (vitamin B12a) detoxifies cyanide by binding with it to form the renally excreted, non-toxic cyanocobalamin or vitamin B12. The cyanocobalamin is then excreted in the urine and bile (it may be reabsorbed by the intrinsic factor mechanism in the ileum allowing effective recirculation of vitamin B12). Because hydroxocobalamin binds without forming methemoglobin, it can be used to treat patients without compromising the oxygen-carrying capacity of haemoglobin (FSANZ 2005; Hamel 2011).

A further detoxification pathway is a reaction of L-cystine with cyanide through the putative intermediate β-thiocyanoalanine to 2-amino-2-thiazoline-4-carboxylic acid (ATCA), where this pathway accounts for about 15–20% of cyanide metabolism

(Mitchell *et al.* 2013; Schrenk *et al.* 2019). Thiocyanate and ATCA are chemically stable metabolites that are not further metabolized but excreted with the urine (Schrenk *et al.* 2019) or excreted in the saliva and urine (FSANZ 2005).

In another detoxification pathway, α-ketoglutarate reacts with cyanide to form α-ketoglutarate cyanohydrin (α-KGCN) and this pathway is assumed to become important when the thiocyanate and ATCA pathways are overwhelmed (Schrenk *et al.* 2019).

Cyanide detoxification also occurs to a lesser extent by mercaptopyruvate (cyanide sulphurtransferase) (Oke 1973). The major substrates for this conversion are thiosulfate and 3-mercaptopyruvate, and the 3-mercaptopyruvate can arise from cysteine via transamination or deamination and this compound can provide sulphur as rapidly as thiosulfate for cyanide detoxification (Oke 1973; Folashade 2013).

Furthermore, cyanide detoxification occurs when methaemoglobin effectively competes with cytochrome oxidase for cyanide, and its formation from hemoglobin affected by sodium nitrile or amyl nitrite is exploited in the treatment of cyanide intoxication (FSANZ 2005).

2.3.5 Human Health Effect of Cyanogenic Glycosides

The toxicological effects of plant toxins after their consumption may range from acute effects of gastroenteritis to more severe toxicities in the central nervous system, leading to death, as seen in cases of poisoning due to cyanide (Taylor 2012). The signs and symptoms of cyanide poisoning reflect the extent of cellular hypoxia and occur when the absorption rate of cyanide exceeds its metabolic detoxification (Schrenk *et al.* 2019).

Consumption of cyanogenic plants such as cassava root, almonds, peaches, plums, apple seeds, apricot or apricot kernels, lima beans, broad beans, or members of the genus sorghum has been reported to cause both acute and sub-acute health problems (depending on dose) such as headache, nausea, vomiting, abdominal pain, dizziness, weakness, diarrhea, anxiety, confusion, decreased consciousness, hypotension (drop in blood pressure), rapid pulse, paralysis, mental confusion, convulsions, cardiac arrest, circulatory and respiratory failure, coma, and in extreme cases death (FAO/WHO 1993; Jones 1998; Francisco and Pinotti 2000; Vetter 2000; Taylor 2012).

When the cyanide level exceeds the limit an individual is able to detoxify, death due to cyanide poisoning can occur (Kwok 2008). The likelihood of cyanide intoxication from the consumption of cyanide-containing food is dependent on body weight. Children are particularly at risk because of their smaller body size (Kwok 2008). It is, therefore, possible that an older person or a child of smaller bodyweight would not be able to detoxify the cyanide resulting from a meal that is inadequately prepared from for example cassava, bamboo shoots, and/or sorghum (Taylor 2012). Approximately 50–60 mg of free cyanide constitutes a lethal dose for an adult man (FAO/WHO 1993; FSANZ 2005).

The onset of signs and symptoms of cyanide poisoning is usually less than 1 minute after inhalation and within a few minutes after ingestion (Hamel 2011). The major determinants of severity and mortality are the source of exposure, the route and the magnitude of exposure, and the effects and the time taken for any treatments that may have been tried (Yen *et al.* 1995; Bolarinwa *et al.* 2013).

Cases of cyanide poison have been recorded in several countries. To know the current situation and cases associated with cyanide, it is recommended to visit the website of the EFSA. Table 2.1 shows some parts of the world where cyanide-induced diseases or cases have been observed.

TABLE 2.1 Cyanide-Induced Diseases Observed in Particular Countries

Country	Food	Disorder
Jamaica	Cassava, lima beans	Peripheral neuritis
Nigeria	Cassava	Goitre, tropical ataxic neuropathy, cyanide poisoning
Mozambique	Cassava	Severe paralyzing illness (konzo)
Zaire	Cassava	Goitre
Turkey	Apricot kernels	Cyanide poisoning
Israel	Apricot kernels	Cyanide poisoning
France	Almond seeds	Cyanide poisoning

2.3.6 Treatment of Cyanide Poisoning

Cyanide poisoning is treatable when quickly recognized and immediately countered with an antidote, and this treatment is based on supportive care with adjunctive antidotal therapy (Schrenk *et al.* 2019). Multiple antidotes exist and are characterized by different antidotal mechanisms such as chelation, formation of stable, less toxic complexes, methemoglobin induction, and sulphur supplementation for detoxification by endogenous rhodanese (Fincham 1990; Borron and Baud 2012).

An ideal cyanide antidote would act rapidly and effectively with little to no treatment-limiting adverse effect (Hamel 2011). Treatment of cyanide poisoning involves the use of a cyanide antidote kit (amyl nitrite + sodium nitrite + thiosulfate) or hydroxocobalamin (Borron et al. 2006). The mechanism of how hydroxocobalamin works is explained under section 3.4: mechanism of cyanide detoxification in humans. A cyanide antidote kit has been reported to provide effective treatment when the antidote kit is combined with supportive treatment such as a 100% O_2 ventilator, vasopressors, and sodium bicarbonate (Bolarinwa *et al.* 2013). The antidotal effect is enhanced by thiosulphate. Hydroxocobalamin antidote is reported to be safer in patients who have pre-existing hypotension or are pregnant (Beasley and Glass 1998; Bolarinwa *et al.* 2013).

2.4 PREVENTING THE EFFECTS OF CYANOGENIC GLYCOSIDES (CYANIDE-INDUCED DISEASES)

Cyanide-induced diseases can be prevented effectively by the removal of cyanogenic compounds in food plants by processing them to safe levels before consumption. The World Health Organization (FAO 1991) for instance, concluded that a level of up to 10 mg/kg HCN in cassava flour is not associated with acute toxicity (Folashade 2013).

Various processing methods are therefore employed to reduce the cyanide content in cyanogenic plants to safe levels. Some of these processing methods are steaming, boiling, roasting, baking, and others also such as peeling, soaking, grating, fermentation, and sun-drying (Kemdirim *et al.* 1995; Tuncel *et al.* 1995; Obilie *et al.* 2004; Cardoso *et al.* 2005).

2.4.1 Effect of Processing on Cyanogenic Glycosides

The major aim of processing crops from cyanogenic plants is to decrease their potential for releasing cyanide upon ingestion as these food items may pose a health risk for consumers if the levels of CNGs are high (Schrenk *et al.* 2019). Several studies have been reported that processing methods such as peeling, drying, grinding, boiling or cooking, soaking, and fermentation caused a significant reduction in the cyanogenic glycosides of processed foods such as roots, tubers, cereals, and leaves (Bolarinwa *et al.* 2016).

These food-processing methods generally utilize the water solubility and degradability of CNGs by endogenous plant enzymes (Schrenk *et al.* 2019) to disintegrate the cyanogens which leads to the production of hydrogen cyanide, and since hydrogen cyanide is volatile, further processing techniques (e.g., boiling, roasting, and drying) will volatilize the remaining hydrogen cyanide to lower levels (Bolarinwa *et al.* 2016; Ndubuisi and Chidiebere 2018). The boiling point of hydrogen cyanide is 26°C, which easily volatilizes during food processing (Montagnac *et al.* 2009; Bolarinwa *et al.* 2015).

Mechanical destruction of the plant cells is mostly achieved by peeling, chopping/slicing, grating, or pounding the raw crops, followed by soaking in water to solubilize the cyanogenic glycosides for extraction and enzymatic degradation (Schrenk *et al.* 2019). The endogenous enzymes that bring about the degradation of CNGs are produced by microorganisms associated with the raw crop or added intentionally during the process of fermentation (Schrenk *et al.* 2019). Occasionally, enzymes such as pectinases are added for the destruction of plant cells. Sun-drying or oven-drying is regularly used to help evaporate the released cyanide as volatile hydrocyanic acid (Schrenk *et al.* 2019).

For example, cassava and bamboo shoots that grow primarily in the tropics contain cyanogenic glycosides, linamarin and taxiphillin, respectively, which break down upon disruption of the plant cells to form hydrogen cyanide (FSANZ 2005). The levels of the hydrogen cyanide produced can be reduced by appropriate preparation of the plant material before consumption as food (Barton 1999; FSANZ 2004). For cassava, peeling and cutting/chopping will disrupt the cell structure of the plant with subsequent liberation of hydrogen cyanide through further processing such as boiling, baking, roasting, or fermentation (Sofyan *et al.* 2018). For bamboo shoots, cutting/chopping into thin strips will liberate hydrogen cyanide, which is removed by boiling (FSANZ 2004).

Some of these processing methods, however, do not reduce the hydrogen cyanide content to acceptable values, thus, care must be taken in the selection of which processing methods to use for in curtailing hydrogen cyanide present in foods. For example, in eastern and southern Africa, cassava is processed into flour by sun-drying the peeled root followed by pounding and sieving or heap fermentation, and this process does not allow enough contact between linamarase and linamarin, thus, making the end product contain up to 59 ppm of HCN equivalents, compared to the WHO safe level of 10 ppm (Montagnac *et al.* 2009).

Several documents regarding the definitions of cyanogenic foods for instance on cassava food commodities and measures to reduce hazards by cassava consumption have been issued by the Codex Alimentarius Commission (Codex) (Schrenk *et al.* 2019). The code of practice for the reduction of hydrocyanic acid (HCN) in cassava and cassava products (CAC/RCP 73–2013) gives guidance on how to produce cassava products with safe concentrations of cyanogenic compounds and advice in support of the reduction of HCN in cassava and lowering uptake of cassava (Schrenk *et al.* 2019).

2.4.1.1 Effect of Soaking and Fermentation on Cyanogenic Glycosides

In western Africa and southern America, cassava parenchyma is grated or crushed into small pieces to disrupt many plant cells and allow good contact between linamarin and linamarase (Mohd Azmi 2019; Nyirenda 2020). The moist mash is then left to ferment for several days in sacks with weights placed on them, making the water-soluble cyanogens squeezed out and the residual HCN gas is removed by roasting. This process reduced the cyanogen content of the product (Gari or Farinha) significantly (Montagnac *et al.* 2009).

Soaking of cassava root has been reported to decrease its total cyanide content by 13–52% after 24 hours, 73–75% after 48 hours, and 90% after 72 hours (Kemdirim *et al.* 1995). Soaking and fermentation of bitter apricot kernels decreased cyanogen levels by about 70% (Tuncel *et al.* 1998). Fermentation of cassava pulp or dough for 4–5 days has been reported to decrease its total cyanide by 52–63% (Kemdirim *et al.* 1995; Obilie *et al.* 2004). The cyanide content of cocoyam flour produced from fermented cocoyam was reported to reduce by 98.6% (Igbadul *et al.* 2014). Prasad and Dhanya (2011) reported an 84.6% reduction in the cyanide content of fermented sorghum leaves.

2.4.1.2 Effect of Drying on Cyanogenic Glycosides

Drying conditions of a plant/food crop such as its moisture content and the rate of moisture loss, together with the extent of tissue disruption of the plant tissue are the main factors that determine the efficiency of cyanide removal (Tivana 2007). When the period of drying a product (e.g., cassava) with higher moisture levels is extended, linamarin breakdown would be enhanced thus, explaining the fact that fast drying rates result in lower detoxification, while slower rates result in higher cyanogen removal (Bolarinwa *et al.* 2016). In 2005, a simple wetting method was developed that reduced the total cyanide content of cassava flour 3–6-fold. The method involved spreading wet flour in a thin layer and standing in the shade for 5 hours to allow the evolution of hydrogen cyanide gas (Bradbury and Denton 2010).

2.4.1.3 Effect of Cooking on Cyanogenic Glycosides

Cyanogenic glycosides are generally water-soluble thus, during cooking, significant amounts of cyanogens are leached into cooking water (Bolarinwa *et al.* 2016). However, care must be taken when the cooking method being used is heating at low moisture content or heating under dry heat, because the removal of the cyanogen is limited to only small amounts (Bolarinwa *et al.* 2016). Several studies have reported an increased reduction of cyanide in cooked products. A 74–80% reduction in total cyanide levels resulted from the steaming process of a cassava product known as Akyeke. (Obilie *et al.* 2004). The process of making another cassava product called Gari resulted in a 90–93% reduction in total cyanide content (Agbor-Egbe and Mbome 2006). A 97% reduction in cyanide levels was reported to result from the optimal cooking conditions of bamboo shoots (98°C–102°C for 148–180 minutes) (Ferreira 1995; Okoye and Uwhen 2016).

2.5 DISEASES CAUSED BY CYANIDE

Consumption of improperly or insufficiently processed cyanogenic plants have been associated with some long-term toxicity of cyanide-induced diseases or disorders.

2.5.1 Iodine Deficiency Diseases (Goitre and Cretinism)

Iodine deficiency diseases are mostly goitre (enlargement of the thyroid gland), and, in its most severe form, cretinism (shortness of stature and severe mental impairment) (Ermans *et al.* 1983; FSANZ 2004). Goitre and cretinism are common diseases in developing countries due to low intake of iodine (<100 µg/day), and these diseases are particularly common in

Africa because of their over-dependence on cyanogenic plants such as cassava as a staple food (Bolarinwa *et al.* 2016). Ingested cyanide from cyanogenic plants is converted in the body to thiocyanate, and this thiocyanate competes with or inhibits the uptake of iodine by the thyroid gland (i.e., uptake of iodide into the thyroid follicle cells via the sodium iodide symporter) (FSANZ 2005; Gbadebo and Oyesanya 2005; Eisenbrand and Gelbke 2016). Populations with very low iodine intake and high thiocyanate levels from consumption of cassava showed severe endemic goiter, which decreased with iodine supplementation (Rosling 1994). Consumption of cyanogenic glycosides even at a very low concentration can also cause iodine deficiency leading to goiter (Odo *et al.* 2014).

2.5.2 Konzo

Konzo is an upper motor neuron disease of acute onset due to continuous large intake of cyanogenic glycosides from insufficiently processed bitter cassava or cassava flour in combination with a low intake of sulphur-containing amino acids, also resulting in thiamine deficiency from inactivation of thiamine when the sulphur in thiamine is utilized for detoxification of cyanide in the human body, thereby causing an irreversible paralysis of the lower limbs in children and women of child-bearing age (Tylleskar *et al.* 1992; Ernesto *et al.* 2002; Adamolekun 2010a; Schrenk *et al.* 2019).

2.5.3 Tropical Ataxic Neuropathy (TAN)

TAN is used to describe several neurological syndromes attributed to toxiconutritional causes (Selvan 2013; Ndubuisi 2018). Dietary exposure to cyanide from the monotonous consumption of inadequately processed cassava products over years results in chronic thiamine deficiency from inactivation of thiamine by the cyanogenic glycosides (Adamolekun 2010b). TAN has occurred mainly in Africa, particularly Nigeria, and is common among older people, 40 years and above. The disease is characterized by unsteady walking, sore tongue, loss of sensation in the hands and feet, blindness, deafness, and weakness (Oluwole *et al.* 2000; FSANZ 2004).

2.5.4 Growth Retardation

Growth retardation is a common health challenge especially among children in developing countries and the contributing factor to this health problem is exposure to cyanogenic glycosides. Growth retardation is particularly a serious problem in populations consuming foods with inadequate proteins especially diets that are low in sulphur-containing amino acids (methionine and cysteine) because detoxification of cyanide in the human body requires sulphur donors from sulphur-containing amino acids. Thus, dietary exposure to cyanide is a contributing factor to growth retardation (Rosling 1994; Banea-Mayambu 2000).

2.5.5 Cyanide Poisoning

Cyanide toxicity occurs when cytochrome oxidase a3 inhibits the terminal enzyme in the respiratory chain and halts electron transport and oxidative phosphorylation (important in adenosine triphosphate (ATP) synthesis and the continuation of cellular respiration) (Beasley and Glass 1998; Hamel 2011). Cyanide poisoning occurs as a result of the consumption of bitter cassava, almond kernels, or apricot kernels and their products without proper processing (Folashade 2013; Bolarinwa *et al.* 2016). Cases of cyanide poisoning after the consumption of drinks produced from the blends of apricot kernels and orange juice have been reported (Atkinson 2006).

2.6 QUANTIFICATION OF CYANOGENIC GLYCOSIDES

Analytical methods used in the quantification of cyanogenic glycosides in plants include either indirect methods (by determining the amount of hydrogen cyanide released after hydrolysis) or direct methods (by determining the intact form) (Bolarinwa *et al.* 2015). The indirect method mostly involves enzymatic hydrolysis followed by colorimetric determination of total cyanide (Bradbury *et al.* 1991; Santamour 1998; Bolarinwa *et al.* 2015). The direct method includes liquid chromatography with refractive index detection (Sornyotha *et al.* 2007), gas chromatography/mass spectrometry (Chassagne *et al.* 1996), and high-performance liquid chromatography with UV detection (HPLC-UV) (Bolarinwa *et al.* 2014).

HPLC-UV or with diode-array detection (HPLC-DAD) has been widely applied to quantify CNGs in food samples after extraction (Schrenk *et al.* 2019). More recently, solid-phase extraction along with liquid chromatography-tandem mass spectrometry (LC-MS/MS) analysis has been applied, improving both sensitivity and selectivity of the analyses (Schrenk *et al.* 2019). Besides liquid chromatography-based techniques, gas chromatography-mass spectrometry (GC-MS), as well as enzyme-linked immunosorbent assays (ELISAs), have been applied less frequently to quantify CNGs in food (EFSA CONTAM Panel 2016; Schrenk *et al.* 2019). No validated methods are available for the quantification of CNGs in food items (Schrenk *et al.* 2019).

2.7 CONCLUSION AND RECOMMENDATIONS

Cyanogenic glycosides are abundant in edible plants and are not toxic on their own. However, when cell structures of plants are disrupted, cyanogenic glycosides are brought together with the corresponding enzyme to produce hydrogen cyanide which has potential health risks. Consumption of improperly processed cyanogenic plants can lead to chronic and acute health problems.

That notwithstanding, cyanogenic glycosides food plants can be safely consumed if suitable measures are taken. The amount of foods containing cyanogenic glycosides ingested that are capable of causing food poisoning depends on many factors, such as individual susceptibility, the processing methods, and the amount of toxin in the plant which may vary according to the species and geographical environment.

Understanding the appropriate processing methods for specific cyanogenic plants will help in significantly reducing the potential health risk of cyanide toxicity. By adequate processing, both the cyanogenic glycosides and or the hydrogen cyanide produced from it can be removed or reduced before consumption.

It is recommended that cyanogenic plants should be cut into smaller pieces and cooked or processed thoroughly to release toxic hydrogen cyanide before consumption in order to curtail the level of the toxin. It is also recommended to limit intake of cyanogenic plants especially when the cooking method chosen is heating under dry heat or at low moisture contents. Seeds of fruits to be processed into juice should also be removed before crushing to avoid cyanide poisoning.

It is advised to take precautions in limiting the amount of intake and also observe safety measures for toxin reduction before consumption especially for children and the elderly. For individuals with illness or with poor health conditions, it is recommended they consult their doctors for further advice before intake.

REFERENCES

Adamolekun, B. (2010a). Etiology of konzo, epidemic spastic paraparesis associated with cyanogenic glycosides in cassava: Role of thiamine deficiency? *Journal of the Neurological Sciences*, 296(1–2), 30–33.

Adamolekun, B. (2010b). Thiamine deficiency and the etiology of tropical ataxic neuropathy. *International Health*, 2(1), 17–21.

Adeparusi, E. O. (2001). Effect of processing on the nutrients and anti-nutrients of lima bean (Phaseolus lunatus L.). *Nahrung/Food*, 45, 94–96.

Agbor-Egbe, T. and Mbome, I. L. (2006). The effects of processing techniques in reducing cyanogen levels during the production of some cameroonian cassava foods. *Journal of Food Composition and Analysis*, 19(4), 354–363.

Atkinson, K. (2006). Apricot kernels carry risk of cyanide poisoning. Available from: http://www.nzherald.co.nz/organisation/story.cfm?o_id=324&objectid=10379786

Bak, S., Kahn, R. A., Nielsen, H. L., Møller, B. L. and Halkier, B. A. (1998). Cloning of three A-type cytochromes P450, CYP71E1, CYP98, and CYP99 from sorghum bicolor (L.) Moench by a PCR approach and identification by expression in escherichia coli of CYP71E1 as a multifunctional cytochrome P450 in the biosynthesis of the cyanogenic glucoside dhurrin. *Plant Molecular Biology*, 36(3), 393–405.

Banea-Mayambu, J. P., Tylleskar, T., Tylleskar, K., Gebre-Medhin, M. and Rosling, H. (2000). Dietary cyanide from insufficiently processed cassava and growth retardation in children in the democratic republic of Congo (formerly zaire). *Ann Trop Pediatr*, 20, 34–40.

Banea, M. J. P., Nahimana, G., Mandombi, C., Bradbury, J. H., Denton, I. C. and Kuwa, N. (2012). Control of konzo in DRC using the wetting method on cassava flour. *Food Chem Toxicol*, 50, 1517–1523.

Barthlet, V. J. and Bacala, R. (2010). Development of optimized extraction methodology for cyanogenic glycosides from flaxseed (Linum usitatissimum). *Journal of AOAC International*, 93, 478–484.

Barton, D., Nakanishi, K. and Meth-Cohn, O. (1999). Polyketides and other secondary metabolites including fatty acids and their derivatives. Barton, D. H. R., Nakanishi, K., & Meth-Cohn, O. *Comprehensive natural products chemistry*, 1st ed. 1–1007.

Beasley, D. M. G. and Glass, W. I. (1998). Cyanide poisoning: Pathophysiology and treatment recommendations. *Occupational Medicine*, 48(7), 427–431.

Bhargava, A., Kumbhare, V., Srivastava, A. and Sahai, A. (1996). Bamboo parts and seeds for additional source nutrition. *J Food Sci Technol*, 32, 145–146.

Bolarinwa, I. F., Orfila, C. and Morgan, M. R. A. (2014). Amygdalin content of seeds kernels and food products commercially-available in the UK. *Food Chemistry*, 152, 1333–139.

Bolarinwa, I. F., Orfila, C. and Morgan, M. R. A. (2015). Determination of amygdalin in apple seeds, fresh apples and processed apple juices. *Food Chemistry*, 170, 437–442. 10.1016/j.foodchem.2014.08.083

Bolarinwa, I. F., Oke, M. O., Olaniyan, S. A. and Ajala, A. S. (2016). A review of cyanogenic glycosides in edible plants. *Toxicology –New aspects to this scientific conundrum*, November. 10.5772/64886

Bolarinwa, I. F., Olaniyan, S. A., Olatunde, S. J., Ayandokun, F. T. and Olaifa, I. A. (2016). Effect of processing on amygdalin and cyanide contents of some nigerian foods. *Journal of Chemical and Pharmaceutical Research*, 8(2), 106–113.

Bolarinwa, I. F. and Orfila, C. (2017). Natural plant toxins – Cyanogenic glycosides in plant foods. pp. 2–3.

Borron, S. W., Stonerook, M., & Reid, F. (2006). Efficacy of Hydroxocobalamin for the Treatment of Acute Cyanide Poisoning in Adult Beagle Dogs. *Clinical Toxicology*, 44, sup1, 5–15. 10.1080/15563650600811672

Borron, S. W. and Baud, F. J. (2012). Antidotes for acute cyanide poisoning. *Current Pharmaceutical Biotechnology*, 13, 1940–1948.

Bradbury, J. H. and Denton, I. C. (2010). Rapid wetting method to reduce cyanogen content of cassava flour. *Food Chemistry*, 121(2), 591–594.

Bradbury, J. H., Egan, S. V. and Lynch, M. J. (1991). Analysis of cyanide in cassava using 344 acid hydrolysis of cyanogenic glucosides. *Journal of Science of Food and Agriculture*, 55, 277–290.

Bradbury, J. H. and Holloway, W. (1988). Antinutritional factors in root crops. In *Chemistry of tropical root crops: Significance for nutrition and agriculture in the Pacific*. Canberra: ACIAR, p. 201.

Brimer, L. 2010). Cyanogenic glycosides in food, feeding stuffs and green medicine. In Bernhoft, A. (Ed.). *Bioactive compounds in plants – Benefits and risks for man and animals*. Oslo: The Norwegian Academy of Science and Letters, pp. 125–143. ISBN 978-82-7099-583-7

Busk, P. K. and Møller, B. L. (2002). Dhurrin synthesis in sorghum is regulated at the transcriptional level and induced by nitrogen fertilization in older plant. *Plant Physiol*, 129, 1222–1231.

Canadian Food Inspection Agency. Cyanides in bitter apricot kernels. (2009). Available from: http://www.hc-sc.gc.ca/fn-an/pubs/securit/2009-apricots-abricots/indexeng.

Cardoso, A. P., Mirione, E., Ernesto, M., Massaza, F., Cliff, J., Haque, M. R. and Bradbury, J. H. (2005). Processing of cassava roots to remove cyanogens. *Journal of Food Composition and Analysis*, 18(5), 451–460.

Castada, H. Z., Liu, J., Ann Barringer, S. and Huang, X. (2020). Cyanogenesis in macadamia and direct analysis of hydrogen cyanide in macadamia flowers, leaves, husks, and nuts using selected ion flow tube–mass spectrometry. *Foods*, 9(2), 174.

CFS (Centre for Food Safety) (2007). Natural toxins in food plants. Risk assessment studies. Food and Environmental Hygiene Department, the Government of the Hong Kong Special Administrative Region. Report No. 27.

Chandra, A. K., Ghosh, D., Mukhopadhyay, S. and Tripathy, S. (2004). Effect of bamboo shoot, bambusa arundinacea (Retz.) willd. On thyroid status under conditions of varying iodine intake in rats. http://nopr.niscpr.res.in/handle/123456789/23555

Chassagne, D., Crouzet, J. C., Bayonove, C. L. and Baumes, R. L. (1996). Identification and quantification of passion fruit cyanogenic glycosides. *Journal of Agriculture and Food Chemistry*, 44, 3817–3820.

Codex Committee on Contaminants in Foods (2008). *Discussion paper on cyanogenic glycosides. CX/CF 09/3/11. 2008.* Rome: FAO/WHO.

Conn, E. E. (1979). Biosynthesis of cyanogenic glycosides. *Naturwissenschaften*, 66(1), 28–34.

Donald, G. B. (2009). Cyanogenic foods (cassava, fruit kernels, and cycad seeds). *Medical Toxicology of Natural Substances*, 55, 336–352.

European Food Safety Authority (EFSA) (2004). Opinion of the scientific panel on food additives, flavourings, processing aids and materials in contact with food (AFC) on hydrocyanic acid in flavourings and other food ingredients with flavouring properties. *EFSA Journal*, 2(11), 105.

EFSA CONTAM Panel (EFSA Panel on Contaminants in the Food Chain) (2016). Scientific opinion on the acute health risks related to the presence of cyanogenic glycosides in raw apricot kernels and products derived from raw apricot kernels. *EFSA Journal*, 14(4), 4424. 47 pp. 10.2903/j.efsa.2016.4424. http://www.efsa. europa.eu/sites/default/files/scientific_output/files/main_documents/4424.pdf

Eisenbrand, G. and Gelbke, H. P. (2016). Assessing the potential impact on the thyroid axis of environmentally relevant food constituents/contaminants in humans. *Archives of Toxicology*, 90, 1841–1857. 10.1007/s00204-016-1735-6

Environmental Protection Agency (EPA) (1990). Summary review of health effects associated with hydrogen cyanide, health issue assessment environmental criteria and assessment office, office of health and environmental assessment office of research and development, U.S. environmental protection agency research triangle park, NC 27711.

Ermans, A. M., Bourdoux, P., Kintheart, J., Lagasse, R., Luvivila, K., Mafuata, M., Thilly, C. M. and Delange, F. (1983). Role of cassava in the etiology of endemic goitre and cretinism. In Delange, F. and Ahluwalia, R. (Eds.). *Cassava toxicity and thyroid: Research and public health issues.* Ottawa: International Development Research Centre, pp. 9–16.

Ernesto, M., Cardoso, A. P., Nicala, D., Mirione, E., Massaza, F., Cliff, J., Hague, M. R. and Bradbury, J. H. (2002). Persistent konzo and cyanogens toxicity from cassava in northern mozambique. *Acta Trop*, 82, 357–362.

FAO/WHO (Food and Agricultural Organization/World Health Organization) (1993). Toxicological evaluation of certain food additives and natural occurring toxicants. Report of the 39th meeting of the Joint FAO/WHO Experts Committee on Food Additives (JECFA). Food Additives Series No. 30. World Health Organization, Geneva. pp. 299–337.

Feng, D., Shen, Y. and Chavez, E. R. (2003). Effectiveness of different processing methods in reducing hydrogen cyanide context of flaxseeds. *Science of Food and Agriculture*, 83, 836–841. 10.1002/jsfa.1412

Ferreira, V. L. P., Yotsuyanagi, K. and Carvalho, C. R. L. (1995). Elimination of cyanogenic compounds from bamboo shoots dendrocalamus giganteus munro. *Trop Sci*, 35, 342–346.

Fincham, S. M., Hill, G. B., Hanson, J. and Wijayasinghe, C. (1990). Epidemiology of prostatic cancer: A case-control study. *The Prostate*, 17(3), 189–206.

Folashade I. (May 2013). *Cyanogenic glycosides in plant foods.* University of Leeds, pp. 1–254. http://etheses.whiterose.ac.uk/5002/1/ISLAMIYAT BOLARINWA THESIS 2013.pdf

Food and Agricultural Organisation (FAO) (1991). Roots, tubers, plantain and bananas in human nutrition. *Food and Agricultural Organisation of the United Nation*, 7(1), 11–21.

Food Standards Australia New Zealand (FSANZ). (2004). Final assessment report proposal P257. Advice on the preparation of cassava and bamboo shoots. Report Number 2–04. Canberra.

Food Standards Australia New Zealand (FSANZ) (2005). Cassava and bamboo shoots. *A Human Health Risk Assessment*, 28, 9.

Francisco, I. A. and Pinotti, M. H. P. (2000). Cyanogenic glycosides in plants. *Brazilian Archives of Biology and Technology*, 43, 487–492.

Ganjewala, D., Kuma, S., Asha, D. S. and Ambika, K. (2010). Advances in cyanogenic glycosides biosynthesis and analyses in plants: A review. *Acta Biol Szeged*, 54, 1–14.

Gbadebo, A. M. and Oyesanya, T. M. (2005). Assessment of iodine deficiency and goitre incidence in parts of yewa area of ogun state, southwestern Nigeria. *Environmental Geochemistry and Health*, 27(5-6), 491.

Gleadow, R. M. and Møller, B. L. (2014). Cyanogenic glycosides: Synthesis, physiology, and phenotypic plasticity. *Annual Review of Plant Biology*, 65, 155–185. 10.1146/annurev-arplant-050213-040027

Gleadow, R. M. and Woodrow, I. E. (2002). Constraints on effectiveness of cyanogenic glycosides in herbivore defense. *Journal of Chemical Ecology*, 28, 1301–1313.

Gruhnert, C., Biehl, B. and Selmar, D. (1994). Compartmentation of cyanogenic glucosides and their degrading enzymes. *Planta*, 195(1), 36–42.

Haisman, D. R., Knight, D. J. and Ellis, M. J. (1967). The electrophoretic separation of the β-glucosidases of almond "emulsin". *Phytochemistry*, 6(11), 1501–1505.

Hall, A. H., Dart, R. and Bogdan, G. (2007). Sodium thiosulfate or hydroxocobalamin for the empiric treatment of cyanide poisoning? *Annals of Emergency Medicine*, 49(6), 806–813.

Hamel, J. (2011). A review of acute cyanide poisoning with a treatment update. *Critical care nurse*, 31(1), 72–82.

Haque, M. R. and Bradbury, J. H. (2002). Total cyanide determination of plants and foods using the picrate and acid hydrolysis methods. *Food Chemistry*, 77, 107–114.

Holzbecher, M. D., Moss, M. A. and Ellenberger, H. A. (1984). The cyanide content of laetrile preparations, apricot, peach and apple seeds. *Clin Toxicol*, 22, 341–347.

Huang, M. T. and Ferraro, T. (1992). Phenolics compounds in food and cancer prevention. In Huang, M. T., Ho, C. T. and Lee, C. Y. (Eds.). *Phenolic compounds in food and their effects on health*. II; Washington D.C.: American Chemical Society. 8–34.

Igbadul, B. D., Amoye, J. and Twadue, I. (2014). Effect of fermentation on the proximate composition, antinutritional factors and functional properties of cocoyam (colocasia esculenta) flour. *Afr J Food Sci Technol*, 5(3), 67–74.

Jones, D. A. (1998). Why are so many food plants cyanogenic? *Phytochemistry*, 47(2), 155–162.

Jones, P. R., Møller, B. L. and Høj, P. B. (1999). The UDP-glucose: P-hydroxymandelonitrile-O-glucosyltransferase that catalyzes the last step in synthesis of the cyanogenic glucoside dhurrin in sorghum bicolor: Isolation, cloning, heterologous expression, and substrate specificity. *Journal of Biological Chemistry*, 274(50), 35483–35491.

Kahn, R. A., Bak, S., Svendsen, I., Halkier, B. A. and Moller, B. L. (1997). Isolation and reconstitution of cytochrome P450ox and in vitro reconstitution of the entire biosynthetic pathway of the cyanogenic glucoside dhurrin from sorghum. *Plant Physiology*, 115(4), 1661–1670.

Kemdirim, O. C., Chukwu, O. A. and Achinewhu, S. C. (1995). Effect of traditional processing of cassava on the cyanide content of Gari and cassava flour. *Plant Foods for Human Nutrition*, 48(4), 335–339.

Klopfenstein, C. F. and Hoseney, R. C. (1995). Nutritional properties of sorghum and millets. In Dendy, D. A. V. (Ed.). *Sorghum and millets chemistry and technology.* 125 St Paul, Minn: American Association of Cereal Chemistry.

Kolind-Hansen, L. and Brimer, L. (2010). The retail market for fresh cassava root tubers in the european union (EU): The case of Copenhagen, Denmark – A chemical food safety issue? *Journal of the Science of Food and Agriculture*, 90, 252–256.

Kulamarva, A. G., Sosle, V. R. and Raghavan, G. S. V. (2009). Nutritional and rheological properties of sorghum. *International Journal of Food Properties*, 12(1), 55–69. 10.1080/10942910802252148

Kwok, J. Cyanide poisoning and cassava. *Food Safety Focus.* (19th issue February, 2008). Incident focus. (2008). http://www.cfs.gov.hk/english/multimedia

Lang, K. (1933). Die rhodanbilung im tierkörper. *Biochem Z*, 259, 243–256.

McMahon, J. M. and Sayre, R. T. (1995). Cyanogenic glycosides: Physiology and regulation of synthesis. 112–121. American Society of Plant Physiologists.

Mitchell, B. L., Bhandari, R. K., Bebarta, V. S., Rockwood, G. A., Boss, G. R. and Logue, B. A. (2013). toxicokinetic profiles of α-ketoglutarate cyanohydrin, a cyanide detoxification product, following exposure to potassium cyanide. *Toxicology Letters*, 222(1), 83–89.

Mohd Azmi, A. F. (2019). *Cyanogenic glycosides in cassava*, Doctoral dissertation University of Leeds.

Møller, B. L. (2010). Functional diversifications of cyanogenic glucosides. *Curr Opin Plant Biol*, 13, 338–347. 10.1016/j.pbi.2010.01.009

Moller, B. L. and Seigler, D. S. (1999). Biosynthesis of cyanogenic glycosides, cyanolipids and related compounds. In Singh, B. K. (Ed.). *Plant amino acids biochemistry and biotechnology.* Marcel Dekker, pp. 563–609.

Montagnac, J. A., Davis, C. R. and Tanumihardjo, S. A. (2009). Processing techniques to reduce toxicity and antinutrients of cassava for use as a stable food. *Comprehensive Reviews in Food Science and Food Safety*, 8, 17–27.

Morant, A. V., Jørgensen, K.0., Jørgensen, C., Paquette, S. M., Sanchez-Perez, R., Møller, B. L. and Bak, S. (2008). Beta- glucosidases as detonators of plant chemical defense. *Phytochemistry*, 69, 1795–1813. 10.1016/j.phytochem.2008.03.006

Ndubuisi, N. and Chidiebere, A. (2018). Cyanide in cassava a review. *International Journal of Genomics and Data Mining.*

Ngudi, D. D. and Kuo, Y. H. and Lambein, F. (2002). Food safety and amino acid balance in processed cassava "cossettes". *Journal of Agricultural and Food Chemistry*, 50(10), 3042–3049.

Nhassico, D., Muquingue, H., Cliff, J., Cumbana, A. and Bradbury, J. H. (2008). Rising african cassava production, diseases due to high cyanide intake and control measures. *J Sci Food Agric*, 88, 2043–2049.

Nyirenda, K. K. (2020). Toxicity potential of cyanogenic glycosides in edible plants. *Medical toxicology*, IntechOpen, pp. 1–19.

Obilie, E. M., Tano-Debrah, K. and Amoa-Awua, W. K. (2004). Souring and breakdown of cyanogenic glucosides during the processing of cassava into Akyeke. *International Journal of Food Microbiology*, 93(1), 115–121.

Odo, C. E., Enechi, O. C. and Oburu, C. S. (2014). Concentrations of anti-nutritional factors in raw edible cocoyam (colocasia esculenta) leaves. *J Pharm Res*, 8(1), 38–40.

Oke, O. L. (1973). Mode of cyanide detoxication. *Chronic cassava toxicity.* Ottawa, CA: IDRC. https://idl-bnc-idrc.dspacedirect.org/bitstream/handle/10625/18796/IDL-18796.pdf?sequence=1

Okoye, N. F. and Uwhen, O. E. (2016). Use of bentonite clay for the reduction of cyanide poisoning after consumption of cassava waste water by wistar albino rat. *Journal of Applied Sciences and Environmental Management*, 20(3), 490–495.

Oluwole, O. S. A., Onabolu, A. O., Link, H. and Rosling, H. (2000). Persistence of tropical ataxic neuropathy in a nigerian community. *J Neurol Neurosurg Psychiatry*, 69, 96–101.

Oluwole, O. S. A., Onabolu, A. O., Mtunda, K. and Mlingi, N. (2007). Characterization of cassava (manihot esculenta Crantz) varieties in Nigeria and Tanzania, and farmers' perception of toxicity of cassava. *Journal of Food Composition and Analysis*, 20(7), 559–567.

Omaye, S. T. (2004). *Food and nutritional toxicology*. CRC press.

Padmaja, G., Balagopalan, C., Moorthy, S. N. and Potty, V. P. (1996). Yuca rava and yuca porridge: The functional properties and quality of two novel cassava food products1. *Cassava Flour And Starch: Progress in Research and Development*, (271), 323.

Poulton, J. E. (1990). Cyanogenesis in plants. *Plant Physiology*, 94(2), 401–405.

Prasad, S. and Dhanya, M. S. (2011). Determination and detoxification of cyanide content in sorghum for ethanol production using saccharomyces cerevisiae strain. *J Metabol Syst Biol*, 2(1), 10–14.

Preedy, V. R. and Watson, R. R. (Eds.). (2020). *Nuts and seeds in health and disease prevention*. Academic press.

Ratnavathi, C. V. and Tonapi, V. A. (2020). Functional characteristics and nutraceuticals of grain sorghum. In Tonapi, V. A., Talwar, H. S., Are, A. K., Bhat, B. V., Reddy, C. R., & Dalton, T. J. (eds.), *Sorghum in the 21st century: Food–fodder–feed–fuel for a rapidly changing world*. Singapore: Springer, pp. 839–858. https://doi.org/10.1007/978-981-15-8249-3_33

Rosling, H. (1994). Measuring effects in humans of dietary cyanide exposure from cassava. *Acta Horticulture*, 375, 271–283.

Santamour, F. S. (1998). Amygdalin in Prunus leaves. *Phytochemistry*, 47, 1537–1538.

Seigler, D. S. (1975). Isolation and characterization of naturally occurring cyanogenic compounds. *Phytochemistry*, 14(1), 9–29.

Seigler, D. S. (1991). Cyanide and cyanogenic glycosides. In Rosenthal, G. A. and Berenbaum, M. R. (Eds.). *Herbivores. Their Interaction with secondary plant metabolites*. San Diego, CA: Academic Press, vol. 1, pp. 35–77.

Selvan, V. A. (2013). A case of tropical ataxic neuropathy. *Apollo Medicine*, 10(3), 223–225.

Schrenk, D., Bignami, M., Bodin, L., Chipman, J. K., del Mazo, J., Grasl-Kraupp, B., Hogstrand, C., Hoogenboom, L. (Ron), Leblanc, J. C., Nebbia, C. S., Nielsen, E., Ntzani, E., Petersen, A., Sand, S., Vleminckx, C., Wallace, H., Benford, D., Brimer, L., Mancini, F. R. and Schwerdtle, T. (2019). Evaluation of the health risks related to the presence of cyanogenic glycosides in foods other than raw apricot kernels. *EFSA Journal*, 17(4), 10.2903/j.efsa.2019.5662

Schulz, V., Gross, R., Pasch, T., Busse, J. and Loeschcke, G. (1982). Cyanide toxicity of sodium nitroprusside in therapeutic use with and without sodium thiosulfate. *Klinische Wochenschrift*, 60, 1393–1400.

Shahidi, F. and Naczk, M. (1995). PhenolicWajants in cereals and legumes. In Shahidi, F. (Ed.). *Food phenolics*. Lancaster, Pa.: Technomic Publishing, pp. 9–52.

Sofyan, S., Rasulu, H., Hasbullah and Hasan, S. (2018). Chemical Properties of high-quality cassava flour (HQCF) from several varieties of cassava. *Pakistan Journal of Nutrition*, 17, 615–621.

Sornyotha, S., Kyu, K. L. and Ratanakhanokchai, K. (2007). Purification and detection of linamarin from cassava root cortex by high performance liquid chromatography. *Food Chemistry*, 104, 1750–1754.

Speijers, G. (1993). Cyanogenic glycosides. *WHO Food Additives Series*, 30, 299–337.

Tamer, C. E., Suna, S. and Özcan-Sinir, G. (2019). Toxicological aspects of ingredients used in nonalcoholic beverages. In Grumezescu, A. M. , & Holban, A. M. (eds.), *Nonalcoholic beverages*, Woodhead Publishing, pp. 441–481. ISBN 9780128152706, https://doi.org/10.1016/B978-0-12-815270-6.00014-1

Taylor, S. L. (2012). *Natural toxins in food*. Encyclopedia of Lifestyle Medicine & Health, p. 27. 10.4135/9781412994149.n227

Tivana, L. D., Bvochora, J., Mutukumira, A. N., Owens, J. D. and Zvauya, R. (2007). Heap fermentation of cassava (manihot esculenta crantz) in Nampula Province, Mozambique. Proceedings of the 13th ISTRC Symposium, pp. 445–450.

Tokpohozin, S. E., Fischer, S., Sacher, B. and Becker, T. (2016). b-D-glucosidase as "key enzyme" for sorghum cyanogenic glucoside (dhurrin) removal and beer bio-flavouring. *Food and Chemical Toxicology*, 97, 217–223.

Toomey, G. (1988). Sorghum as substitute. Food enterprises for indian woman. *Int. Develop. Res. Centre Reports*, 17(3), 20–21.

Tuncel, G., Nout, M. J. R. and Brimer, L. (1995). The effects of grinding, soaking and cooking on the degradation of amygdalin of bitter apricot seeds. *Food Chemistry*, 53, 447–451.

Tuncel, G., Nout, M. J. R. and Brimer, L. (1998). Degradation of cyanogenic glycosides of bitter apricot seeds (Prunus armeniaca) by endogenous and added enzymes as affected by heat treatments and particle size. *Food Chemistry*, 63(1), 65–69.

Tylleskar, T., Banae, M., Bikangi, N., Cooke, R. D., Poulter, N. H. and Rosling, H. (1992). Cassava cyanogens and konzo, an upper motorneuron disease found in Africa. *Lancet*, 339, 208–211.

Vetter, J. (2000). Plant cyanogenic glycosides. *Toxicon*, 38, 11–36.

Wajant, H., Riedel, D., Benz, S. and Mundry, K. W. (1994). Immunocytological localization of hydroxynitrile lyases from sorghum bicolor L. and linum usitatissimum L. *Plant Science*, 103(2), 145–154.

Yamashita, T., Sano, T., Hashimoto, T. and Kanzawa, K. (2007). Development of a method to remove cyanogenic glycosides from flaxseed meal. *International Journal of Food Science and Technology*, 42, 70–75.

Yen, D., Tsai, J., Wang, L. M., Kao, W. F., Hu, S. C., Lee, C. H., Deng, J. F. (1995). The clinical experience of acute cyanide poisoning. *The American Kournal of Emergency Medicine*, 13(5), 524–528.

Zagrobelny, M., Bak, S. and Møller, B. L. (2008). Cyanogenesis in plants and arthropods. *Phytochemistry*, 69(7), 1457–1468.

Zagrobelny, M., Bak, S., Rasmussen, A. V., Jørgensen, B., Naumann, C. M. and Møller, B. L. (2004). Cyanogenic glucosides and plant–insect interactions. *Phytochemistry*, 65(3), 293–306.

ABBREVIATIONS

ARfD	Acute reference dose
ATCA	2-amino-2-thiazoline-4-carboxylic acid
CONTAM Panel	EFSA Panel on Contaminants in the Food Chain
CN	Cyanide

CN-	Cyanide ion
CNGs	Cyanogenic glycosides
Codex	Codex Alimentarius Commission
CYP	Cytochrome P450
ELISAs	Enzyme-linked immunosorbent assays
FAO/WHO	Food and Agriculture Organization of the United Nations/World Health Organization
FSANZ	Food Standards Australia New Zealand
GC-MS	Gas Chromatography-Mass Spectrometry
HCN	Hydrocyanic acid
HPLC	High-performance liquid chromatography
HPLC-DAD	High performance liquid chromatography high performance liquid chromatographic method with diode-array detection
HPLC-MS	High performance liquid chromatography-mass spectrometry
HPLC-UV	High performance liquid chromatography with UV detection
α-KGCN	a-ketoglutarate cyanhydrin
LC-MS/MS	Liquid chromatography-tandem mass spectrometry
MS	Mass spectrometry
UV	Ultraviolet
WHO	World Health Organization

CHAPTER 3

Pyrrolizidine Alkaloids as Food Toxins

Javed Ahamad[1], Mohd. Javed Naim[2], Subasini Uthirapathy[3], and Showkat R. Mir[4]

[1]Department of Pharmacognosy, Faculty of Pharmacy, Tishk International University, Erbil, Kurdistan Region, Iraq
[2]Department of Pharmaceutical Chemistry, Faculty of Pharmacy, Tishk International University, Erbil, Kurdistan Region, Iraq
[3]Department of Pharmacology, Faculty of Pharmacy, Tishk International University, Erbil, Kurdistan Region, Iraq
[4]Department of Pharmacognosy, School of Pharmaceutical Education and Research, Jamia Hamdard, New Delhi, India

CONTENTS

3.1 Introduction	53
3.2 Sources of Pyrrolizidine Alkaloids	54
3.3 Chemistry of Pyrrolizidine Alkaloids	55
3.4 Pharmacological Actions of Pyrrolizidine Alkaloids	56
3.4.1 Anti-Inflammatory Activity	57
3.4.2 Antimicrobial Activity	58
3.4.3 Anticancer Activity	59
3.4.4 Anti-HIV Activity	59
3.4.5 Antiulcer Activity	59
3.4.6 Acetylcholinesterase Inhibitor Activity	59
3.5 Toxicological Effects of Pyrrolizidine Alkaloids	60
3.5.1 Mutagenic Effects of PAs	60
3.5.2 Carcinogenicity Produced by PAs	61
3.5.3 Hepatotoxic Effects of PAs	62
3.6 Conclusions	62
References	63

3.1 INTRODUCTION

Medicinal plants are major components in herbal medicines, dietary supplements, cosmetic products, and they are also used as pharmaceutical excipients. In recent years, there has been increasing use of herbal medicines and dietary supplements to treat various chronic diseases and to promote health (Ahamad et al. 2019). There have been several

DOI: 10.1201/9781003222194-4

outbreaks of human poisoning as a result of ingestion of contaminated grains as well as case reports of poisoning caused by intentional ingestion of herbal medicines and dietary supplements containing toxic phytochemicals such as pyrrolizidine alkaloids (PAs), cyanogenetic glycosides, glycoalkaloids, furanocoumarins, and lactones (Chen et al. 2010).

Pyrrolizidine alkaloids (PAs) and their N-oxides are composed of necine and its derivative, and about 660 PAs and their N-oxides have been identified in over 6,000 plant species. The most important families that contain PAs are Boraginaceae, Fabaceae, and Asteraceae families (Roeder 2000). PAs are predominantly found in the genera of *Crotalaria, Senecio, Heliotropium, Echium, Symphytum,* and *Cynoglossum* (Stegelmeier et al. 1999). Several plant species are known to contain toxic PAs, such as *Liparis nervosa, Heliotropium digynum, Senecio jacobaea, S. brasiliensis, Heliotropium indicum, Castanospermum australe, Alexa leiopetala, Solenanthus lanatus,* and *Echium confusum* (Moreira et al. 2018).

Plants containing PAs are the most prevalent toxic phytochemicals that affect humans, livestock, and wildlife (Arzt and Mount 1999). PAs are also the most common plant toxic phytochemicals linked to several diseases in humans, and most of the common sources of the contamination are staple foods, honey, milk, herbal teas, and herbal medicines (Arseculeratne et al. 1981, Bach et al. 1989; Bah et al. 1994; Deinzer et al. 1977; Edgar et al. 2002; Culvenor et al. 1981). The majority of the PAs found are known to be genotoxic, tumorigenic, and hepatotoxic (Arzt and Mount 1999). The mutagenicity of PAs and their N-oxides has been studied extensively in human beings and grazing livestock (Griffin and Segall 1986; Gimmler-Luz et al. 1990; Fu et al. 2004; Frei et al. 1992; Green and Muriel 1975). Although the harmful effects of PAs are well documented, and their contents in relevant food and feed matrices have been monitored for years, regulatory limits in the European Union have yet to be established. The European Food Safety Authority (EFSA) assessed the risk of PAs exposure using a margin of exposure (MOE) approach based on a Benchmark Dose Lower Confidence Limit 10% (BMDL10) of 237 g/kg body weight/day (BW/d) for the PAs and their N-oxides sum (EFSA 2017). Until now, Directive 2002/32/EC only permitted a maximum content of 100 mg/kg of PA and their N-oxides containing *Crotalaria* spp. in animal feed materials and compound feed (Directive 2002/32/EC). PAs are also reported to possess several beneficial bioactivities such as anti-inflammatory (Ghosh and Singh 1974), antimicrobial (Neto et al. 2016), anticancer (Appadurai 2014), anti-HIV (Vlietinck 1997; Pal et al. 1989), and antiulcer (Toma et al. 2004a,b). In this book chapter, sources, chemistry, pharmacological, and toxicological actions of pyrrolizidine alkaloids will be extensively studied and compiled.

3.2 SOURCES OF PYRROLIZIDINE ALKALOIDS

Pyrrolizidine alkaloids (PAs) are a diverse group of chemicals and distributed in several plant families. Approximately 660 PAs (including N-oxides) have been identified in 6,000 different plant species (The Plant List 2013). The most common families containing PAs are Asteraceae (Compositae), Boraginaceae, and Fabaceae (Leguminosae). Fabaceae (950 genera and 24,000 species) has only four genera (900 species) that contain PAs, and the most common tribe Crotalarieae, of which genus *Crotalaria* (700 species) is most important. In the Asteraceae family (60 genera and 3200 species), only 10% of species are reported to contain PAs. Most of the genera have been placed in the tribe Senecioneae, and *Senecio* genus with 1,600 species have been recognized as most important source of PAs (Langel et al. 2011). Boraginaceae (2,700 species) is a smaller family in comparison

TABLE 3.1   Limits of PAs/N-oxides in Food Products and Spices

S. no.	Products	Category (ingredients)	Limits of PAs/N-oxides
1.	Herbal infusions	Rooibos, anise, lemon balm, chamomile, thyme, peppermint, lemon verbena (dried product)	400 µg/kg
2.	Other herbal infusions	Dried product	200 µg/kg
3.	Tea	Tea (*Camellia sinensis*) and flavoured tea (dried product)	150 µg/kg
		Tea and herbal infusions for infants and young children (dried product)	75 µg/kg
		Tea and herbal infusions for infants and young children (liquid)	1.0 µg/kg
4.	Food supplements	Containing herbal ingredients	400 µg/kg
5.	Pollen-based food supplements	Pollen and pollen products	500 µg/kg
6.	Herbs	Dried	400 µg/kg
7.	Borage, lovage, marjoram, and oregano	Dried	1000 µg/kg
8.	Borage	Fresh, frozen	750 µg/kg
9.	Cumin seeds	Seed spice	400 µg/kg

Source: https://webarchive.nationalarchives.gov.uk/ukgwa/20211201132024/ https://www.food.gov.uk/news-alerts/consultations/june-2019-stakeholder-update-on-rapidly-developing-policy-on-food-contaminants.

with the previous two families, but it contains a relatively high number of PA-producing plants (1,700 species, 40 genera). The Heliotropioideae subfamily (160 species) contains the genus *Heliotropium*, while the Boraginoideae subfamily includes genera like *Echium*, *Symphytum*, and *Cynoglossum* as important genera that contain PAs (El-Shazly and Wink 2014). PA concentrations are known to vary significantly between species and between parts of the plant. The content of PAs in Crotalarieae, Senecioneae, and Heliotropioideae families may range from 0.1 mg/kg to 100 mg/kg d.w. (e.g., in seeds), while concentrations in Eupatorieae and Boraginoideae species may range from 0.1 mg/kg to 10 mg/kg d.w. (Cramer et al. 2014; Kowalczyk et al. 2018). Table 3.1 comprises permissible limits of PAs/N-oxides in food products and spices.

3.3 CHEMISTRY OF PYRROLIZIDINE ALKALOIDS

Pyrrolizidine alkaloids (PAs) are made up of a necine base and one or two necine bases linked together by ester linkages. The necine base can either be completely saturated or 1,2-unsaturated (Figure 3.1). Although the number of different necine bases found in nature is limited, when combined with the pool of necine acids available, a significant number of diverse structures are produced (Bruneton 2008; Moreira et al. 2018). PAs are usually found in the form of tertiary bases or pyrrolizidine alkaloids N-oxides. Necic acids are aliphatic carboxylic acids that can be simple (angelic and tiglic acids), monocarboxylic acids (trachelantic and viridifloric acids), or dicarboxylic acids (C-8 or C-10 acids) with

FIGURE 3.1 Chemical structure of basic moiety in pyrrolizidine alkaloids (PAs) and their different forms, where, R_1 and R_2 correspond to different necic acids.

hydroxyisopropylbutanoic structures at C-7 (senecic and isatinecic acids) (Hartmann and Witte 1995; Valese et al. 2016). In Figure 3.2, different types of necic acids are presented.

The pyrrolizidine alkaloids (PAs) can be classified into four major groups: retronecine, heliotridine, otonecine, and platynecine based on types of necine moiety present in it (Figure 3.3). Retronecine, heliotridine, and otonecine are unsaturated bases, whereas platynecine type PAs are saturated. Retronecine and heliotridine are diastereomers, with a distinct orientation of OH group at position C-7 (Xia et al. 2008). The example of retronecine-type PAs includes riddelliine, retrorsine, monocrotaline, and symphytine. Lasiocarpine and heliotrine are heliotridine-type, and senkirkine and petasitenine are otonecine-type PAs (These et al. 2013).

Isoleucine plays an important role in the biosynthesis of pyrrolizidine alkaloids (PAs) as these give rise to putrescine which works as starting molecule for the synthesis of necine acids, and homospermidine is the central precursor for many necine bases (Langel et al. 2011). Regardless of the initial steps, the reaction is catalyzed by homospermidine synthase and the result is the symmetrical intermediate homospermidine. Subsequently, homospermidine is cyclized to the iminium ion as intermediate, which is reduced and cyclized to trachelanthamidine, and also gives isorectronecanole (Böttcher et al. 1994; Hartmann 1999). Trachelanthamidine further leads to retronecine, and it gives otonecine through intermediate. The possible biosynthesis of pyrrolizidine alkaloids is summarized in Figure 3.4.

3.4 PHARMACOLOGICAL ACTIONS OF PYRROLIZIDINE ALKALOIDS

Pyrrolizidine alkaloids (PA) are found in a wide range of plants around the world, including many species that are consumed by humans. PAs are known for their broad range of pharmacological properties that can be used in drug development programs, as well as

FIGURE 3.2 Chemical structure of different types of necic acids present in pyrrolizidine alkaloids (PAs).

FIGURE 3.3 Major types of pyrrolizidine alkaloids (PAs): (a) retronecine, (b) heliotridine, (c) otonecine, and (d) platynecine.

the toxicity that these compounds can produce in humans and animals. Pyrrolizidine alkaloids (PAs) have a diverse range of biological features that can be used in drug discovery and development.

3.4.1 Anti-Inflammatory Activity

The activity of pyrrolizidine alkaloids (PAs), such as crotalaburnine against improved edema and vascular permeability caused by carrageenan, bradykinin, 5-hydroxytryptaminea, and

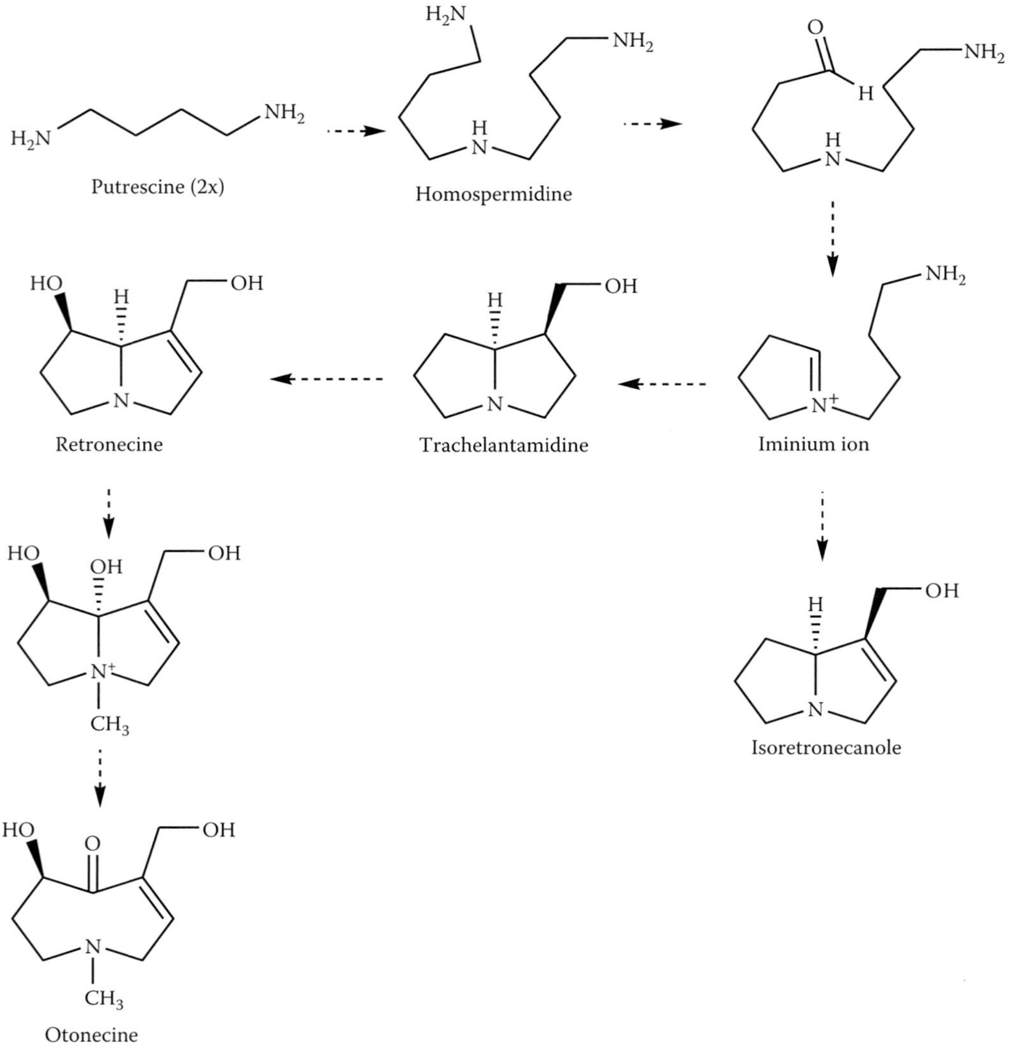

FIGURE 3.4 Possible biosynthesis of pyrrolizidine alkaloids and their *N*-oxides.

prostaglandin, was investigated by Ghosh and Singh (1974). In this study, crotalaburnine was studied in rats to see if it could prevent the cotton-pellet formation in tissues and results were compared with the standard drug hydrocortisone. The result showed that crotalaburnine was more efficacious against acute edema produced by carrageenan and hyaluronidase at a dose of 10 mg/kg. Crotalaburnine was also found to be two times as effective as hydrocortisone in a cotton-pellet granuloma test (Ghosh and Singh 1974).

3.4.2 Antimicrobial Activity

Many alkaloids have been reported as potent antimicrobials, which is consistent with this class of secondary metabolites' defense function in plants (Macel 2011). Usaramine, monocrotaline, and azido-retronecine have been shown to have antimicrobial efficacy against *Pseudomonas aeruginosa, Staphylococcus epidermidis,* and *Trichomonas vaginalis* (Neto et al. 2016). Usaramine was tested for its ability to prevent biofilm development in *S. epidermidis* and *P. aeruginosa.* The PA group of chemicals, which includes senecionine, seneciphylline, jacobine, and jaconine, was significantly effective, although high doses were required, with each PA's effective dose ranging from 0.33 mM to 3.33 mM for the most sensitive fungus genus *Trichoderma* (Hol and Van 2002).

3.4.3 Anticancer Activity

PAs have been reported as potent inhibitors of human cancer cell lines in *in-vitro* studies such as cervical squamous, prostate, lung, and breast (Yu et al. 2020). Indicine N-oxide isolated from *Heliotropium indicum* reduced the proliferation of previously documented cancer cell lines with an IC_{50} value of 100 µM. Cell cycle arrest was detected at the mitosis stage, with no noticeable changes in spindle or interphase microtubule architecture (Appadurai and Rathinasamy 2014). The anticancer effects of Nervosine VII isolated from *Liparis nervosa* were studied for possible autophagy and apoptosis in HCT116 human colorectal cancer cells. It showed anticancer effects by activation of MAPKs signaling pathway including JNK, ERK1/2, and p38 and suppressing the p53 signaling pathway (Huang et al. 2020).

3.4.4 Anti-HIV Activity

Polyhydroxylated PA has been shown to interact with the activity of the human immunodeficiency virus (HIV). *Castanospermum australe* contains australine and alexine, and these PAs inhibit cell union with virions and, as a result, syncytium formation (Vlietinck 1997; Pal et al. 1989). Alexine, as well as the other PAs from *Alexa leiopetala* and *C. australe,* showed inhibitory activity against HIV. The potent actions were from 7,7a-diepialexine, which had an IC_{50} of 0.38 mM. The inhibition of pig kidney α-glucosidase 1 and the reduced cleavage of the precursor HIV-1 glycoprotein (gp)160 were found to be associated with anti-HIV action (Taylor et al. 1992).

3.4.5 Antiulcer Activity

PAs have been reported as a potent ganglionic inhibitor as well as reported beneficial actions in the treatment of ulcers. The anti-ulcer activity of *Senecio brasiliensis* PAs such as integerrimine, retrorsine, senecionine, usaramine, and seneciphylline was studied in rats and mice. These PAs showed potent antiulcerogenic activity in both acute and chronic gastric ulcers (Toma et al. 2004a). In another study by Toma et al. (2004b), the authors found preventive effects of PAs from S. *brasiliensis* on a gastric and duodenal induced ulcer in mice and rats.

3.4.6 Acetylcholinesterase Inhibitor Activity

Acetylcholinesterase (AchE) is a parasympathetic neurotransmitter protein that catalyzes the breakdown of acetylcholine (ACh). It is typically located in the central and peripheral nervous systems' synaptic gaps, where it is in charge of terminating nerve impulses. It plays a vital role in brain activity. Over-stimulation of ACh receptors can result in mental illnesses such as depression. When present in low levels, however, additional diseases, like myasthenia and Alzheimer's disease can occur. As a result, these enzyme inhibitors are being investigated as potential pharmacological targets (Khan et al. 2018; Nair et al. 2004). Several PAs and their N-oxides were isolated from *Solenanthus lanatus*, such as acetylheliosupine-n-oxide, heliosupine-n-oxide, and angeloylechinatine-n-oxide. All of these compounds inhibited AChE with IC_{50} values ranging from 0.53 to 0.60 mM (Benamar et al. 2016). In another study by Benamar et al. (2017), they isolated several compounds from *Echium confusum* such as echimidine-n-oxide, echimidine, and 7-O-angeloyllycopsamien-n-oxide and 7-O-angeloylretronecine. These compounds showed significant AChE inhibition with IC_{50} values ranging from 0.275 to 0.769 mM. In Table 3.2, the pharmacological action of plants containing PAs/N-oxides is presented.

3.5 TOXICOLOGICAL EFFECTS OF PYRROLIZIDINE ALKALOIDS

Pyrrolizidine alkaloids (PAs) are known to cause severe toxicities in human beings such as genotoxicity, carcinogenicity, hepatotoxicity, and mutagenicity. Half of the known PAs have been reported to be toxic. Even PAs causes mild to severe toxicity to livestocks and grazing animals. The most common toxic effects due to PAs include weight loss, jaundice, depression, diarrhea, abdominal pain circling, and blindness (Wei et al. 2021). The most predominant PAs that cause toxicities are riddelliine, retrorsine, monocrotaline, symphytine (retronecine-type); lasiocarpine and heliotrine (heliotridine-type); and senkirkine and petasitenine (otonecine-type) (Ruan et al. 2014). These PAs can be metabolized into highly reactive pyrrole esters by cytochrome-P450 monooxygenase enzymes in the liver, which can bind to proteins and DNA, causing protein dysfunction and genotoxicity (Robertson and Stevens 2017). The hepatic sinusoidal obstruction syndrome, formerly known as veno-occlusive disease, is the most common adverse health outcome of acute PAs toxicity. Chronic diseases such as liver cirrhosis, cancer, and pulmonary arterial hypertension have been linked to long-term ingestion of small quantities of PAs and their N-oxides (Mattocks 1986; Neuman et al. 2015; Edgar et al. 2011; Kakar et al. 2010). The PAs and their N-oxides cause several types of genotoxicities such as DNA binding, DNA cross-linking, DNA-protein cross-linking, sister chromatid exchange, chromosomal aberrations, etc. (Fu et al. 2004).

3.5.1 Mutagenic Effects of PAs

The mutagenic effects of PAs/N-oxides are studied extensively by many researchers and are well documented (Griffin and Segall 1986; Gimmler-Luz et al. 1990; Fu et al. 2004; Frei et al. 1992; Green and Muriel 1975). PAs/N-oxides mutagenic effects are also reported in grazing livestock (de Lanux-Van Gorder 2000; Fletcher et al. 2009). For assessing the mutagenicity of PAs, the *Salmonella typhimurium/mammalian microsome* assay has been widely employed. In the presence of S9, retrorsine was assessed using TA98, TA100, TA1535, and TA1537 tester strains, and the PA was mutagenic for TA1535 and TA1537, implying that it caused both base-pair substitution and frameshift mutations (Wehner et al. 1979).

TABLE 3.2 Pharmacological Action of Plants Containing PAs/N-oxides

S. no.	Plants	PAs/N-oxides	*In-vitro/in-vivo* studies	Reference
1.	*Liparis nervosa*	Nervosine I, nervosine II, nervosine III, nervosine IV, nervosine V, nervosine VI	Anti-inflammatory activity	Ghosh and Singh 1974
2.	*Heliotropium digynum*	Heliotrine-*n*-oxide, heliotrine, 7-angelyolsincamidine-*n*-oxide, and europine	*In-vitro* antioxidant activity	Ghosh and Singh 1974
3.	*Senecio jacobaea*	Jacobine, jaconine, senecionine, and seneciphylline	Antimicrobial activity	Macel 2011
		Usaramine, monocrotaline, and azido-retronecine	Antimicrobial activity	Neto et al. 2016; Hol and Van 2002
4.	*Senecio brasiliensis*	Usaramine, seneciphylline, senecionine, integerrimine and retrorsine	Anti-ulcer activity	Toma, et al. 2004a
5.	*Heliotropium indicum*	Indicine N-oxide	Anticancer activity	Appadurai 2014
6.	*Castanospermum australe*	Australine and alexine	Anti-HIV activity	Vlietinck 1997
7.	*Alexa leiopetala* and *C. australe*	Alexine	Anti-HIV activity	Pal et al. 1989; Taylor et al. 1992
8.	*Solenanthus lanatus*	Heliosupine-n-oxide, 30-o-acetylheliosupine-*n*-oxide, and 7-0-angeloylechinatine-n-oxide	Acetylcholinesterase activity	Khan et al. 2018; Nair et al. 2004
9.	*Echium confusum*	7-O-angeloyllycopsamine-N-oxide, echimidine-N-oxide, echimidine, and 7-O-angeloylretronecine	Acetylcholinesterase activity	Benamar et al. 2016; Benamar, et al. 2017

3.5.2 Carcinogenicity Produced by PAs

Several clinical studies show the carcinogenic potential of PAs and their *N*-oxides (Mattocks et al. 1986). The liver is the most common carcinogenic target for PAs. Other organs also affected by PAs were the kidney, bladder, lung, and spinal cord (Chan et al. 1994; Hirono et al. 1983; Mori et al. 1984; Peterson et al. 1983). Roeder (1999) investigated PAs mutagenic potential and found that senkirkine had the highest mutagenic potential.

TABLE 3.3 Toxicological Effects of PAs/N-oxides

S. no.	Plant sources	PA compounds	Toxic effects	References
1.	*S. longilobus* and *T. farfara*	Monocrotaline, lasiocarpine, heliotrine, clivorine, and riddelline	Mutagenicity and carcinogenicity	Stewart et al. 2001
2.	*H. indicum*	Indicine, acetyl indicine, indicinini	Geno toxicity	Zhao et al. 2011
3.	*H. europaeum*	Dehydropyrrolizidine alkaloids	Geno toxicity	Yang et al. 2001
4.	*Tussilago farfara*	Senecionine, senkirkine	Carcinogenicity	Lebada et al. 2000
5.	*Senecio longilobus*	Riddelliine, retrorsine N-oxide	Mutagenic potential	Roeder 1999
6.	*Cynoglossum officinale*	Heliotridine, retronecine, and otonecine	Genotoxicity and tumourigenicity	El-Shazly et al. 1996
7.	*Crotalaria assamica*	Trichodesmine and monocrotaline	Neurotoxicity	Huxtable et al. 1996

3.5.3 Hepatotoxic Effects of PAs

Wilmot and Robertson (1920) studied the hepatotoxic effect of PAs on humans. It was discovered in the 1960s that these chemicals can cause chronic lung toxicity (Mclean 1970), and Mattocks (1986) claimed that animal experiments proved that several of them are hepatotoxic. Many manifestations of PA toxicity, which have been observed in both animals and humans, do not appear for weeks or months after the initial exposure. Pyrrole metabolites are the most common cause of liver damage, which results in parenchymal cell megalocytosis and liver cirrhosis. Animals with acute hepatotoxicity have a loss of sinusoidal endothelial cells, as well as dilated and congested sinusoids. Chronic toxicity, including centrilobular parenchymal cell necrosis, hepatocyte megalocytosis, liver cirrhosis, bile duct proliferation, fibrosis, vascular damage, and metabolic function loss, as well as tumor formation can occur. Hepatotoxicity and neurotoxicity are all possible side effects of PAs and their N-oxides (Hanumegowda et al. 2003). The most prominent toxicological effects of PAs/N-oxides are described in Table 3.3.

3.6 CONCLUSIONS

In recent decades, a number of outbreaks of human poisoning cases have been reported due to ingestion of contaminated food grains and herbal medicines. PAs and their N-oxides are known to cause mild to severe toxicities such as genotoxicity, carcinogenicity, and mutagenicity in human beings and animals. Thus, the quality and safety of food items and herbal medicines should be checked, and the content of PAs and their N-oxides in such materials analyzed. The finding of the present study provides comprehensive data on PAs and their N-oxides sources, chemistry, and pharmacological and toxicological effects. The present study expands the knowledge about PAs' toxicological effects, and helps in designing strategies to prevent contamination of food items and herbal medicines with PAs and their N-oxides.

REFERENCES

Ahamad, J., Toufeeq, I., Khan, M. A., Ameen, M. S. M., Anwer, E. T., et. al. (2019). Oleuropein: A natural antioxidant molecule in the treatment of metabolic syndrome. *Phytother Res*, 33(12), 3112–3128.

Appadurai, P. and Rathinasamy, K. (2014). Indicine N-oxide binds to tubulin at a distinct site and inhibits the assembly of microtubules: A mechanism for its cytotoxic activity. *Toxicol Lett*, 225, 66–77.

Arseculeratne, S. N., Gunatilaka, A. A. and Panabokke, R. G. (1981). Studies on medicinal plants of Sri Lanka: Occurrence of pyrrolizidine alkaloids and hepatotoxic properties in some traditional medicinal herbs. *J Ethnopharmacol*, 4, 159–177.

Arzt, J. and Mount, M. E. (1999). Hepatotoxicity associated with pyrrolizidine alkaloid (*Crotalaria* spp) ingestion in a horse on Easter Island. *Vet Hum Toxicol*, 41, 96–99.

Bach, N., Thung, S. N. and Schaffner, F. (1989). Comfrey herb tea-induced hepatic veno-occlusive disease. *Am J Med*, 87, 97–99.

Bah, M., Bye, R. and Pereda-Miranda, R. (1994). Hepatotoxic pyrrolizidine alkaloids in the Mexican medicinal plant *Packera candidissima* (Asteraceae: Senecioneae). *J Ethnopharmacol*, 43, 19–30.

Benamar, H., Tomassini, L., Venditti, A., Marouf, A., Bennaceur, M. and Nicoletti, M. (2016). Pyrrolizidine alkaloids from *Solenanthus lanatus* DC. with acetylcholinesterase inhibitory activity. *Nat Prod Res*, 30, 2567–2574.

Benamar, H., Tomassini, L., Venditti, A., Marouf, A., Bennaceur, M., Serafini, M. and Nicoletti, M. (2017). Acetylcholinesterase inhibitory activity of pyrrolizidine alkaloids from *Echium confusum* coincy. *Nat Prod Res*, 31, 1277–1285.

Böttcher, F., Ober, D. and Hartmann, T. (1994). Biosynthesis of pyrrolizidine alkaloids: Putrescine and spermidine are essential substrates of enzymatic homospermidine formation. *Can J Chem*, 72, 80–85.

Bruneton, J. (2008). *Farmacognosia*, 2nd ed., Acribia: Zaragoza Spain, ISBN 978-1-84585-006-7.

Chan, P. C., Mahler, J., Bucher, J. R., Travlos, G. S. and Reid, J. B. (1994). Toxicity and carcinogenicity of riddelliine following 13 weeks of treatment to rats and mice. *Toxicon*, 32, 891–908.

Chen, T., Mei, N. and Fu, P. P. (2010). Genotoxicity of pyrrolizidine alkaloids. *J Appl Toxicol*, 30(3), 183–196.

Cramer, L., Schiebel, H. M., Ernst, L. and Beuerle, T. (2014). Pyrrolizidine alkaloids in the food chain: Development, validation, and application of a new HPLC-ESI-MS/MS sum parameter method. *J. Agric. Food Chem*, 61, 11382–11391.

Culvenor, C. C., Edgar, J. A. and Smith, L. W. (1981). Pyrrolizidine alkaloids in honey from *Echium plantagineum* L. *J Agric Food Chem*, 29, 958–960.

Deinzer, M. L., Thomson, P. A., Burgett, D. M. and Isaacson, D. L. (1977). Pyrrolizidine alkaloids: Their occurrence in honey from tansy ragwort (*Senecio jacobaea* L.). *Science*, 195, 497–499.

de Lanux-Van Gorder, V. (2000). Tansy ragwort poisoning in a horse in southern Ontario. *Can Vet J*, 41, 409–410.

Directive 2002/32/EC of the European Parliament and of the Council of 7 May 2002 on undesirable substances in animal feed. *Off. J. Eur. Communities L.* 2002, 140, 10–22.

Edgar, J. A., Colegate, S. M., Boppré, M. and Molyneux, R. J. (2011). Pyrrolizidine alkaloids in food: A spectrum of potential health consequences. *Food Addit Contam A Chem*, 28, 308–324.

Edgar, J. A., Roeder, E. and Molyneux, R. J. (2002). Honey from plants containing pyrrolizidine alkaloids: A potential threat to health. *J Agric Food Chem*, 50, 2719–2730.

EFSA (European Food Safety Authority) (2017). Risks for human health related to the presence of pyrrolizidine alkaloids in honey, tea, herbal infusions and food supplements. *EFSA Journal*, 15.

El-Shazly, A., Sarg, T., Ateya, A., Aziz, E. A., Witte, L. and Wink, M. (1996). Pyrrolizidine alkaloids of *cynoglossum officinale* and *cynoglossum amabile (*family *boraginaceae*). *Biochem Syst Ecol*, 24, 415–421.

El-Shazly, A. and Wink, M. (2014). Diversity of pyrrolizidine alkaloids in the Boraginaceae structures, distribution, and biological properties. *Diversity*, 6, 188–282.

Fletcher, M. T., McKenzie, R. A., Blaney, B. J. and Reichmann, K. G. (2009). Pyrrolizidine alkaloids in *Crotalaria* taxa from northern Australia: Risk to grazing livestock. *J Agric Food Chem*, 57, 311–319.

Frei, H., Luthy, J., Brauchli, J., Zweifel, U., Wurgler, F. E. and Schlatter, C. (1992). Structure/activity relationships of the genotoxic potencies of sixteen pyrrolizidine alkaloids assayed for the induction of somatic mutation and recombination in wing cells of *Drosophila melanogaster*. *Chem Biol Interact*, 83, 1–22.

Fu, P. P., Xia, Q., Lin, G. and Chou, M. W. (2004). Pyrrolizidine alkaloids-genotoxicity, metabolism enzymes, metabolic activation, and mechanisms. *Drug Metab, Rev*, 36, 1–55.

Ghosh, M. and Singh, H. (1974). Inhibitory effect of a pyrrolidine alkaloid, crotalaburnine, on rat paw oedema and cotton pellet granuloma. *Br J Pharmacol*, 51, 503–508.

Gimmler-Luz, M. C., Erdtmann, B. and Balbueno, R. A. (1990). The effect of the pyrrolizidine alkaloid integerrimine on the chromosomes of mouse bone marrow cells. *Mutat Res*, 241, 297–304.

Green, M. H. and Muriel, W. J. (1975). Use of repair-deficient strains of Escherichia coli and liver microsomes to detect and characterise DNA damage caused by pyrrolizidine alkaloids heliotrine and monocrotaline. *Mutat Res*, 28, 331–336.

Griffin, D. S. and Segall, H. J. (1986). Genotoxicity and cytotoxicity of selected pyrrolizidine alkaloids, a possible alkenal metabolite of the alkaloids, and related alkenals. *Toxicol. Appl. Pharmacol*, 86, 227–234.

Hanumegowda, U. M., Copple, B. L., Shibuya, M., Malle, E., Ganey, P. E. and Roth, R. A. (2003). Basement membrane and matrix metalloproteinase sinmonocrotaline-induced liver injury. *Toxicological Sciences*, 76(1), 237–246.

Hartmann, T. (1999). Chemical ecology of pyrrolizidine alkaloids. *Planta*, 207, 483–495.

Hartmann, T. and Witte, L. (1995). Chemistry, biology and chemoecology of the pyrrolizidine alkaloids. *Alkaloids. Chem Biol Perspect*, 9, 155–233.

Hirono, I., Ueno, I., Aiso, S., Yamaji, T. and Haga, M. (1983). Carcinogenic activity of *farfugium japonicum* and *senecio cannabifolius*. *Cancer Lett*, 20, 191–198.

Hol, W. and Van, V. J. (2002). Pyrrolizidine alkaloids from Senecio jacobaea affect fungal growth. *J Chem Ecol*, 28, 1763–1772.

Huang, S., Zhao, S. M., Shan, L. H. and Zhou, X. L. (2020). Antitumor activity of nervosine VII, and the crosstalk between apoptosis and autophagy in HCT116 human colorectal cancer cells. *Chin J Nat Med*, 18, 81–89.

Huxtable, R., Yan, C., Wild, S., Maxwell, S. and Cooper, R. (1996). Physicochemical and metabolic basis for the differing neurotoxicity of the pyrrolizidine alkaloids, trichodesmine and monocrotaline. *Neurochem Res*, 21, 141–146.

Kakar, F., Akbarian, Z., Leslie, T., Mustafa, M. L., Watson, J., Van Egmond, H. P., et al. (2010). An outbreak of hepatic veno-occlusive disease in Western Afghanistan associated with exposure to wheat flour contaminated with pyrrolizidine alkaloids. *J Toxicol*, 313280.

Khan, H., Amin, S., Kamal, M. A. and Patel, S. (2018). Flavonoids as acetylcholinesterase inhibitors: Current therapeutic standing and future prospects. *Biomed Pharmacother*, 101, 860–870.

Kowalczyk, E., Sieradzki, Z. and Kwiatek, K. (2018). Determination of pyrrolizidine alkaloids in honey with sensitive gas chromatography-mass spectrometry method. *Food Anal Meth*, 11, 1345–1355.

Langel, D., Ober, D. and Pelser, P. B. (2011). The evolution of pyrrolizidine alkaloid biosynthesis and diversity in the Senecioneae. *Phytochem Rev*, 10, 3–74.

Lebada, R., Schreier, A., Scherz, S., Resch, C., Krenn, L. and Kopp, B. (2000). Quantitative analysis of the pyrrolizidine alkaloids senkirkine and senecionine in *Tussilago farfara* L. by capillary electrophoresis. *Phytochem Anal*, 11, 366–369.

Macel, M. (2011). Attract and deter: A dual role for pyrrolizidine alkaloids in plant-insect interactions. *Phytochem Rev*, 10, 75–82.

Mattocks, A. R., Driver, H. E., Barbour, R. H. and Robins, D. J. (1986). Metabolism and toxicity of synthetic analogue of macrocyclic diesterpyrrolizidine alkaloids. *Chemico-Biological Interactions*, 56(1), 95–108.

Mclean, E. K. (1970). The toxic actions of pyrrolizidine (senecio) alkaloids. *Pharmacol Rev*, 22(4), 429–483.

Moreira, R., Pereira, D. M., Valentão, P. and Andrade, P. B. (2018). Pyrrolizidine alkaloids: Chemistry, pharmacology, toxicology and food safety. *Int J Mol Sci*, 19(6), 1668. 10.3390/ijms19061668

Mori, H., Kawai, K., Ohbayashi, F., Bunai, Y., Yamada, K. and Hirono, I. (1984). Some toxic properties of a carcinogenic pyrrolizidine alkaloid, petasitenine. *J Toxicol Sci*, 9, 143–149.

Nair, V. P. and Hunter, J. M. (2004). Anticholinesterases and anticholinergic drugs. *Contin. Educ. Anaesth. Crit. Care Pain*, 4, 164–168.

Neto, TdSN, Gardner, D., Hallwass, F., Leite, A. J. M., de Almeida, C. G., Silva, L. N., et al. (2016). Activity of pyrrolizidine alkaloids against biofilm formation and *trichomonas vaginalis*. *Biomed Pharmacother*, 83, 323–329.

Neuman, M. G., Cohen, L. B., Opris, M., Nanau, R. and Jeong, H. (2015). Hepatotoxicity of pyrrolizidine alkaloids. *J Pharm Pharm Sci*, 18, 825–843.

Pal, R., Hoke, G. M. and Sarngadharan, M. (1989). Role of oligosaccharides in the processing and maturation of envelope glycoproteins of human immunodeficiency virus type 1. *Proc Natl Acad Sci, USA*, 86, 3384–3388.

Peterson, J. E., Jago, M. V., Reddy, J. K. and Jarrett, R. G. (1983). Neoplasia and chronic disease associated with the prolonged administration of dehydroheliotridine to rats. *J Natl Cancer Inst*, 70, 381–386.

Robertson, J. and Stevens, K. (2017). Pyrrolizidine alkaloids: Occurrence, biology, and chemical synthesis. *Nat Prod Rep*, 34, 62–89.

Roeder, E. (1999). Analysis of pyrrolizidine alkaloids. *Curr Org Chem*, 3, 557–576.

Roeder, E. (2000). Medicinal plants in China containing pyrrolizidine alkaloids. *Pharmazie*, 55, 711–726.

Ruan, J., Yang, M., Fu, P., Ye, Y. and Lin, G. (2014). Metabolic activation of pyrrolizidine alkaloids: Insights into the structural and enzymatic basis. *Chem Res Toxicol*, 27, 1030–1039.

Stegelmeier, B. L., Edgar, J. A., Colegate, S. M., Gardner, D. R., Schoch, T. K., Coulombe, R. A. and Molyneux, R. J. (1999). Pyrrolizidine alkaloid plants, metabolism and toxicity. *J Nat Toxins*, 8, 95–116.

Stewart, M. J. and Steenkamp, V. (2001). Pyrrolizidine poisoning: A neglected are in human toxicology. *Therapeutic Drug Monitoring*, 23(6), 698–7078.

Taylor, D., Nash, R., Fellows, L., Kang, M. and Tyms, A. (1992). Naturally occurring pyrrolizidines: Inhibition of α-glucosidase 1 and anti-HIV activity of one stereoisomer. *Antivir Chem Chemother*, 3, 273–277.

The Plant List (2013). Version 1.1, released September 2013. http://www.theplantlist.org.

These, A., Bodi, D., Ronczka, S., Lahrssen-Wiederholt, M. and Preiss-Weigert, A. (2013). Structural screening by multiple reaction monitoring as a new approach for tandem mass spectrometry: Presented for the determination of pyrrolizidine alkaloids in plants. *Anal Bioanal Chem*, 405, 9375–9383.

Toma, W., Trigo, J. R., Bensuaski de Paula, A. C. and Monteiro Souza Brito, A. R. (2004a). Modulation of gastrin and epidermal growth factor by pyrrolizidine alkaloids obtained from *Senecio brasiliensis* in acute and chronic induced gastric ulcers. *Can J Physiol Pharmacol*, 82, 319–325.

Toma, W., Trigo, J. R., de Paula, A. C. B. and Brito, A. R. M. S. (2004b). Preventive activity of pyrrolizidine alkaloids from *Senecio brasiliensis* (Asteraceae) on gastric and duodenal induced ulcer on mice and rats. *J Ethnopharmacol*, 95, 345–351.

Valese, A. C., Molognoni, L., de Sá Ploêncio, L. A., de Lima, F. G., Gonzaga, L. V., Górniak, S. L., et al. (2016). A fast and simple LC-ESI-MS/MS method for detecting pyrrolizidine alkaloids in honey with full validation and measurement uncertainty. *Food Control*, 67, 183–191.

Vlietinck, A., De Bruyne, T. and Vanden Berghe, D. (1997). Plant substances as antiviral agents. *Curr Org Chem*, 1, 307–344.

Wehner, F. C., Thiel, P. G. and van Rensburg, S. J. (1979). Mutagenicity of alkaloids in the *Salmonella*/microsome system. *Mutat Res*, 66, 187–190.

Wei, X., Ruan, W. and Vrieling, K. (2021). Current knowledge and perspectives of pyrrolizidine alkaloids in pharmacological applications: A mini-review. *Molecules*, 26, 1970. 10.3390/molecules26071970

Wilmot, F. C. and Robertson, G. W. (1920). *Senecio* disease or cirrhosis of the liver due to *Senecio* poisoning. *Lancet*, 23, 848–849.

Xia, Q., Yan, J., Chou, M. W. and Fu, P. P. (2008). Formation of DHP-derived DNA adducts from metabolic activation of the prototype heliotridine-type pyrrolizidine alkaloid, heliotrine. *Toxicol Lett*, 178, 77–82.

Yang, Y. C., Yan, J., Doerge, D. R., Chan, P. C., Fu, P. P. and Chou, M. W. (2001). Metabolic activation of the tumorigenic pyrrolizidine alkaloid, riddelliine, leading to DNA adduct formation *in vivo*. *Chem Res Toxicol*, 14, 101–109.

Yu, Z., Guo, G. and Wang, B. (2020). Lycopsamine inhibits the proliferation of human lung cancer cells *via* induction of apoptosis and autophagy and suppression of interleukin-2. *J Buon*, 25, 2358–2363.

Zhao, Y., Xia, Q., Yin, J. J., Lin, G. and Fu, P. P. (2011). Photo irradiation of dehydropyrrolizidine alkaloids—Formation of reactive oxygen species and induction of lipid peroxidation. *Toxicol Lett*, 205, 302–309.

CHAPTER 4

Furanocoumarins and Lectins as Food Toxins

Kavita Munjal[1], Saima Amin[2], Showkat R. Mir[2], Vinod Kumar Gauttam[3], and Sumeet Gupta[4]

[1]Department of Pharmacognosy, M. M. College of Pharmacy, M. M. (Deemed to be University), Mullana, Ambala, Haryana, India
[2]Department of Pharmacognosy & Phytochemistry, School of Pharmaceutical Education and Research, Jamia Hamdard, New Delhi, India
[3]IES Institute of Pharmacy, Bhopal, India
[4]Department of Pharmacology, M. M. College of Pharmacy, M. M. (Deemed to be University), Mullana, Ambala, Haryana, India

CONTENTS

4.1	Introduction	67
4.2	Sources of Furanocoumarins and Lectins	70
4.3	Chemistry of Furanocoumarins and Lectins	72
4.4	Pharmacological Actions of Furanocoumarins and Lectins	73
4.5	Toxicological Reports of Furanocoumarins and Lectins	75
4.6	Conclusions	75
References		75

4.1 INTRODUCTION

Coumarins are phenolic compounds found in a wide range of plants. Coumarins, which are characterized by their benzopyran-2-one nucleus, have been isolated from a variety of plants belonging to the Apiaceae, Rutaceae, Fucaceae, and Leguminosae families [1]. Furanocoumarins are a subclass of phenolic compounds that belong to coumarin derivatives. They are secondary metabolites that have a role in the plant's defense against insects, diseases, and other organism infestations. They can be subdivided into two parts: linear, commonly known as psoralens; and angular, commonly known as angelicins. Psoralen, xanthotoxin, bergapten, and iso-pimpinellin are the most abundant linear furanocoumarins in higher plants, while angelicin, pimpinellin, sphondin, and iso-bergapten are the most common angular furanocoumarins. Psoralen is the most well-known furanocoumarin. Linear furanocoumarins are found largely in the Rutaceae, Moraceae, Leguminosae, and Apiaceae families of higher plants, whereas angular furanocoumarins are found primarily in the Apiaceae and Leguminosae families [2].

DOI: 10.1201/9781003222194-5

Furanocoumarins are tricyclic aromatic compounds that have a furan ring bonded to α-benzopyrone system (coumarin), which contributes the ketone functional group to its structure. Because the furan can be connected different positions, there are many isomers and derivatives with different atoms attached to the core structure, making furanocoumarins an immensely diverse and variable chemical family. The linear type furanocoumarins are formed when the furan ring is joined at C-7 and C-8 positions of the aromatic ring. The angular type furanocoumarins are formed when the furan ring is attached at the C-6 and C-7 positions [3]. These isomers dominate the distribution of these variants by a large margin in comparison to other isomeric forms. These two isomeric variants of furanocoumarins are psoralen and angelicin, respectively. In nature, furanocoumarins are often found as a complex mixture of isomers and derivatives, with some clearly dominating over the others. Important structures of furanocoumarins are depicted in Figure 4.1.

Furanocoumarins are biologically active with a variety of therapeutic and clinical applications. They can be found in all parts of the plants, but they are more concentrated in fruits, leaves, and roots. They are mostly found with resins and essential oil. Their distribution varies significantly between plant species, as well as by geographic location and climate [4].

Furanocoumarins' most notable attribute is their remarkable capacity to sensitive skin to light, in particular near ultraviolet light (UV-A, 320–400 nm) acting as plant-derived photosensitizers. They have been utilized to treat a variety of skin and autoimmune illnesses when combined with UV-A (PUVA therapy and photophoresis). In addition to its side effects (erythema, edema, hyperpigmentation, and premature aging of skin), furanocoumarin photochemotherapy has a number of adverse effects [5]. In various plants (e.g., parsley, celery, carrots, and limes), the furanocoumarins react with UV light via its parent chemical, psoralen, to cause skin eruptions. The eruption normally starts 24 hours after exposure and peaks 48–72 hours afterwards and may result in strong toxicity, mutagenicity, and possibly carcinogenicity later on. The long-wave ultraviolet A (320–400 nm) radiation causes phyto-photodermatitis, which is a cutaneous phototoxic inflammatory condition characterized by rashes, itching, eruption, etc. Many of the furanocoumarins can be categorized as photosensitizing chemicals as their temporal use may sensitize the skin to exposure to light [6].

The mechanism of action is widely understood. When a photon of the right wavelength strikes a furanocoumarin, the energy is absorbed, elevating the molecule from its ground state to a triple excited state. Energy is released in the form of heat, fluorescence, and/or phosphorescence upon return to the ground state, and a photoproduct gets formed. In phytophotodermatitis, two different photochemical reactions have been identified, which occur independently. Type I reaction occurs in the absence of oxygen, while a type II reaction occurs in its presence. These photochemical reactions, the light-activated furanocoumarin causes DNA interstrand cross-linking between the furan ring and the pyrimidine bases, particularly thymine of DNA, causing DNA damage by intercalation into the double-helical structure of DNA through molecular complexing. During the type I oxygen-independent reaction, the nuclear DNA attaches to the ultraviolet-activated furanocoumarins. Likewise, activated furanocoumarins also cause cell membrane damage and edema due to oxygen-dependent processes. This causes arachidonic acid metabolic pathways to be activated, ensuing cell death (damaged cells and apoptotic keratinocytes). These changes may result in erythema, blistering, epidermal necrosis, and eventually epidermal desquamation clinically [4,7–9].

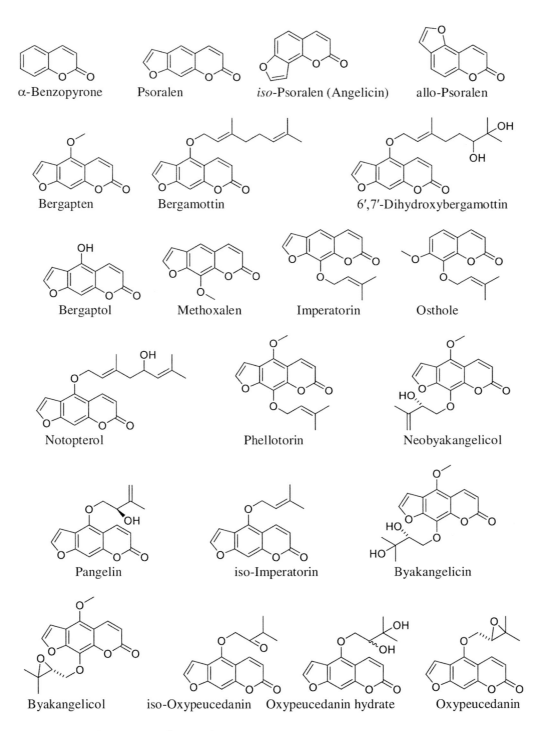

FIGURE 4.1 Structures of some furanocoumarins.

Furanocoumarins have been to treat skin conditions and are incorporated to sunscreen preparations. The notion was that by increasing melanin production, they would be able to defend themselves from UV rays. While this may have helped in the short term, it has been shown to raise cancer in the long run, even when the furanocoumarin exposure is stopped [10]. The citrus oils are frequently employed as a significant component of many cosmetics. Citrus fruits, particularly grapes and lemon, having a number of compounds in their peels that may interact negatively with medications. Citrus fruit juice is usually made from the entire fruit, including the peel. Bergamottin (also known as bergamot) is a natural furanocoumarin found in the peel that has been shown to inhibit particular isoforms of the cytochrome P450 enzyme (CYP3A4). Inhibition of this enzyme inhibits certain medications from undergoing metabolism, resulting in a higher concentration of the drugs in the bloodstream. When exposed to sunlight, bergamottin and other compounds in citrus oils (e.g., lime, grapefruit, orange, and lemon) are phototoxic, causing substantial skin damage. When exposed to UV radiation, 5-methoxypsoralen, the main phototoxic component of bergamot oil, has displayed carcinogenic activity in bacterial assays as well as clastogenic effects in mammalian cells in culture [11,12].

Applying a phototoxin to the skin can prove to be hazardous. As a result, laws governing the maximum amount of furanocoumarins allowed in commercial cosmetics have been enacted and implemented. Many government regulatory organizations, including the European Commission, have set a furanocoumarin limit of 1 ppm in cosmetic items. It is desired that the use of cosmetics from legal manufacturers should not contribute to health problems in the majority of the population [13,14].

Lectins are a class of protein that have the ability to recognize and bind to specific carbohydrates. They can bind to monosaccharides, oligosaccharides, lipopolysaccharides, glycosides, glycoproteins, glycolipids, or other glycol-conjugates. These interactions are mostly reversible and non-enzymatic in nature. The carbohydrate-binding properties of lectins are key to many of their biological activities such as cell-molecule and cell-cell interactions in a variety of biological systems. The specificity of these interactions have been exploited for their purification and characterization. Lectins possess the ability to agglutinate erythrocytes. Thus, they are also referred to as hemagglutinins or phytoagglutinins, because they were initially found in plant extracts only. Traditionally, they have been used as histology and blood transfusion reagents. Recently their role has been extended to include detection and isolation of glyco-conjuagtes, histochemistry of cells and tissues, and for following changes on cell surfaces during physiological and pathological processes, from cell differentiation to cancer.

4.2 SOURCES OF FURANOCOUMARINS AND LECTINS

Citrus plants have long been known to produce coumarins and furanocoumarins as defense chemicals to cope with herbivorous insects and pathogens. In humans, these chemical compounds are strong photosensitizers and can interact with medications, leading to the 'grapefruit juice effect'. Removing coumarins and furanocoumarins from food and cosmetics imply additional costs and might alter product quality. Thus, the selection of Citrus cultivars displaying low coumarin and furanocoumarin contents constitutes a valuable alternative. In a study, the contents of these compounds within the peel and the pulp of 61 Citrus species was undertaken using UPLC-MS analysis. It was found that Citrus peel contained larger diversity and higher concentrations of coumarin/furanocoumarin than the pulp of the same fruits. On the basis of analysis, the Citrus species were classified into four groups that corresponded to four ancestral taxa namely pummelos, mandarins, citrons, and papedas. Three of the four ancestral taxa (pummelos, citrons, and papedas) were found to synthesize high amounts of these compounds, whereas mandarins appeared practically devoid of them [15].

Grapefruits (*Citrus paradisi*) are medium-sized fruit trees, belonging to the Rutaceae family. Grapefruit, considered to be an accidental cross between of pomelos (*C. maxima*) and sweet orange (*C. sinensis*), was first discovered in the 18th century. Different grapefruit varieties vary in color from white to red depending on the presence or absence of lycopenes [16]. According to data from the Food and Agricultural Organization of the United Nations, China and the United States are the world's largest grapefruit producers. In China, 5.01 million tons of grapefruit were reportedly produced in 2019 [17]. The foremost kinds of grapefruit encompass pink, ruby pink, superstar ruby, Thompson, and white Marsh. Because of several bioactive phytochemicals, including flavonoids, carotenoids, coumarins, and organic acids, the grapefruits possess several beneficial effects [18–20]. Flavonoids are considered to be the most important bioactive components present in grapefruits. The major flavonoids found in grapefruits, including hesperetin, naringenin, narirutin, and didymin, have been extensively studied both in-vitro and in-vivo to confirm their role in benefiting human health [21,22]. Although these phytochemicals exhibit diverse bioactive activities, some compounds have been shown to interact with numerous medications causing adverse effects due to the delay in metabolism of drugs. Furanocoumarins and flavanones are the major culprits responsible for grapefruit juice causing these drug interactions. Several studies have reported that furanocoumarins present in grapefruit interact with medications by interfering with the haptic and intestinal enzyme cytochrome P450 [23]. Several studies have developed different processing methods for the removal of furanocoumarins from grapefruits such as heat treatment, UV irradiation, and mixing with autoclaved fungi [24]. Numerous comprehensive reviews summarized the grapefruit juice drug interaction mechanisms mainly focusing on the role of furanocoumarins. Although furanocoumarins can cause undesirable effects because of interactions with certain medications, recent evidence has emerged from several in-vitro and in-vivo studies suggesting that furanocoumarins possess additional biological activities, such as antioxidative, antiproliferative, antiinflammatory, and bone health promoting effects [25–27]. As grapefruit juice is one of the most popular fruit juices worldwide, grapefruit furanocoumarins deserve more attention with regard to their health benefits.

Psoralea corylifolia is the most abundant source of furanocoumarins with a total content of more than 2.5% by weight; two main compounds have been identified as psoralen and iso-phoralen [28]. The composition pattern observed, with two dominant compounds, is reported in the majority of relevant literature data [29]. According to data in the literature, minor constituents of *P. corylifolia* include bakuchicin and psoralidin [30].

The seeds of *Cnidium monnieri* and the roots of *Angelica archangelica* contain a considerable amount of furanocoumarins and several of them have been identified successfully with different identification tools. Osthole was reported as the most important coumarin derivative related to furanocoumarins in *C. monnieri*. Other constituents were iso-pimpinellin, auraptenol, and morinzin. Imperatorin become diagnosed as the primary furanocoumarin of *A. archangelica* in the literature; other compounds included pimpinellin and phellopterin [31–33].

Lectins are mostly of plant origin; however, many of them have been reported from animals such as eel, snail, horseshoe crab, etc. In plants, lectins are abundant in seeds, grains, and storage tissues. Ricin from *Ricinus communis* and abrin from *Abrus precatorius* are one of the earliest representatives of lectins. Lectins from soya beans, green peas, wheat germ, and mushrooms were among the first isolated in their pure form. Invertebrates and lower vertebrates were the first sources of lectins from animals. The first lectin from the animal source was isolated and characterized from the eel. It was shown to be specific for L-fucose. Conventionally, the lectin families and lectin domains are named after the first lectin of the family that was characterized for its molecular structure and carbohydrate-binding properties. Twelve families have been identified for the plant-derived lectins based on the presence of typical lectin domain. Lectins have been also reported from mushrooms, algae, fungi, and sponges.

4.3 CHEMISTRY OF FURANOCOUMARINS AND LECTINS

Furanocoumarins are tricyclic aromatic compounds composed of a furan ring fused to a α-benzopyrone (coumarin) system. Even though the furan moiety can be in either a 2,3- or a 3,2-arrangement at the c, f, g, or h bonds of the coumarin, most naturally occurring derivatives belong to the psoralen (furo [3,2-g][1]benzopyran-7-one), allo-psoralen (furo [2,3-f][1]benzopyran-7-one), angelicin (furo[2,3-h][1]benzopyran-7-one), and furo[3,2-c] coumarin (furo[3,2-c][1]benzopyran-7-one) subclasses (Figure 4.1). The whole class can be structurally divided into angular and linear furanocoumarins, with the latter being compounds with the furan ring attached at 6,7 positions on the aromatic ring (e.g., angelicin), whereas in the former, the furan ring is attached to 7,8 positions (e.g., psoralen). In higher plants, the most abundant linear furanocoumarins are psoralen, xanthotoxin, bergapten, and iso-pimpinellin, while the most relevant angular furanocoumarins are angelicin, pimpinellin, sphondin, and iso-bergapten [34–36].

More than 50 natural furanocoumarins are known, and several screenings have revealed a consistent qualitative and quantitative heterogeneity at both intra-generic and intra-specific levels [37]. As a result, a large variability in biological activities has been observed for plants containing furanocoumarins and different degrees of toxicity and drug interactions may be encountered with those used as foods [38]. The literature reports a wide variation in the content of furanocoumarins in plants, which can be explained by a combination of natural and genetic factors, influence of storage or processing, and by constraints at both sample treatment and analytical protocols [39]. Unfortunately, pharmacological studies have rarely provided an adequate chemical characterization of the plant material and, at the same time, the evaluation of the dietary intake for furanocoumarins is considered extremely difficult [40]. The data on their average human consumption from food plants are scarce and could be more than the published estimates [41].

From a chemotaxonomical standpoint, furanocoumarins have been isolated from few genera within a limited group of plant families, including Apiaceae, Fabaceae, Moraceae, and Rutaceae and, to a minor degree in Amaranthaceae, Asteraceae, Cyperaceae, Meliaceae, Pittosporaceae, Rosaceae, and Solanaceae [42]. Apiaceae and Rutaceae provide the larger number of species and the most valuable sources of furanocoumarins for pharmacological and toxicological purposes. In terms of food and medicinal plants, Apiaceae usually harbor, on a fresh-weight basis, a higher content than Rutaceae [43].

Inhibitory action of furanocoumarins against insects is understood to take place through a cytochrome P450-mediated metabolism with CYP6B enzymes especially for rapid and precise furanocoumarins metabolization [44]. On this regard, the natural capability of some furanocoumarins to interact, inactivate, and influence microsomal enzymes is the foundation of their relevance in drug–herb interactions and of their role in the modulation of pharmacokinetic properties of multiple substances in vivo [45].

In plant life, furanocoumarins play a key function as phytoalexins and they are highly regarded for their involvement in both constitutive and triggered plant protection. They provide protection against microorganisms, nematodes, phytophagous insects, herbivores, and plant competitors [46,47]. The biosynthesis of furanocoumarins can therefore be prompted and more suitable via direct exposure to microorganisms, insects, and fungi, as well as via abiotic elicitors, including UV radiation and bodily harm, with specific intensities resulting in distinctive accumulation tiers. Their toxicity in nature is based totally on direct contact and photoactivation and it is dependent on their potential to create DNA adducts under the effect of UV-A, as discussed previously. It results in a robust cytotoxicity and acute infection in animals, making furanocoumarin-containing flowers a category characterized via excessive bioactive and toxic capability [48]. In this regard, unique toxicities are described for linear and angular furanocoumarins, such as psoralene-type linear furanocoumarins, like xanthotoxin and bergapten, which display sturdy photosensitizing consequences in comparison to angelicin-type angular forms, whose phototoxic effect is weaker.

Lectins are carbohydrate-binding proteins. They yield lectin-carbohydrate complexes without changing the carbohydrates, thus are termed as non-enzymatic proteins. The interactions sites are restricted to limited polypeptide segments. These are called carbohydrate recognition domains. The specificity of lectins to particular a carbohydrate or glycol-conjugate is due to this domain. It is interesting to note that multiple lectin domains can recognize similar carbohydrates. Lectin-monosaccahride interactions are weaker than the lectin-complex carbohydrates as more binding sites are available for interaction. Taking advantage of the specific interaction between lectins and carbohydrates, affinity chromatography has been successfully used for the purification of lectins.

4.4 PHARMACOLOGICAL ACTIONS OF FURANOCOUMARINS AND LECTINS

Pharmacological and biochemical studies have indicated that several furanocoumarins (iso-imperatorin, oxypeucedanin, notopterol, and bergapten) have anti-inflammatory, analgesic, anti-cancer, and anti-coagulant potential [49–51]. In 2015, Uto et al. evaluated the anti-inflammatory activities of constituents present in either aerial or root parts of *Angelica acutiloba* (Siebold and Zucc.) Kitag. on a RAW264.7 cell model treated with lipopolysaccharide and found a modest anti-inflammatory activity of bergaptol, through the attenuation of the production of nitrite, prostaglandin E2, interleukin-6, and TNF-alpha [52]. This anti-inflammatory activity of bergaptol was found to be due to the up-regulation of heme oxygenase-1 expression, which regulates antioxidative activity. Oxypeucedanin has also shown anti-inflammatory activity similar to that of bergaptol, acting mainly on nitrite production and in the expression of inducible nitric oxide synthase on LPS-treated RAW 264.7 cells [53,54].

The furanocoumarins has been found out to affect the brain activity. Imperatonin is reported to cause irreversible inactivation of gama-aminobutyric acid (GABA)-transaminase, resulting in increased GABA quantity within the synaptic clefts of neurons and within the brain [55,56]. Moreover, imperatorin showed dose dependent increase in the maximal electro-convulsions in mice that reached the peak of the maximum anticonvulsant effect 30 minutes after systemic administration. Furthermore, imperatorin at sub-threshold doses increase the anticonvulsant effects of co-administered antiepileptic drugs such as carbamazepine, phenytoin, and phenobarbital, but not those of valproate. Eventually, a higher carbamazepine concentration was achieved inside the brain in experimental animals, but not for phenytoin and phenobarbital [57]. The mechanism by which the co-administration of imperatorin induces these effects seems to be that the furanocoumarin enhances the diffusion of carbamazepine via the blood–brain barrier by way of modulating its permeability and/or inhibiting the multi-drug resistance proteins or the P-glycoproteins [53,57].

Few studies have mentioned the bone-health-promoting property of furanocoumarins, particularly bergapten in *in-vitro* and *in-vivo* studies. Meng et al. identified the proliferation-stimulating fraction from crude extract of *Cnidium monnieri* (L.) Cusson using osteoblast-like UMR106 cells. It was observed that bergapten drastically improved the cell proliferation of UMR 106 cells in a dose- and time-dependent way [58]. It also improved the level of alkaline phosphatase (ALP), bone nodule formation, type I collagen synthesis, and bone morphogenetic protein-2 (BMP-2) gene expression [59]. Modulation of key enzymes in bone health promotion by bergapten is not restricted to the above-defined proteins; however, it involves additionally the expression of Runt-associated transcription thing 2 (RUNX2), modulation of the Wnt/ff-catenin pathway, and osteocalcin (non-collagenous protein secreted by means of osteoblasts) [60,61]. Oral administration of bergapten for three months in an ovariectomized mouse model (one of the principal animal models for analyzing postmenopausal osteoporosis) resulted in an exceptional decrease in delaying bone loss, as shown through evaluation of the bone mineral density, separation, and number of trabeculae [62,63]. Bergapten restricts the increase of osteoclastic cell, responsible for bone resorption, through activation of the apoptotic pathway in LPS-induced osteoblast precursor RAW 264.7 cells. These facts provide experimental evidences on the modulation of bone strength in rats by bergapten, the main constituent of *C. paradisi* [64–66].

Research related to lectins has picked up recently. Concanavalin A from *Canavalia* spp. and Phytohemagglutinins from *Phaseolus vulgaris* have been extensively studied. They have been studied for pro-inflammatory, anticancer, and mitogenic activities on lymphocytes [67,68]. Concanavalin A was found to affect apoptosis and autophagy by modulating cellular-signaling pathways in most of the cancer cell lines. It was also shown to disintegrate the ability of mitochondrial membrane in human cancer A375 cells, triggering the release of the cytochrome C and the caspase activation, both concerned in cell apoptosis. Concanavalin A additionally triggered apoptosis in ovarian cancers SKOV3 cells with the aid of modulating the expression of apoptosis-related proteins cyclooxygenase 2 (COX-2), B-cellular lymphoma 2 (Bcl-2), and serine-threonine protein kinase AKT and activating the Foxola-Bim signaling pathways. *Phaseolus vulgaris* lectin was demonstrated to inhibit the proliferation of nasopharyngeal carcinoma HONE-1 cells [69]. Moreover, plant lectins have been also used as polyclonal activators of T-cells to examine the *in-vitro* lymphocyte features. Concanavalin A and phytohemagglutinins revealed to bind membrane carbohydrates expressed on T-lymphocytes, and then to activate mitogenic response of lymphocytes and the production of cytokines [70].

The algae *Ticocarpus crinitus* and the mussel *Mytilus trossulus* lectins were confirmed to enhance the synthesis of cytokines. *T. crinitus* lectin exhibited a growth in a dose-dependent manner of tumor necrosis component α (TNFα), interleukin 6, and interferon γ (INFγ) in human whole-blood cells [71]. The *M. trossulus* lectin, at a concentration of 5 and 80 μg/mL, showed a stimulating effect on spontaneous and lipopolysaccharides (LPS)-triggered production of TNFα with the aid of human peripheral blood cells.

However, a stimulating activity on spontaneous and induced production of IFNγ become most effective observed at a concentration of 80 µg/mL. The increase of lymphocytes turned into also stimulated via the mussel *M. trossulus* lectin [72]. Several lectins are pursued for their potential therapeutic effects against cancer, diabetes, arthritis, etc. Wheat germ agglutinin has the ability to preferentially agglutinate malignant cells. Certain lectins are also used as carriers for the delivery of chemotherapeutic agents. They are also used in investigations related to bacteria, fungi, and bacteriophages.

4.5 TOXICOLOGICAL REPORTS OF FURANOCOUMARINS AND LECTINS

Single oral treatment of rats with 5,000 mg/kg body weight of hydro-ethanolic leaf extract of *Flabellaria paniculata* (HLE-FP) produced no loss of life. The behavioral manifestations observed include dyspnoea, restlessness/agitation, and generalized body tremor, feed, and water refusal within 24 hours of administration. These manifestations regularly subsided after 24 hours. The LD50 was presumed to be more than 5,000 mg/kg. Single intraperitoneal treatment with HLE-FP at doses of 250, 500, 1,000, 2,000, and 4,000 mg/kg caused 0, 0, 0, 20, and 80% mortality, respectively, in the treated mice within 24 hours' post-treatment. Using log dose-probit analysis, the i.p. LD50 was estimated to be 2,951.2 mg/kg. Behavioral manifestations on i.p. administration of HLE-FP includes restlessness and writing within the first 15 minutes, and calmness and anorexia within 48 hours' post-treatment for surviving mice [73].

4.6 CONCLUSIONS

Furanocoumarins, the most important ingredients of *B. gaudichaudii* roots, are photo-toxic materials that the plant makes use of as a protection mechanism in opposition to phytopathogenic microorganisms and herbivores. The mechanism of movement is intercalation inside the double helix of the DNA structure and in molecular complexation, and when activated through light, they react with pyrimidine bases, particularly with thymine, which can sell toxic, mutagenic, and carcinogenic outcomes [74].

Acute and long-term consumption of lectins can be problematic. Lectins bind to the surface carbohydrates in gut, resulting in damage to their lining and may cause colitis, IBS, etc. Their role have been established in peptic ulcer, nephropathy, insulin-dependent diabetes, rheumatoid arthritis, and other diseases. Lectins also resist digestion in rodents and result in their poor performance especially after the ingestion of lectins from kidney beans. Lectin feeding has been reported to result in enlargement of intestine, liver, and pancreas in several experimental animals.

REFERENCES

1. Stefanachi, A., et al. (2018). Coumarin: A natural, privileged and versatile scaffold for bioactive compounds. *Molecules (Basel, Switzerland)*, 23(2), 250.
2. Río, J. A. D., et al. (2014). Furanocoumarins: Biomolecules of therapeutic interest. In Attaur, R. (ed.). *Studies in Natural Products Chemistry*. Elsevier: Amsterdam, pp. 145–195.

3. Bruni, R., Barreca, D., Protti, M., Brighenti, V., Righetti, L., Anceschi, L., Mercolini, L., Benvenuti, S., Gattuso, G., Pellati, F. (2019). Botanical sources, chemistry, analysis, and biological activity of furanocoumarins of pharmaceutical interest. *Molecules*, 24(11). 2163. https://doi.org/10.3390/molecules24112163

4. Christensen, L. P. (2018). Polyphenols and polyphenol-derived compounds from plants and contact dermatitis. In Watson, R. R., Preedy, V. R. and Zibadi, S. (ed.). *Polyphenols: Prevention and Treatment of Human Disease* (Second Edition). Academic Press: Cambridge, pp. 349–384.

5. Potapenko, A., Malakhov, M. V. and Kiagova, A. A. (2004). Photobiophysics of furanocoumarins. *Biofizika*, 49(2), 322–338.

6. Parsons, B. J. (1980). Psoralen photochemistry. *Photochem Photobiol*, 32(6), 813–821.

7. Diekmann, J., et al. (2019). The Photoaddition of a psoralen to DNA proceeds via the triplet state. *Journal of the American Chemical Society*, 141(34), 13643–13653.

8. Gupta, A. K. and Anderson, T. F. (1987). Psoralen photochemotherapy. *J Am Acad Dermatol*, 17(5 Pt 1), 703–734.

9. Nevezhin, E. V., et al. (2015). On the mechanism of erythrocyte hemolysis induced by photooxidized psoralen. *Biochemistry (Mosc)*, 80(6), 763–768.

10. Melough, M. M., Cho, E. and Chun, O. K. (2018). Furocoumarins: A review of biochemical activities, dietary sources and intake, and potential health risks. *Food Chem Toxicol*, 113, 99–107.

11. Seden, K., et al. (2010). Grapefruit-drug interactions. *Drugs*, 70(18), 2373–2407.

12. Girennavar, B., et al. (2006). Furocoumarins from grapefruit juice and their effect on human CYP 3A4 and CYP 1B1 isoenzymes. *Bioorg Med Chem*, 14(8), 2606–2612.

13. Frérot, E. and Decorzant, E. (2004). Quantification of total Furocoumarins in citrus oils by HPLC coupled with UV, fluorescence, and mass detection. *Journal of Agricultural and Food Chemistry*, 52(23), 6879–6886.

14. Giménez-Arnau, A. M. (2016). Opinion of the scientific committee on consumer safety (SCCS) – opinion on the safety of the use of methylisothiazolinone (MI) (P94), in cosmetic products (sensitisation only). *Regul Toxicol Pharmacol*, 76, 211–212.

15. Dugrand-Judek, A., Olry, A., Hehn, A., Costantino, G., Ollitrault, P., Froelicher, Y. and Bourgaud, F. (2015). The distribution of coumarins and furanocoumarins in citrus species Closely Matches citrus phylogeny and reflects the organization of Biosynthetic pathways. *PLoS One*, 10(11), e0142757.

16. Peterson, J. J., Beecher, G. R., Bhagwat, S. A., Dwyer, J. T., Gebhardt, S. E., Haytowitz, D. B. and Holden, J. M. (2006). Flavanones in grapefruit, lemons, and limes: A compilation and review of the data from the analytical literature. *Food Compos Anal*, 19, S74e80.

17. Food and Agriculture Organization of United Nations STAT (FAOSTAT). http://www.fao.org/faostat/en/#data/QC/compare [Accessed on 30 November 2021].

18. Madrigal-Bujaidar, E., Roaro, L. M., Garcí'a-Aguirre, K., GarciaMedina, S. and Alvarez-Gonzalez, I. (2013). Grapefruit juice suppresses azoxymethane-induced colon aberrant crypt formation and induces antioxidant capacity in mice. *Asian Pac. J. Cancer Pre*, 14, 6851e6.

19. Mahgoub, A. A. (2002). Grapefruit juice potentiates the antiinflammatory effects of diclofenac on the carrageenaninduced rat's paw oedema. *Pharmacol. Res*, 45, 1e4.

20. Chudnovskiy, R., Thompson, A., Tharp, K., Hellerstein, M., Napoli, J. L. and Stahl, A. (2014). Consumption of clarified grapefruit juice ameliorates high-fat diet induced insulin resistance and weight gain in mice. *PloS One*, 9, e108408.
21. Tripoli, E., La Guardia, M., Giammanco, S., Di Majo, D. and Giammanco, M. (2007). Citrus flavonoids: molecular structure, biological activity and nutritional properties: A review. *Food Chem*, 104, 466e79.
22. Benavente-Garcia, O. and Castillo, J. (2008). Update on uses and properties of citrus flavonoids: New findings in anticancer, cardiovascular, and anti-inflammatory activity. *J. Agric. Food Chem*, 56, 6185e205.
23. Guo, L. Q., Fukuda, K., Ohta, T. and Yamazoe, Y. (2000). Role of furanocoumarin derivatives on grapefruit juice- mediated inhibition of human CYP3A activity. *Drug Metab Disposition*, 28, 766e71.
24. Uesawa, Y. and Mohri, K. (2006). The use of heat treatment to eliminate drug interactions due to grapefruit juice. *Biol Pharm Bull*, 29, 2274e8.
25. Bailey, D. G., Malcolm, J., Arnold, O. and David Spence, J. (1998). Grapefruit juice-drug interactions. *Br J Clin Pharmacol*, 46, 101e10.
26. Bailey, D. G., Dresser, G. and Arnold, J. M. O. (2013). Grapefruitemedication interactions: Forbidden fruit or avoidable consequences? *Can Med Assoc J*, 185, 309e16.
27. Guo, L. Q. and Yamazoe, Y. (2004). Inhibition of cytochrome P450 by furanocoumarins in grapefruit juice and herbal medicines. *Acta Pharmacol Sin*, 25, 129e36.
28. Wang, X., Wang, Y., Yuan, J., Sun, Q., Liu, J. and Zheng, C. (2004). An e_cient new method for extraction, separation and purification of psoralen and iso-psoralen from Fructus Psoraleae by supercritical fluid extraction and high-speed counter-current chromatography. *J. Chromatogr. A*, 1055, 135–140.
29. Zhang, X., Zhao, W., Wang, Y., Lu, J. and Chen, X. (2016). The chemical constituents and bioactivities of Psoralea corylifolia linn.: A review. *Am. J. Chin. Med.*, 44, 35–60.
30. Alalaiwe, A., Hung, C. F., Leu, Y. L., Tahara, K., Chen, H. H., Hu, K. Y. and Fang, J. Y. (2018). The active compounds derived from *Psoralea corylifolia* for photochemotherapy against psoriasis-like lesions: The relationship between structure and percutaneous absorption. *Eur. J. Pharm. Sci.*, 124, 114–126.
31. Wang, Z. C., Feng, D. Q. and Ke, C. H. (2013). Coumarins from the herb cnidium monnieri and chemically modified derivatives as antifoulants against *Balanus albicostatus* and *Bugula neritina* larvae. *Int. J. Mol. Sci.*, 14, 1197–1206.
32. Wu, X., Gao, X., Zhu, X., Zhang, S., Liu, X., Yang, H., Song, H. and Chen, Q. (2019). Fingerprint analysis of cnidium monnieri (L.) cusson by high-speed counter current chromatography. *Molecules*, 24, 4496.
33. Waksmundzka-Hajnos, M., Petruczynik, A., Dragan, A., Wianowska, D. and Dawidowicz, A. L. (2004). Effect of extraction method on the yield of furocoumarins from fruits of *archangelica officinalis* hoffm. *Phytochem. Anal.*, 15, 313–319.
34. Stanjek, V., Piel, J. and Boland, W. (1999). Synthesis of furanocoumarins: Mevalonate-independent prenylation of umbelliferone in *Apium graveolens* (Apiaceae). *Phytochemistry*, 50, 1141–1145.
35. Hamerski, D. and Matern, U. (1988). Elicitor-induced biosynthesis of psoralens in Ammi Majus L suspension-cultures: Microsomal conversion of demethylsuberosin into (+)-marmesin and psoralen. *Eur. J. Biochem*, 171, 369–375.

36. Murray, R. H., Mendez, J. and Brown, S. A. (1982). *The Natural Coumarins: Occurrence, Chemistry and Biochemistry*. Chichester, UK: Johns Wiley & Sons.
37. Peroutka, R., Schulzová, V., Botek, P. and Hajšlová, J. (2007). Analysis of furanocoumarins in vegetables (Apiaceae) and citrus fruits (Rutaceae). *J. Sci. Food Agric.*, 87, 2152–2163.
38. De Castro, W. V., Mertens-Talcott, S., Rubner, A., Butterweck, V. and Derendorf, H. (2006). Variation of flavonoids and furanocoumarins in grapefruit juices: A potential source of variability in grapefruit juice drug interaction studies. *J. Agric. Food Chem.*, 54, 249–255.
39. Schulzová, V., Hajšlová, J., Botek, P. and Peroutka, R. (2007). Furanocoumarins in vegetables: Influence of farming system and other factors on levels of toxicants. *J. Sci. Food Agric.*, 87, 2763–2767.
40. Zamora-Ros, R., Knaze, V., Rothwell, J. A., Hémon, B., Moskal, A., Overvad, K. and Touillaud, M. (2016). Dietary polyphenol intake in Europe: The european prospective investigation into cancer and nutrition (EPIC) study. *Eur. J. Nutr.*, 55, 1359–1375.
41. Schlatter, J., Zimmerli, B., Dick, R., Panizzon, R. and Schlatter, C. H. (1991). Dietary intake and risk assessment of phototoxic furocoumarins in humans. *Food Chem. Toxicol.*, 29, 523–530.
42. Meleoug, M. M., Cho, E. and Chun, O. K. (2018). Furocoumarins: A review of biolochemical activities, dietary sources and intake, and potential health risks. *Food Chem. Toxicol.*, 113, 99–107.
43. Bourgaud, F., Olry, A. and Hehn, A. (2014). Recent advances in molecular genetics of furanocoumarin synthesis in higher plants. In Jacob, C., Kirsch, G., Slusarenko, A. J., Winyard, P. G. and Burkholz, T. (eds.). *Recent Advances in Redox Active Plant and Microbial Products*. Dordrecht, The Netherlands: Springer, pp. 363–375.
44. Heidel-Fischer, H. M. and Vogel, H. (2015). Molecular mechanisms of insect adaptation to plant secondary compounds curr. *Opin. Insect Sci.*, 8, 8–14.
45. Fujita, K. I. (2004). Food-drug interactions via human cytochrome P450 3A (CYP3A). *Drug Metab. Drug Interact*, 20, 195–218.
46. Fracarolli, L., Rodrigues, G. B., Pereira, A. C., Massola Júnior, N. S., Silva-Junior, G. J., Bachmann, L., Wainwright, M., Bastos, J. K. and Braga, G. U. L. (2016). Inactivation of plant-pathogenic fungus *Colletotrichum acutatum* with natural plant-produced photosensitizers under solar radiation. *J. Photochem. Photobiol. B Biol.*, 162, 402–411.
47. Caboni, P., Saba, M., Oplos, C., Aissani, N., Maxia, A., Menkissoglu-Spiroudi, U. and Ntalli, N. (2015). Nematicidal activity of furanocoumarins from parsley against meloidogyne spp. *Pest Manag. Sci.*, 71, 1099–1105.
48. Meiners, T. (2015). Chemical ecology and evolution of plant–insect interactions: A multitrophic perspective. *Curr. Opin. Insect Sci.*, 8, 22–28.
49. Bai, Y., Li, D., Zhou, T., Qin, N., Li, Z., Yu, Z. and Hua, H. (2016). Coumarins from the roots of *angelica dahurica* with antioxidant and antiproliferative activities. *J. Funct. Foods*, 20, 453–462.
50. Qian, G.-S., Wang, Q., Leung, K.S.-Y., Qin, Y., Zhao, Z. and Jiang, Z.-H. (2007). Quality assessment of rhizoma et radix notopterygii by HPTLC and HPLC fingerprinting and HPLC quantitative analysis. *J. Pharm. Biomed. Anal.*, 44, 812–817.
51. Pan, M. H., Lai, C. S. and Ho, C. T. A. (2010). Anti-inflammatory activity of natural dietary flavonoids. *Food Funct*, 1, 15–31.

52. Uto, T., Tung, N. H., Taniyama, R., Miyanowaki, T., Morinaga, O. and Shoyama, Y. (2015). Anti-inflammatory activity of constituents isolated from aerial part of *angelica acutiloba* kitagawa. *Phytother. Res.*, 29, 1956–1963.

53. Wang, C. C., Lai, J. E., Chen, L. G., Yen, K. Y. and Yang, L. L. (2000). Inducible nitric oxide synthase inhibitors of chinese herbs. Part 2: Naturally occurring furanocoumarins. *Bioorg. Med. Chem.*, 8, 2701–2707.

54. Choi, S. Y., Ahn, E. M., Song, M. C., Kim, D. W., Kang, J. H., Kwon, O. S., Kang, T. C. and Baek, N. I. (2005). In vitro GABA-transaminase inhibitory compounds from the root of *angelica dahurica*. *Phytother. Res.*, 19, 839–845.

55. Łuszczki, J. J., Głowniak, K. and Czuczwar, S. J. (2007). Imperatorin enhances the protective activity of conventional antiepileptic drugs against maximal electroshock-induced seizures in mice. *Eur. J. Pharmacol.*, 574, 133–139.

56. Łuszczki, J. J., Głowniak, K. and Czuczwar, S. J. (2007). Time-course and dose-response relationships of imperatorin in the mouse maximal electroshock seizure threshold model. *Neurosci. Res.*, 59, 18–22.

57. Brandt, C., Bethmann, K., Gastens, A. M. and Löscher, W. (2006). The multidrug transporter hypothesis of drug resistance in epilepsy: Proof-of-principle in a rat model of temporal lobe epilepsy. *Neurobiol. Dis.*, 24, 202–211.

58. Meng, F. H., Xiong, Z. L., Sun, Y. and Li, F. (2004). Coumarins from *Cnidium monnieri* (L.) and their proliferation stimulating activity on osteoblast-like UMR106 cells. *Pharmazie*, 59, 643–645.

59. Tang, C. H., Yang, R. S., Chien, M. Y., Chen, C. C. and Fu, W.-M. (2008). Enhancement of bone morphogenetic protein-2 expression and bone formation by coumarin derivatives via p38 and ERK-dependent pathway in osteoblasts. *Eur. J. Pharmacol.*, 579, 40–49.

60. Franceschi, R. T. and Xiao, G. (2003). Regulation of the osteoblast-specific transcription factor, Runx2: Responsiveness to multiple signal transduction pathways. *J. Cell Biochem.*, 88, 446–454.

61. Hoang, Q. Q., Sicheri, F., Howard, A. J. and Yang, D. S. (2003). Bone recognition mechanism of porcine osteocalcin from crystal structure. *Nature*, 425, 977–980.

62. Xiao, J. J., Zhao, W. J., Zhang, X. T., Zhao, W. L., Wang, X. X., Yin, S. H., Jiang, F., Zhao, Y. X., Chen, F. N. and Li, S. L. (2015). Bergapten promotes bone marrow stromal cell differentiation into osteoblasts in vitro and in vivo. *Mol. Cell. Biochem.*, 409, 113–122.

63. Zheng, M., Ge, Y., Li, H., Yan, M., Zhou, J. and Zhang, Y. (2014). Bergapten prevents lipopolysaccharide mediated osteoclast formation, bone resorption and osteoclast survival. *Int. Orthop.*, 38, 627–634.

64. Deyhim, F., Garica, K., Lopez, E., Gonzalez, J., Ino, S., Garcia, M. and Patil, B. S. (2006). Citrus juice modulates bone strength in male senescent rat model of osteoporosis. *Nutrition*, 22, 559–563.

65. Deyhim, F., Mandadi, K., Faraji, B. and Patil, B. S. (2008). Grapefruit juice modulates bone quality in rats. *J. Med. Food.*, 11, 99–104.

66. Deyhim, F., Mandadi, K., Patil, B. S. and Faraji, B. (2008). Grapefruit pulp increases antioxidant status and improves bone quality in orchidectomized rats. *Nutrition*, 24, 1039–1044.

67. Sumner, J. B. and Howel, S. F. (1936). Identification of hemagglutinin of jack bean with Concanavalin A. *J. Bacteriol.*, 32, 227–237.

68. Hamblin, J. and Kent, S. P. (1973). Possible role of phytohemagglutinin in *Phaseolus vulgaris* L. *Nat. New Biol.*, 245, 28–30.

69. Ang, A. S., Cheung, R. C., Dan, X., Chan, Y. S., Pan, W. and Ng, T. B. (2014). Purification and characterization of a glucosamine-binding antifungal lectin from *Phaseolus vulgaris* cv. Chinese pinto beans with antiproliferative activity towards nasopharyngeal carcinoma cells. *Appl. Biochem. Biotechnol.*, 172, 672–686.

70. Jiang, Q. L., Zhang, S., Tian, M., Zhang, S. Y., Xie, T., Chen, D. Y., Chen, Y. J., He, J., Liu, J., Ouyang, L., et al. (2015). Plant lectins, from ancient sugar-binding proteins to emerging anti-cancer drugs in apoptosis and autophagy. *Cell Prolif*, 48, 17–28.

71. Molchanova, V. I., Chernikov, O. V., Chikalovets, I. V. and Lukyanov, P. A. (2010). Purification and partial characterization of the lectin from the marine red alga *Tichocarpus crinitus* (Gmelin) Rupr. (rhodophyta). *Bot. Mar.*, 53, 69–78.

72. Chikalovets, I. V., Kondrashina, A. S., Chernikov, O. V., Molchanova, V. I. and Lukyanov, P. A. (2013). Isolation and general characteristics of lectin from the mussel *Mytilus trossulus*. *Chem. Nat. Comp.*, 48, 1058–1061.

73. Akindele, A. J., Adeneye, A. A., Salau, O. S., Sofidiya, M. O. and Benebo, A. S. (2014). Dose and time-dependent sub-chronic toxicity study of hydroethanolic leaf extract of *Flabellaria paniculata Cav.* (Malpighiaceae) in rodent. *Front Pharmacol*, 5, 78.

74. Marques, G., Gutiérrez, A. and Del Río, J. C. (2007). Chemical characterization of lignin and lipophilic fractions from leaf fibers of curaua (*Ananas erectifolius*). *J. Agric. Food Chem*, 55, 1327–1336.

CHAPTER 5

Glycoalkaloids as Food Toxins

Javed Ahamad[1], Subasini Uthirapathy[2], Esra T. Anwer[3], Mohd. Javed Naim[4], and Showkat R. Mir[5]

[1]Department of Pharmacognosy, Faculty of Pharmacy, Tishk International University, Erbil, Kurdistan Region, Iraq
[2]Department of Pharmacology, Faculty of Pharmacy, Tishk International University, Kurdistan Region, Iraq
[3]Department of Pharmaceutics, Faculty of Pharmacy, Tishk International University, Kurdistan Region, Iraq
[4]Department of Pharmaceutical Chemistry, Faculty of Pharmacy, Tishk International University, Kurdistan Region, Iraq
[5]Department of Pharmacognosy, School of Pharmaceutical Education and Research, Jamia Hamdard, New Delhi, India

CONTENTS

5.1 Introduction	81
5.2 Sources of Glycoalkaloids	82
5.3 Chemistry of Glycoalkaloids	83
5.4 Pharmacological Actions of Glycoalkaloids	85
5.5 Toxicological Activities of Glycoalkaloids	86
5.6 Conclusions	90
References	90

5.1 INTRODUCTION

Glycoalkaloids (GAs) are naturally occurring plant-protective compounds found in a variety of regularly consumed plants. GAs are most predominantly found in the Solanaceae family, especially in potatoes (*Solanum tuberosum L.*), tomatoes (*Solanum lycopersicum L.*), and eggplant (*Solanum melongena L.*). Potatoes are considered the most common source of food GAs, and α-solanine and α-chaconine are the most prevalent types of GAs present in them (Friedman et al. 1997). Tomatoes (e.g., α-tomatine and dehydrotomatine), and eggplant (e.g., solasonine and solamargine) are also major sources of GAs (Blankemeyer et al. 1998; Friedman 2013). Excessive accumulation of GAs during growth, harvesting, and postharvest procedures are the major factors that lead to accumulation of GAs in these food items. Ingestion of food items with excess GAs could cause human health problems because of their dose-dependent toxicity (Morris and Petermann 1985). Hence, cultivation, postharvesting, and storage conditions and technologies should be developed to reduce production and accumulation of

DOI: 10.1201/9781003222194-6

GAs in food crops to reduce toxicities due to consumption of these food items (Mondy and Ponnawpalam 1985).

GAs, especially α-solanine and α-chaconine, are known to cause gastrointestinal problems such as gastritis, gastrointestinal disturbance, nausea, vomiting, diarrhea, fever, low blood pressure, and, in high doses, cause a fast pulse rate along with neurological and occasional death in humans and livestock (Langkilde et al. 2009). These GAs, on the other hand, could be useful as insecticides, antimicrobials, and antifungal agents if extracted from the source. In recent years, the anticancer activity of GAs has been studied extensively, and several preclinical and *in-vitro* studies revealed their anticancer potential (Lv et al. 2014; Friedman et al. 2005; Ito et al. 2007; Huang et al. 2015). This book chapter is intended to review and discuss the source, chemistry, pharmacological, and toxicological actions of the major GAs present in food items.

5.2 SOURCES OF GLYCOALKALOIDS

Glycoalkaloids (GAs) are exclusively found in the Solanaceae family plants, especially in potatoes (*Solanum tuberosum*), tomatoes (*Solanum lycopersicum*), and brinjal (*Solanum melongena*). These GAs are mainly produced in the edible parts of plants such as leaves, flowers, fruits, sprouts, roots, and skin, etc. The GAs' content in different plants is summarized Table 5.1.

Potatoes (*Solanum tuberosum*) are the most common source of GAs, and they contain solanidane-type glycoalkaloids. α-solanine and α-chaconine are the most prevalent types of GAs present in potatoes (Friedman et al. 1997). Potatoes also contain other GAs such as β-chaconine, γ-chaconine, β1-solanine, β2-solanine, and γ-solanine (Friedman 2006). The wild variety of potato, *S. chacoense,* also contains leptines and leptidines. These are steroidal alkaloids in addition to the α-solanine and α-chaconine (Chen and Miller 2000).

Tomatoes (*Solanum lycopersicum*) are also reported to contain about 100 GAs. GAs are reported in all parts of the tomato plant, but mostly in leaves, stems, roots, flowers, and green fruits (Kozukue et al. 2004). The GAs of tomatoes are spirosolane-type glycosides and the most predominant ones are α-tomatine and dehydrotomatine (Friedman 2013). Fujiwara et al. (2004) also isolated esculeoside, a spirosolane-type glycoside from tomatoes.

Eggplant or brinjal (*Solanum melongena*) also contains a significant amount of GAs, and mainly fruits contain these alkaloids. The most predominant GAs in eggplant are

TABLE 5.1   Content of Glycoalkaloids (GAs) in Potatoes, Tomatoes, and Eggplant

Source plants	Major GAs		Reference
Potato	α-Solanine 12 to 5000 mg/kg	α-Chaconine	Wood and Young 1974
Tomato	α-Tomatine 144 to 4900 mg/kg	Dehydrotomatine 14 to 330 mg/kg	Milner et al. 2011
Eggplant	Solamargine 6.25 to 20.5 mg/100 g	Solasonine	Mennella et al. 2012

solamargine and solasonine (Blankemeyer et al. 1998). Bajaj et al. (1979) studied the chemical composition of 21 varieties of *S. melongena* and found the GA content ranges from 6.25 to 20.5 mg/100 g of fresh weight of fruits. Other species of eggplant are also reported to contain GAs such as *S. aethiopicum, S. integrifolium,* and *S. sodomaeum* (Mennella et al. 2012).

5.3 CHEMISTRY OF GLYCOALKALOIDS

Glycoalkaloids (GAs) are nitrogen-containing steroidal alkaloids such as aglycon and attached with oligosaccharides. Usually GAs contain six-membered heterocyclic rings with nitrogen. These GAs are produced in Solanaceae family plants such as potatoes, tomatoes, and eggplant. The GAs are mainly spirosolane-type present in tomatoes (α-tomatine and dehydrotomatine), and eggplant (solasonine and solamargine), or solanidane-type like α-solanine and α-chaconine present in potatoes (Friedman and McDonald 1997; Friedman 2015) (Figure 5.1).

The GAs are produced in plants in response to the defense system of these plants. GAs in potatoes and tomatoes are synthesized *via* the mevalonate pathway. The starting material in the biosynthesis of these glycoalkaloids is acetyl-CoA, which leads to the synthesis of mevalonate, and further, it ultimately leads to the synthesis of cholesterol. Further cholesterol works as a base for the biosynthesis of different GAs (Heftmann 1983; Akiyama et al. 2021). The possible biosynthetic pathway of spirosolane type of GAs in potatoes and spirosolane-type GAs in tomatoes are presented in Figure 5.2a and Figure 5.2b, respectively.

16DOX: 22,26-hydroxycholesterol 16α-hydroxylase; PGA: potato glycoalkaloid biosynthesis; DPS: dioxygenase for potato solanidane synthesis; 3βHSD1: 3β-hydroxysteroid dehydrogenase 1; S5αR$_2$: steroid 5α-reductase; UGT: uridine diphosphate-dependent glycosyltransferase; Glc: glucose; Gal: galactose; Rha: rhamnose; and Xyl: xylose.

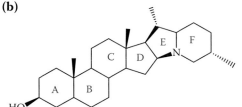

FIGURE 5.1 (a) Spirosolane-type GAs in tomatoes and eggplant, and (b) solanidane-type in potatoes.

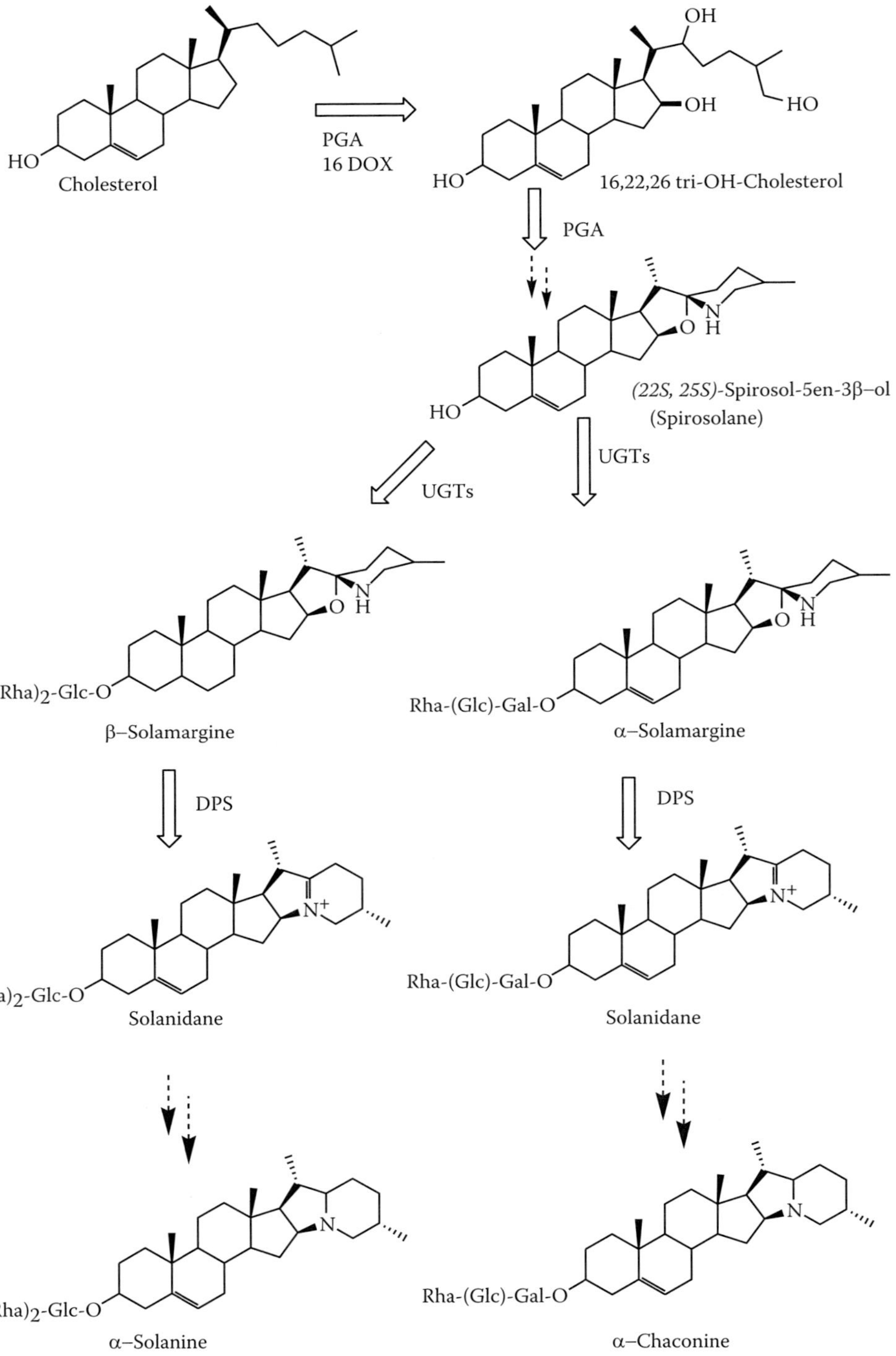

FIGURE 5.2A Possible biosynthesis of spirosolane type of glycoalkaloids present in potatoes (α-solanine and α-chaconine).

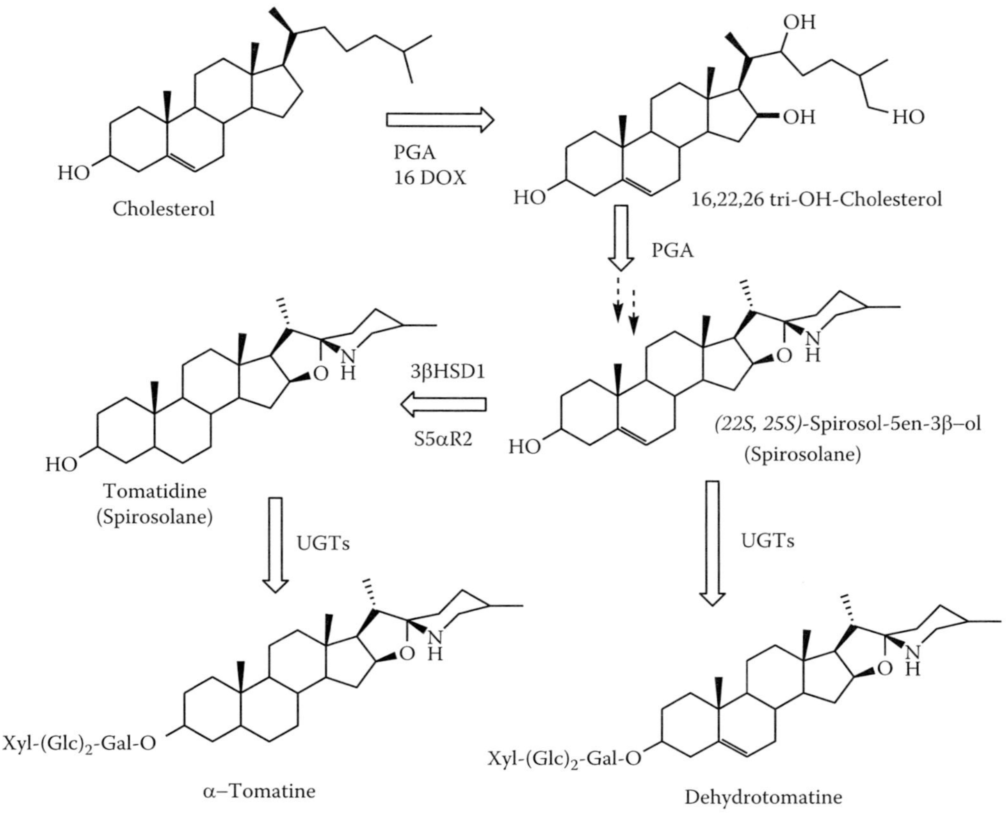

FIGURE 5.2B Possible biosynthesis of spirosolane-type of glycoalkaloids present in tomatoes (α-tomatine and dehydrotomatine).

5.4 PHARMACOLOGICAL ACTIONS OF GLYCOALKALOIDS

Potatoes (α-solanine, α-chaconine), tomatoes (α-tomatine, dehydrotomatine), and eggplant (solasonine, solamargine) are all *Solanum* species that contain GAs. In potatoes, its tubers, roots, sprouts, and leaves mostly contain GAs (Jadhav et al. 1981; Friedman 2006). The tomato plant is thought to utilize these GAs to defend itself against bacteria, fungi, viruses, and insects. Glycoalkaloids from potatoes like α-solanine and α-chaconine, and tomatoes such as tomatine and tomatidine, inhibit cancer cell growth as well as tumor growth *in-vivo*. Apoptosis is a process of cell destruction that comprises many signaling pathways and biomarkers. These GAs can be used to inhibit cancer cell growth (Lee 2006). The hydroxycinnamic acid conjugates chlorogenic and dicaffeoylquinic acids make up the majority of the antioxidant phenolic compounds present in eggplant fruit (Whitaker and Stommel 2003; Patil et al. 1972).

Glycoalkaloids containing foods also contain a high level of non-toxic phytochemicals, which have beneficial effects on humans. Most of the GAs have been reported to possess antipyretic, anti-inflammatory, antihyperglycemic, antiallergenic, and antibiotic activities (Friedman et al. 2003).

For the better utilization of GAs in plants, a thorough understanding of their biological characteristics is required, which recognizes not only the complexity of live cells, but also the potential for unique interrelationships between some or all of the components within each cell. Because solanine toxicity is not traditionally associated with a cholinergic syndrome, the relevance of plasma cholinesterase inhibition to human solanine toxicity is uncertain. α-solanine and α-chaconine have also been shown to be cytotoxic in experimental animals (Phillips et al. 1996; Roddick et al. 1988). Both α-solanine and α-chaconine are reversible inhibitors of human plasma cholinesterase, according to *in-vitro* research (butyrylcholinesterase). Intraperitoneal dosages of solanine caused mild to moderate inhibition of both butryl cholinesterases and cholinesterases in rabbits (Maga 1980; Sinden et al. 1984). However, additional research is needed to understand the mechanism of these GAs and their toxic effects on normal living cells before they may be used in cancer treatment. Furthermore, better extraction and isolation techniques should be applied for these GAs from their natural sources, and for the determination of biological activities, the safety and efficacy of these GAs need further long-term studies (Chao et al. 2012). GAs are also reported as having insecticidal, antimicrobial, and antifungal activities (Hameed et al. 2017). GAs have both toxic and beneficial bioactivities, and both types of actions of GAs should be examined and explored. Table 5.2 summarizes the beneficial pharmacological activities of the different GAs present in food items.

5.5 TOXICOLOGICAL ACTIVITIES OF GLYCOALKALOIDS

Almost all nightshades were reported to contain toxic glycoalkaloids such as solanine, tomatine, and solasonine (e.g. eggplant, potato, tomato). These bitter-tasting phytochemicals are most likely responsible for plant defense against insects, fungi, and bacteria. Green tomatoes are high in toxic GAs like α-tomatine and dehydrotomatin (Kozukue et al. 2004; Phillips et al. 1996). Even modest concentrations of these glycoalkaloids might cause gastrointestinal tract problems and pain (Drossman et al. 2006). Tomatine embryotoxicity is another pharmacologic and toxicological aspect, due to its high affinity for cholesterol and ability to break cell membranes (Siener et al. 2016).

Arthritis is an inflammatory disorder in which the walls of blood vessels become weakened. This sickness causes swollen and painful joints as a result of a decrease in the bloodstream in the organs. Some researchers feel that eating habits and certain foods are the most important factors influencing the start or progression of arthritis (Slavin and Ailani 2017). In a study including over 1,000 arthritis patients, tomatoes and other nightshade vegetables like potatoes and eggplant were identified as foods that may increase inflammation and pain concerns. Dietary restriction and avoidance of particular foods, such as tomatoes and other nightshade vegetables, are highly advised for those suffering from arthritis and discomfort (Awasthi et al. 2011). Despite the noted health benefits of tomatoes, there have been some unfavorable side effects associated with it, and tomato-based goods consumption (Peksa et al. 2006).

Glycoalkaloids ingestion can lead to gastrointestinal problems such as nausea, diarrhea, and stomach pain. Higher doses of GAs cause acute intoxication, with severe symptoms such as neurological problems, rapid pulse, low blood pressure, and in the worst-case scenario, coma or death (Dale and Mackay 2007). The hazardous dose is

TABLE 5.2 Pharmacological Activities of Glycoalkaloids *in-vitro* and *in-vivo*

Glycoalkaloids compound	Pharmacological action (*in-vitro/in-vivo*)	Observations	Reference
Tomatine	Anticancer effects (AIF–nucleus expression methods)	Tomatine caused apoptosis	Chao et al. 2012
α-tomatine	Breast cancer (MCF-7 breast adenocarcinoma cell line)	Inhibited human breast adenocarcinoma cell lines	Sucha et al. 2013
α-tomatine	Anticancer effects (lung cancer NCI-H460)	Inhibited migration, and suppressed the cell adhesion in lung cancer cells	Shieh et al. 2011
α-chaconine	Anticancer effects (Hep G2 cell line)	Produced cell cytotoxicity of Hep G2 cell line	Lee et al. 2004
γ-chaconine	Anticancer effects (colon cell line)	γ-Chaconine showed activity against the colon cells and liver cells	Lee et al. 2004
α-solanine	Anticancer effects (Pancreatic cancer cells)	α-Solanine inhibited metastasis	Lv et al. 2014
α- solanine	Anticancer effects (Bcl-2 expression method by mitochondrial pathway)	α-Solanine stimulated mitochondrial pathway *via* p53 and Bcl-2 expression of apoptosis	Mohsenikia et al. 2013
α-solanine	*In-vitro* and *in-vivo* anticancer activity (P38-MAPK signaling pathway)	α-Solanine induced prostate cancer cell inhibition through P38 MAPK pathway	Pan et al. 2016
Steroidal GAs, solamargine	*In-vitro* anticancer effects (HT29,T47D and hepatoma cell lines)	Inhibited the growth of human cancer cells such as colon, prostate, breast and hepatoma cells	Hu et al. 1999; Zhou et al. 2014
Solamargine	*In-vitro* anticancer effects (MAPK- STAT3 pathway)	Solamargine suppressed the proliferation of lung cancer *via* P38 by inducing Stat3	Zhou et al. 2014
Solasodine	*In-vivo* anticancer effects (Murine sarcoma 180 cell lines)	Solasodine inhibited murine sarcoma cell lines	Cham 2013
Solanine and chaconine	*In-vivo* anticancer effects (Liver and stomach cancer cell)	Inhibited liver and stomach cancer cell growth	Friedman et al. 2005
Lycopene and α-tomatine	*In-vivo* and *in-vitro* anticancer effects and CVD	Inhibited human cancer cells and prevention therapy for cardiovascular diseases	Friedman 2013

(*Continued*)

TABLE 5.2 *(Continued)*

Glycoalkaloids compound	Pharmacological action (*in-vitro/in-vivo*)	Observations	Reference
α-chaconine and α-solanine	Antifungal activities	α-Chaconine and α-solanine are found potent against the pathogenic strains of *Trichomonas*	Fewell and Roddick 1993, 1994
α-solanine and α-chaconine	Antibacterial activities	Potato peel, and α-chaconine and α-solanine inhibited the bacterial growth	Amanpour et al. 2015
α-chaconine and α-solanine	Anti-inflammatory activity	α-chaconine and α-solanine showed reduction of inflammation and interleukin 2 and 8	Kenny et al. 2013
Tomatidine	Anti-inflammatory activity	Tomatidine blocked the NF-KB and JNK signaling pathway	Sandrock and VanEtten 1998
α-tomatine	Insecticidal activity	α-Tomatine shows antifungal activity and resistant pathogens	Chen et al. 2010
Tomatine, chaconine, solanine, solasonine, and solamargine	Antimalarial activity	GAs produced dose dependent action against malarial parasite	Satoh 1967
Solanine	*In-vivo* hyperglycemia effect	Solanine increased the blood glucose level by stimulation of adrenal gland	Akinnuga et al. 2010
Tomatine, tomatidine, and dehydrotomatine	Antidiabetic activity	Reduced the blood glucose level *in-vivo* model	Akinnuga et al. 2010; Krits et al. 2007

predicted to be 2–5 mg/kg of body weight, whereas the deadly dose is considered to be 3–6 mg/kg (Nowacki 2009). Glycoalkaloids with a concentration of more than 14 mg/kg cause bitterness, while those with a concentration of more than 20 mg/kg cause a burning sensation in the throat and mouth (Bushway et al. 1983). Based on an anticipated 158 g of potatoes per capita, a study predicted a total daily intake of glycoalkaloids of between 0.4 and 1.7 mg/person/day. A daily dose of 3.6–8 mg per person would be obtained by eating these potatoes with peels (St-Jules et al. 2016).

Glycoalkaloids are being researched because of their potential health effects when consumed through food (Claringbold et al. 1982). The neurological system is harmed by glycoalkaloids because they interfere with the ability to manage acetylcholine, which is vital in nerve impulse transmission. Solanine produces toxicity such as altering the cell membranes and GIT disturbances (Harvey et al. 1985). The toxicity could be related to the glycoalkaloids' anticholinesterase effect on the central nervous system, as well as cell membrane disturbances impacting the GIT and other systemic organs (Lachman et al. 2001).

These poisonous glycoalkaloids have been shown to accumulate in the body, particularly when glycoalkaloid-containing foods are ingested daily. Glycoalkaloids have been demonstrated to stay in the bloodstream for up to 24 hours after ingestion (Peksa et al. 2006). As a result, a cumulative safety risk among regular or frequent potato and potato product consumers may be possible in the long run. Because the symptoms of toxicity are similar to those of other gastrointestinal disorders, poisoning is more difficult to diagnose (Tajner-Czopek et al. 2012). It's crucial to think about how much glycoalkaloid buildup from food is controlled by body metabolism. When two people stopped eating potatoes and other products of potatoes, glycoalkaloid levels in their blood dropped dramatically, eventually becoming negligible by the next week. Once in the bloodstream, the elimination level appears to be very low. This indicates that these substances may accumulate in the body's numerous organs, especially in the liver (Tajner-Czopek et al. 2012; Porter 1972).

Total glycoalkaloids levels of 2–5 mg/kg are hazardous to humans because they elicit systemic and gastrointestinal effects as well as downregulation of acetylcholinesterase (Langkilde et al. 2009). These glycoalkaloids cause toxicity by forming a combination with the sterol in cellular membranes, causing cells to break down (Roddick et al. 1988). Weakness, depression, coma, convulsions, partial paralysis, and mental disorientation are some of the neurological indications and symptoms of glycoalkaloids intoxication (Smith et al. 1996). Some glycoalkaloids from *Solanum* species, such as α-solasonine, α-solamargine, and aglycone solasodine, have been shown to have anticancer activity against a variety of tumor cell lines (Korpan et al. 2004).

Potatoes produce toxic effects in human beings when the concentrations of solanine exceed 0.02% (Nishie et al. 1971). In a study conducted on a Greek family of eight people, they ate a dinner of young potato shoots and wide beans in 1932, and the first toxic consequence of solanine poisoning happened. There were no signs of poisoning at first, but within 12 hours, symptoms such as headache, colic pain, hot skin, nausea, and vomiting emerged. Weakness, depression, seizures, diarrhea, stomach discomfort, and difficulty breathing were among the additional symptoms (Willimott 1933). In another study, solanine induces itching in the neck region, dyspnea, sleepiness, and hyperesthesia at a modest dose of 200 mg/per oral. At higher doses, it causes nausea, vomiting, and diarrhea (Slavin and Ailani 2017). The toxicological effects of different compounds of Solanaceae are listed in Table 5.3.

TABLE 5.3 Toxicological Effects of Solanaceae

Glycoalkaloids compounds	Toxic effects	References
Tomatine and solanine	Arthritis and body aches	Friedman et al. 1992; Slavin and Ailani 2017
Tomatine	Migraine	Evans et al. 2015; Nazari and Eghbali 2012
Lycopene	Lycopenodermia	Townsend et al. 2013; Friedlander et al. 2012
Potassium and calcium (macronutrients of tomato and potato	Kidney disorders	Khanna and White 2009; St-Jules et al. 2016
β-Fructofuranosidase and polygalacturonase	Allergies	Foetisch et al. 2001; Kondo et al. 2001
Citric acid and maleic acid and	Burning sensations and urinary problems	Vella et al. 2012; Pang and Ali 2015

5.6 CONCLUSIONS

Glycoalkaloids are present in regularly consumed food items such as potatoes, tomatoes, and eggplant. In recent years, an increasing number of toxicological events were reported by food contaminations with glycoalkaloids. Glycoalkaloids such as α-solanine and α-chaconine (potato); α-tomatine and dehydrotomatine (tomato); and solasonine and solamargine (eggplant) are reported to produce mild to severe complications in human beings as well as grazing animals. The mild complications with ingestion of glycoalkaloids are gastrointestinal problems such as gastritis, gastrointestinal disturbance, nausea, vomiting, diarrhea, fever, and low blood pressure; while with high doses of glycoalkaloids, a fast pulse rate along with neurological and occasional death in humans and farm animals. Thus it is very important to identify, analyze, and characterize different types of glycoalkaloids present in food items. This book chapter expands the knowledge of GAs, particularly their sources and chemistry, and provides comprehensive data on their beneficial and toxicological aspects.

REFERENCES

Akinnuga, A. M., Bamidele, O., Ebunlomo, A. O., Adeniyi, O. S., Adeleyea, G. S., Ebomuche, L. C. (2010). Hypoglycaemic effects of dietary intake of ripe and unripe *Lycopersicon esculentum* [tomatoes] on streptozotocin-induced diabetes mellitus in rats. *J Biolog Sci*, 10(2), 50–53.

Akiyama, R., Watanabe, B., Nakayasu, M., Lee, H. J., Kato, J. and Umemoto, N. (2021). The biosynthetic pathway of potato solanidanes diverged from that of spirosolanes due to evolution of a dioxygenase. *Nature Communications*, 12, 1300. 10.1038/s41467-021-21546-0

Amanpour, R., Abbasi-Maleki, S., Neyriz-Naghadehi, M. and Asadi-Samani, M. (2015). Antibacterial effects of *Solanum tuberosum* peel ethanol extract *in vitro. J Herb Med Pharmacol*, 4, 45–48.

Awasthi, M., Malhotra, S. and Modgil, R. (2011). Dietary habits of kidney stone patients of Kangra district, Himachal Pradesh, North India. *J Human Ecology*, 34, 163–169.

Bajaj, K. L., Kaur, G. and Chadha, M. L. (1979). Glycoalkaloid content and other chemical constituents of the fruits of some eggplant (*Solanum melongena*, L.) varieties. *Journal of Plant Foods*, 3(3), 163–168.

Blankemeyer, J. T., McWilliams, M. L., Rayburn, J. R., Weissenberg, M. and Friedman, M. (1998). Developmental toxicology of solamargine and solasonine glycoalkaloids in frog embryos. *Food and Chemical Toxicol*, 36(5), 383–389.

Bushway, R. J., Bureau, J. L. and McGann, D. F. (1983). Alpha-chaconine and alpha-solanine content of potato peels and potato peel products. *J Food Sci*, 48, 84–86.

Cham, B. E. (2013). Drug therapy: Solamargine and other solasodine rhamnosyl glycosides as anticancer agents. *Modern Chemotherapy*, 2(02), 33.

Chao, M. W., Chen, C. H., Chang, Y. L., Teng, C. M. and Pan, S. L. (2012). α-Tomatine-mediated anti-cancer activity in vitro and in vivo through cell cycle-and caspase independent pathways. *PLoS One*, 7(9), e44093. 10.1371/journal.pone.0044093

Chen, Y., Li, S., Sun, F., Han, H., Zhang, X., Fan, Y., et al. (2010). *In vivo* antimalarial activities of glycoalkaloids isolated from Solanaceae plants. *Pharmaceutical Biology*, 48(9), 1018–1024.

Chen, Z. and Miller, A. R. (2000). Steroidal alkaloids in Solanaceous vegetable crops. *Horticultural Rev*, 25, 171–196.

Claringbold, W. D. B., Few, J. D. and Renwick, J. H. (1982). Kinetics and retention of solanidine in man. *Xenobiotica*, 12, 293–302.

Dale, M. F. B. and Mackay, G. R. (2007). Inheritance of table and processing quality. *Potato Genetics*, 1994, 285–315.

Drossman, D. A. and Dumitrascu, D. L. (2006). Rome iii: New standard for functional gastrointestinal disorders. *J Gastrointestinal Liver Diseases*, 15, 237–241.

Evans, E. W., Lipton, R. B., Peterlin, B. L., Raynor, H. A., Thomas, J. G., O'leary, K. C., et al. (2015). Dietary intake patterns and diet quality in a nationally representative sample of women with and without severe headache or migraine. *Headache: The J Head and Face Pain*, 55, 550–561.

Fewell, A. M. and Roddick, J. G. (1993). Interactive antifungal activity of the glycoalkaloids α-solanine and α-chaconine. *Phytochemistry*, 33(2), 323–328.

Fewell, A. M., Roddick, J. G. and Weissenberg, M. A. (1994). Interactions between the glycoalkaloids solasonine and solamargine in relation to inhibition of fungal growth. *Phytochemistry*, 37(4), 1007–1011.

Foetisch, K., Son, D., Altmann, F., Aulepp, H., Conti, A., Haustein, D. and Vieths, S. (2001). Tomato (*Lycopersicon esculentum*) allergens in pollen-allergic patients. *European Food Research and Technology*, 213, 259–266.

Friedlander, J. I., Shorter, B. and Moldwin, R. M. (2012). Diet and its role in interstitial cystitis/bladder pain syndrome (ic/bps) and comorbid conditions. *BJU International*, 109, 1584–1591.

Friedman, M. (2006). Potato glycoalkaloids and metabolites: Roles in the plant and in the diet. *J Agric Food Chem*, 54(23), 8655–8681.

Friedman, M. (2013). Anticarcinogenic, cardioprotective, and other health benefits of tomato compounds lycopene, α-tomatine, and tomatidine in pure form and in fresh and processed tomatoes. *J Agric Food Chem*, 61(40), 9534–9550.

Friedman, M. (2015). Chemistry and anticarcinogenic mechanisms of glycoalkaloids produced by eggplants, potatoes, and tomatoes. *J Agric Food Chem*, 63, 3323–3337.

Friedman, M., Lee, K. R., Kim, H. J., Lee, I. S. and Kozukue, N. (2005). Anticarcinogenic effects of glycoalkaloids from potatoes against human cervical, liver, lymphoma, and stomach cancer cells. *J Agricultural and Food Chem*, 53(15), 6162–6169.

Friedman, M. and McDonald, G. M. (1997). Potato glycoalkaloids: Chemistry, analysis, safety, and plant physiology. *CRC Crit Rev Plant Sci*, 16, 55–132.

Friedman, M., McDonald, G. M. and Filadelfi-Keszi, M. (1997). Potato glycoalkaloids: Chemistry, analysis, safety, and plant physiology. *Critical Rev in Plant Sci*, 16(1), 55–132.

Friedman, M., Rayburn, J. and Bantle, J. (1992). Structural relationships and development toxicity of Solanum alkaloids in the frog embryo teratogenesis assay-xenopus. *J Agric Food Chem*, 40, 1617–1624.

Friedman, M., Roitman, J. N. and Kozukue, N. (2003). Glycoalkaloid and calystegine contents of eight potato cultivars. *J Agric Food Chem*, 51, 2964–2973.

Fujiwara, Y., Takaki, A., Uehara, Y., Ikeda, T., Okawa, M., Yamauchi, K., et al. (2004). Tomato steroidal alkaloid glycosides, esculeosides A and B, from ripe fruits. *Tetrahedron.*, 60(22), 4915–4920.

Hameed, A., Ijaz, S., Mohammad, I. S., Muhammad, K. S., Akhtar, N. and Khan, H. M. S. (2017). Aglycone solanidine and solasodine derivatives: A natural approach towards cancer. *Biomed Pharmacother*, 94, 446–457.

Harvey, M. H., McMillan, M., Morgan, M. R. A. and Chan, H. W. S. (1985). Solanidine is present in sera of healthy individuals in amounts dependent on their dietary potato consumption. *Human J Toxicol*, 4, 187–194.

Heftmann, E. (1983). Biogenesis of steroids in Solanaceae. *Phytochemistry*, 22, 1843–1860.

Hu, K., Kobayashi, H., Dong, A., Jing, Y., Iwasaki, S. and Yao, X. (1999). Antineoplastic agents III: Steroidal glycosides from *Solanum nigrum. Planta Medica*, 65(01), 035–038.

Huang, H., Chen, X., Li, D., He, Y., Li, Y., Du, Z., et al. (2015). Combination of α-tomatine and curcumin inhibits growth and induces apoptosis in human prostate cancer cells. *PLoS One*, 10(12), e0144293. 10.1371/ journal.pone.0144293

Ito, S. I., Ihara, T., Tamura, H., Tanaka, S., Ikeda, T., Kajihara, H., et al. (2007). α-Tomatine, the major saponin in tomato, induces programmed cell death mediated by reactive oxygen species in the fungal pathogen *Fusarium oxysporum. FEBS Letters*, 581(17), 3217–3222.

Jadhav, S. J., Sharma, R. P. and Salunkhe, D. K. (1981). Naturally occurring toxic alkaloids in foods. *CRC Crit Rev Toxicol*, 9, 21–104.

Kenny, O. M., McCarthy, C. M., Brunton, N. P., Hossain, M. B., Rai, D. K., Collins, S. G., et al. (2013). Anti-inflammatory properties of potato glycoalkaloids in stimulated Jurkat and Raw 264.7 mouse macrophages. *Life Scie*, 92(13), 775–782.

Khanna, A. and White, W. B. (2009). The management of hyperkalemia in patients with cardiovascular disease. *The American J Med*, 122, 215–221.

Kondo, Y., Urisu, A. and Tokuda, R. (2001). Identification and characterization of the allergens in the tomato fruit by immunoblotting. *Inter Archives Allergy Immunol*, 126, 294–299.

Korpan, Y. I., Nazarenko, E. A., Skryshevskaya, I. V., Martelet, C., JaffrezicRenault, N. and Anna, V. (2004). Potato glycoalkaloids: true safety or false sense of security? *Trends Biotechnol*, 22(3), 147–151.

Kozukue, N., Han, J. S., Lee, K. R. and Friedman, M. (2004). Dehydrotomatine and α-tomatine content in tomato fruits and vegetative plant tissues. *J Agric Food Chemi*, 52(7), 2079–2083.

Krits, P., Fogelman, E. and Ginzberg, I. (2007). Potato steroidal glycoalkaloid levels and the expression of key isoprenoid metabolic genes. *Planta*, 227(1), 143–150.

Lachman, J., Hamouz, K., Orsák, M. and Pivec, V. (2001). Potato glycoalkaloids and their significance in plant protection and human nutrition – Review. *Rostlinná Výroba*, 47, 181–191.

Langkilde, S., Mandimika, T., Schrøder, M., Meyer, O., Slob, W., Peijnenburg, A. and Poulsen, M. (2009). A 28-day repeat dose toxicity study of steroidal glycoalkaloids, α-solanine and α-chaconine in the Syrian Golden hamster. *Food Chem Toxicol*, 47(6), 1099–1108.

Lee, M. R. (2006). The Solanaceae: Foods and poisons. *J R Coll Physicians Edinb*, 36, 162–169.

Lee, K. R., Kozukue, N., Han, J. S., Park, J. H., Chang, E. Y., Baek, E. J., et al. (2004). Glycoalkaloids and metabolites inhibit the growth of human colon (HT29) and liver (HepG2) cancer cells. *J Agric Food Chem*, 52(10), 2832–2839.

Lv, C., Kong, H., Dong, G., Liu, L., Tong, K., Sun, H., et al. (2014). Antitumor efficacy of α-solanine against pancreatic cancer *in vitro* and *in vivo*. *PLoS One*, 9(2), e87868. 10.1371/journal.pone.0087868

Maga, J. A. (1980). Potato glycoalkaloids. *CRC Critical Reviews in Food Science and Nutrition*, 12, 371–405.

Mennella, G., Lo Scalzo, R., Fibiani, M., D'Alessandro, A., Francese, G., Toppino, L., et al. (2012). Chemical and bioactive quality traits during fruit ripening in eggplant (*S. melongena* L.) and allied species. *J Agric Food Chem*, 60(47), 11821–11831.

Milner, S. E., Brunton, N. P., Jones, P. W., O'Brien, N. M., Collins, S. G. and Maguire, A. R. (2011). Bioactivities of glycoalkaloids and their aglycones from Solanum species. *J Agric Food Chem*, 59(8), 3454–3484.

Mohsenikia, M., Alizadeh, A. M., Khodayari, S., Khodayari, H., Karimi, A., Zamani, M., et al. (2013). The protective and therapeutic effects of alpha-solanine on mice breast cancer. *Europ J Pharmacol*, 718(1–3), 1–9.

Mondy, N. I. and Ponnawpalam, R. (1985). Effect of magnesium fertilizers on total glycoalkaloids and nitrate-N in Katahdin tubers. *J Food Sci*, 50(2), 535–536.

Morris, S. C. and Petermann, J. B. (1985). Genetic and environmental effects on levels of glycoalkaloids in cultivars of potato (*Solanum tuberosum* L.). *Food Chem*, 18(4), 271–282.

Nazari, F. and Eghbali, M. (2012). Migraine and its relationship with dietary habits in women. *Iranian J Nursing Midwifery Res*, 17, S65.

Nishie, K., Gumbmann, M. and Keyl, A. (1971). Pharmacology of solanine. *Toxicol Appl Pharmacol*, 19(1), 81–92.

Nowacki, W. (2009). Characteristics of native potato cultivars register. *Plant Breeding and Acclimatization*, 1–34.

Pan, B., Zhong, W., Deng, Z., Lai, C., Chu, J., Jiao, G., et al. (2016). Inhibition of prostate cancer growth by solanine requires the suppression of cell cycle proteins and the activation of ROS/P38 signaling pathway. *Cancer Med*, 5(11), 3214–3222.

Pang, R. and Ali, A. (2015). The chinese approach to complementary and alternative medicine treatment for interstitial cystitis/bladder pain syndrome. *Translational Andrology Urology*, 4, 653.

Patil, B. C., Sharma, R. P., Salunkhe, D. K., et al. (1972). Evaluation of solanine toxicity. *Food Cosmet Toxicol*, 10, 395–398.

Peksa, A., Lubowska, G., Anilowski, K., Lisinska, G. and Rytel, E. (2006). Changes of glycoalkaloids and nitrate contents in potatoes during chip processing. *J Food Chem*, 97(1), 151–156. ISSN 0308-8146, https://doi.org/10.1016/j.foodchem.2005.03.035

Phillips, B. J., Hughes, J. A., Phillips, J. C., et al. (1996). A study of the toxic hazard that might be associated with the consumption of green potato tops. *Food Chem Toxicol*, 34, 439–448.

Porter, W. L. (1972). A note on the melting point of α-solanine. *American Potato J*, 49(10), 403–406.

Roddick, J. G., Rijnenberg, A. L. and Osman, S. F. (1988). Synergistic interaction between potato glycoalkaloids α-solanine and α-chaconine in relation to destabilization of cell membranes: Ecological implications. *J Chem Ecol*, 14(3), 889–902.

Sandrock, R. W. and VanEtten, H. D. (1998). Fungal sensitivity to and enzymatic degradation of the phytoanticipin α-tomatine. *Phytopathology*, 88(2), 137–143.

Satoh, T. (1967). Glycemic effects of solanine in rats. *The Japanese J Pharmacol*, 17(4), 652–658.

Shieh, J. M., Cheng, T. H., Shi, M. D., Wu, P. F., Chen, Y., Ko, S. C., et al. (2011). α-Tomatine suppresses invasion and migration of human non-small cell lung cancer NCI-H460 cells through inactivating FAK/PI3K/Akt signaling pathway and reducing binding activity of NF-κB. *Cell Biochem Biophys*, 60(3), 297–310.

Siener, R., Seidler, A., Voss, S. and Hesse, A. (2016). The oxalate content of fruit and vegetable juices, nectars and drinks. *J Food Composition Anal*, 45, 108–112.

Sinden, S. L., Sanford, L. L. and Webb, R. E. (1984). Genetic and environmental control of potato glycoalkaloids. *American Potato J*, 61(3), 141.

Slavin, M. and Ailani, J. (2017). A clinical approach to addressing diet with migraine patients. *Current Neurology Neuroscience Reports*, 17, 17.

Smith, D. B., Roddick, J. G. and Jones, J. L. (1996). Potato glycoalkaloids: Some unanswered questions. *Trends Food Sci Technol*, 7(4), 126–131.

St-Jules, D. E., Goldfarb, D. S. and Sevick, M. A. (2016). Nutrient non-equivalence: Does restricting high-potassium plant foods help to prevent hyperkalemia in hemodialysis patients? *J Renal Nutr*, 26, 282–287.

Sucha, L., Hroch, M., Rezacova, M., Rudolf, E., Havelek, R., Sispera, L., et al. (2013). The cytotoxic effect of α-tomatine in MCF-7 human adenocarcinoma breast cancer cells depends on its interaction with cholesterol in incubation media and does not involve apoptosis induction. *Oncology Reports*, 30(6), 2593–2602.

Tajner-Czopek, A., Rytel, E., Kita, A., Pęksa, A. and Hamouz, K. (2012). The influence of thermal process of coloured potatoes on the content of glycoalkaloids in the potato products. *Food Chem*, 133(4), 1117–1122.

Townsend, M. K., Devore, E. E., Resnick, N. M. and Grodstein, F. (2013). Acidic fruit intake in relation to incidence and progression of urinary incontinence. *Intern Urogynecology J*, 24, 605–612.

Vella, M., Robinson, D. and Cardozo, L. (2012). Painful bladder syndrome. *Obstetrics, Gynaecology & Reproductive Med*, 22, 44–49.

Whitaker, B. D. and Stommel, J. R. (2003). Distribution of hydroxycinnamic acid conjugates in fruit of commercial eggplant (*Solanum melongena* L.) cultivars. *J Agric Food Chem*, 51, 3448–3454.

Willimott, S. (1933). An investigation of solanine poisoning. *Analyst*, 58(689), 431–439.

Wood, F. A. and Young, D. A. (1974). TGA in potatoes. *Canada Department of Agric*, 1533, 1–3.

Zhou, Y., Tang, Q., Zhao, S., Zhang, F., Li, L., Wu, W., et al. (2014). Targeting signal transducer and activator of transcription 3 contributes to the solamargine-inhibited growth and induced apoptosis of human lung cancer cells. *Tumor Biol*, 35(8), 8169–8178.

SECTION II

Analytical Techniques Exploited in Structural Characterization and Identification: *Qualitative Application*

CHAPTER 6

Infrared Spectroscopy

Mohd. Javed Naim[1], Javed Ahamad[2], and Esra T. Anwer[3]

[1]Department of Pharmaceutical Chemistry, Faculty of Pharmacy, Tishk International University, Erbil, Kurdistan Region, Iraq

[2]Department of Pharmacognosy, Faculty of Pharmacy, Tishk International University, Erbil, Kurdistan Region, Iraq

[3]Department of Pharmaceutics, Faculty of Pharmacy, Tishk International University, Erbil, Kurdistan Region, Iraq

CONTENTS

6.1	Introduction	97
6.2	Current Analytical Methods	98
6.3	Infrared Spectroscopy	98
6.4	The Need for Analysis	99
6.5	IR Spectroscopy in Food Toxin Analysis	99
	6.5.1 Glycoalkaloids	99
	6.5.2 Cyanogenic Glycosides	102
	6.5.3 Pyrrolizidine Alkaloids	104
	6.5.4 Furanocoumarins	105
	6.5.5 Lectins	106
6.6	Conclusions	107
References		107

6.1 INTRODUCTION

Naturally occurring toxic molecules/substances are basically generated by plants, fungi, and algae or are the products of metabolism that show harmful effects when consumed in appropriate quantities are known as food toxins. Their presence indicates some specific functions in the plant and are proved to defend plants from insects, microbes, and predators. These natural chemicals show diversity in their chemical structures and nature as well as toxicity level one can exert (Dolan et al. 2010; Kaiser et al. 2020; Reese et al. 2019; Schilter et al. 2014).

Well-known classes are glycoalkaloids (Milner et al. 2011), cyanogenic alkaloids (Vetter 2000), pyrrolizidine alkaloids (Bates et al. 2014; Bates et al. 2016; Roberts et al. 2018; EFSA 2005; EFSA 2011), furanocoumarins (Fracarolli et al. 2016), and lectins (Peumans et al. 1995). The extent of toxicity depends on the levels of toxin present in food followed by the susceptibility of an individual to these toxins. Natural toxins present in food may show a series of clinical symptoms leading to acute illness which when

DOI: 10.1201/9781003222194-8

untreated bring about chronic illness. The initial symptoms may vary from mild gastric upset, neurological disorder, and respiratory paralysis among children and elderly people. Cases of acute poisoning were observed in the intake of wild mushrooms or ineffectively treated ginkgo. Severe toxicity is realized more frequently by alkaloids. Pyrrolizidine alkaloids are observed in liver toxicity when consumed for long periods (EFSA 2005; EFSA 2011).

Analysis of contaminants and toxins in food necessitates the improvement and authentication of available analytical techniques for screening, identification, and quantitative analysis of contaminants. There is no doubt to say that chromatographic techniques can assess a variety of toxins simultaneously with significant selectivity and sensitivity to yield appropriate results but high cost and enormous time consumption in these methods can be seen as a disadvantage. Furthermore, the samples used during analysis are found to be destroyed. As a result, alternate techniques like IR spectroscopy are being progressively established to deliver easy and fast results.

IR spectroscopy is a non-destructive method that can prove to be a boon in the analysis and authentication of samples. Since its workings depend on analyzing the interaction between radiation and samples, it is a suitable technique for analyzing raw samples, manufacturing methodologies, and validation (Levasseur-Garcia 2018; Nawrocka et al. 2013; McMullin et al. 2015). This chapter highlights the advancement and capability of IR spectroscopy as a substitute to prevailing techniques for the analysis of food toxins, as well as IR spectroscopy coupled with chemometric methods.

6.2 CURRENT ANALYTICAL METHODS

Various analytical methods have been in use for analysis of food toxins in food products, plant extracts, and herbs among which some of the well-recognized techniques are UV spectroscopy, LC, HPLC, GC, mass spectrometry, and capillary electrophoresis (Ignat et al. 2001; Lu & Rasco 2012; Bittner et al. 2013). In the past few years, GC coupled with mass spectrometry or LC coupled to diode array detector, as well as tandem mass spectrometry (ESI-MS/MS) are the most extensively used analytical methods (Muñoz-González et al. 2012; Sánchez-Patán et al. 2011; Achamlale et al. 2009). Gas chromatography provides a higher resolution as well as higher sensitivity in comparison to liquid chromatography, but the sample preparation process is a tedious and laborious procedure that involves isolation by several extraction steps followed by derivatization (Alonso Garcıa et al. 2004; Berthod et al. 1999; Bucić-Kojić et al. 2007). Sample preparation is simpler for liquid chromatography and is sensitive enough to give good quantification when coupled with mass spectrometry (Cao et al. 2009; Caridi et al. 2007; Chen et al. 2000; Cimpoiu 2006; Bunaciu et al. 2012; Cozzolino 2015a).

6.3 INFRARED SPECTROSCOPY

Infrared spectroscopy is a speedy, selective, and sensitive analytical method to assess food toxins and to evaluate the bioavailability of various toxins and characterization of phytochemicals (De Jesus Inacio et al. 2020; Ahamad et al. 2013, 2014a,b). Infra-red spectroscopy has been introduced in order to control the quality of food as well as food products in industries for e.g. the superiority or quality of food can be easily monitored by assessing various factors like its moisture content, protein etc. (McGoverin et al. 2010).

Infrared Spectroscopy

TABLE 6.1 Pros and Cons of Near and Mid IR in Comparison to Outdated Traditional Methods

Characteristics	IR (near/mid)	Traditional methods
Sample pre processing	Not required or minimal	Required
Cost	Relatively low	Medium to high
Speed of analysis	High	Low to medium
Need of standards	Not required	Required
Data analysis and interpretation	Chemometrics is needed	Simple
Quantitative analysis	Yes	Yes
Qualitative analysis	Yes	Yes

The unbeatable response to the infrared spectroscopic technique is due to its rapidity, non-destructiveness, and eco-friendly nature as it does not require any toxic solvents and reagents during its procedure. Because of its flexible and adaptable nature, it is being used widely in biology, medicine, soil chemistry, and agri-foods (García-Cañas et al. 2010). Since infrared spectroscopy analyses the interaction between infrared radiation and matter, it appears to be an appropriate technique for analyzing raw materials, analysis of manufacturing procedures, and validation of finished goods. Because of its advantageous outcomes, it has now become an excellent analytical tool in the field of phyto-analytics and can assess the physical and chemical parameters simultaneously (Ignat et al. 2001). Infrared spectroscopy shows a greater degree of precision in the analysis of food toxin and provides qualitative and quantitative information at a very reliable and acceptable cost in no time (Ignat et al. 2001; Lu & Rasco 2012; Bittner et al. 2013, Bunaciu et al. 2012; Cozzolino 2015a). Table 6.1 highlights and relates the cons and pros of employing spectroscopic techniques and outdated traditional methods.

6.4 THE NEED FOR ANALYSIS

The advancement and execution of IR techniques for carrying out analysis of food toxin has been made promising by the progression and implementation of multivariate data (MVA) methodologies in chemometrics. For three decades, chemometrics have attracted the attention of scientists throughout the world. Predictive analytical methods (calibration) like principle component and partial least square (PLS) regression are now extensively employed in the expansion of infrared analytical techniques for the assessment of food toxins in a variety of food materials. Recently, the analysis of toxins in food items using infrared spectroscopy combined with chemometric prediction technique is being extensively employed (Ignat et al. 2001; Lu & Rasco 2012; Bittner et al. 2013, Bunaciu et al. 2012; Cozzolino 2015b).

6.5 IR SPECTROSCOPY IN FOOD TOXIN ANALYSIS

6.5.1 Glycoalkaloids

Complex biochemical pathways are used by plants to produce secondary metabolites in order to handle/manage hostile stimuli from environments like herbivoral damage, nutrient deprivation, and pathogen. These metabolites can be specific to a species or genera and plays an important role in improving plant viability by enhancing their capacity to survive in harsh environmental conditions (Harborne 2014), for protection from pathogens, for UV light absorption, for radical scavenging, and anticancer agents (Kennedy & Wightman 2011). On the basis of their biosynthetic origin, secondary metabolites are categorized as terpenoids, phenolic compounds, and alkaloids (Yazaki 2005). Glycoalkaloids are steroidal alkaloids equipped with a sterol framework in six heterocyclic rings with nitrogen. They protect the plants by forming a defense system in many widely consumed plants belonging to the Solanaceae family like potatoes, tomatoes, and aubergine. Before the discovery of chaconine in the year 1854, solanine was the sole molecule widely existing in potatoes. Tomatine (tomatine + dehydrotomatine) was discovered in the year 1948 in tomatoes followed by the discovery of solasonine and solamargine (glycoalkaloids of aubergine) (Roddick 1996; Milner et al. 2011). Glycoalkaloids basically consist of a basic aglycone part (C-27 cholestane with nitrogen containing rings) and oligosaccharide that overall makes the glycoalkaloids amphiphatic in nature. The aglycone part is further categorized into solanidanes, spirosilanes, epiminocholestanes, α-epiminoyclohemiketals, and 3-aminospirostanes (Makkar et al. 2009; Mazid et al. 2011; Vaananen 2007). Approximately 90 unique steroidal alkaloids were recognized in about 350 species of the Solanaceae family. Nitrogen in glycoalkaloids is identified to be present as a primary secondary or tertiary nitrogen that finally affects the nature of the compound (Tek 2006; Dinan et al. 2001). The oligosaccharide portion of the glycoalkaloids is also found to show significant activity (Roddick 1996). Insufficient data are available for determining the biosynthetic pathway or the regulation factors for glycoalkaloids but there is an assumption that the aglycone part is synthesized via the mevalonate/isoprenoid pathway (Krits et al. 2007). In Figure 6.1a–c, the major glycoalkaloids are presented.

Tilahun et al. (2020) developed a non-destructive VIS/NIR approach for detecting α-solanine and α-chaconine levels in intact potato cultivars. 180 tubers of Atlantic and Trent potato cultivars were tested for VIS/NIS spectra, color values, and the reference substances α-solanine and α-chaconine with replication every 2 weeks until 10 weeks of storage. To cultivate newer prototypes that correlate color and spectral analysis data in potatoes (α-solanine and α-chaconine), partial least square (PLS) regression analysis was used. At a wavelength of 500–100 nm, the transmittance spectral data of undamaged potato tubers were determined. In the detection and analysis of glycoalkaloids, the devised approach was shown to be chemical-free, quick, and cost effective. Segtnan et al. (2006) used NIRS to determine the amount of acrylamide in potato chips. Acrylamide is a carcinogenic component that can be found in a variety of heat-treated foods. For this investigation, potato tuber samples were carved, fried, and pulverized before NIR analysis. The spectrum forecast models were built using a PLS regression analytical technique. The correlation coefficient between predicted and standard acrylamide readings was 0.952, with an RMSECV of 246.88 µg/kg. NIR spectroscopy could be utilized to determine the amount of acrylamide in processed potato chips if the correlation coefficient is higher and the RMSECV is lower.

Pedreschi et al. (2010) used NIR spectroscopy to determine the amount of acrylamide in potato chips. Raw potatoes were hydrogenated with palm oil, cut, and fried at various times to get 60 samples for VIS/NIR analysis. With a correlation coefficient of 0.83 and SEP values of 266 µg/kg, the best model emerged. The results of the investigation revealed that, due to the higher probability of error in analysis, this method is better suited for sample classification rather than content estimation.

Tamasi et al. (2021) determined tomatine content in eight varieties of tomato at different maturity stages by ATR-FT-MIR analyses. The tomatine content varies in green

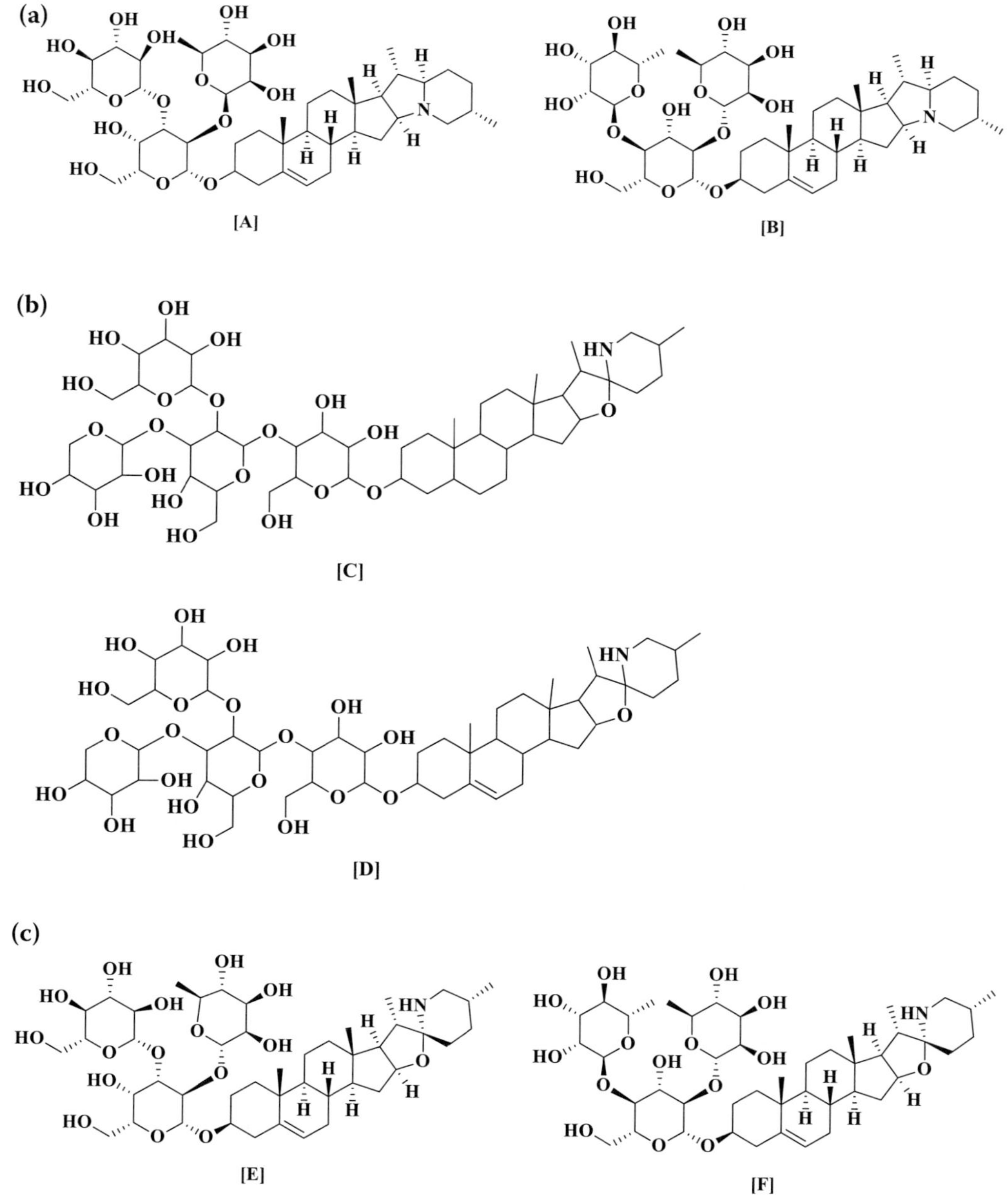

FIGURE 6.1 (a) Structure of α-solanine [A] and α-chaconine [B]. (b) Structure of α-tomatine [C] and dehydrotomatine [D]. (c) Structure of solasonine [E] and so-lamargine [F].

(664–1,874 mg/kg), turning (325–621 mg/kg), and pink (144–280 mg/kg) on the dry weight basis of tomato. Tomatine shows main IR bands in the region between 1,100 and 950 cm^{-1}. The authors compared the IR outcomes with HPLCESI-QqQ-MS/MS and thermogravimetric analysis (TGA) and found that IR is a fast and suitable technique for quantification when combined with chemometry.

6.5.2 Cyanogenic Glycosides

Cyanogenic glycosides are present naturally in plants and associated plant based food products. They are secondary metabolites that exist as glycosides of α-hydroxynitriles (Vetter 2000). They are polar, water-soluble glycosides and found to be gathered in the cell vacuoles of plants. They are O-β-glycosides of α-hydroxynitriles (cyanohydrins) (Seigler 1991). They also are originally derived from amino acid and occurs in almost 2,500 species of plants. The chief cyanogenic glycosides occurring/obtained from various edible parts of plants are amygdalin, dhurrin, linamarin, linustatin, lotaustralin, prunasin, and taxiphyllin (Haque and Bradbury 2002; Codex Committee on Contaminants in foods 2008). Fabaceae, Leguminosae, Compositae, Rosaceae, and Linaceae are some families yielding cyanogenic glycosides whose identification can prove to be a useful tool for informative taxonomic markers (Bolarinwa et al. 2016).

The basic framework of these glycosides comprises of a core carbon that is connected to a cyano (CN) group and R1 (CH_3 or C_6H_5 or p-C_6H_5OH) and R2 (H or CH_3 or C_2H_5) as substituents that are linked to glucose or gentiobiose through a glycosidic linkage. Table 6.2 gives the information regarding the occurrence of major cyanogenic glycosides in foods meant for human consumption (Cressey et al. 2019), which represent potential bases for hydrogen cyanide (World Health Organization food additive series 30, Geneva 1993). The ability of living organisms to produce hydrocyanic acid (HCN) is known as cyanogenesis (Conn 1981). Hydrogen cyanide (HCN) was first identified and isolated in the year 1802 from the leaves of bitter almonds and peaches. Furthermore, Robiquet and Charlard recognized amygdalin from bitter almonds, which on enzymatic hydrolysis, gives α-hydroxynitrile and sugar moiety (Vetter 2000).

Enzymatic hydrolysis of cyanogenic glycosides yields corresponding cyanohydrin followed by its conversion to hydrogen cyanide, a ketone, and an aldehyde by means of endogenous β-glucosidase, by wounding, or by action of gut microflora (Codex Committee on Contaminants in Foods 2008).

Two distinct β-glucosidase enzymes are present in flaxseed for the conversion of cyanogenic glycosides to hydrogen cyanide (Shahidi and Wanasundara 1997). The gentiobiose sugar moiety of linustatin and neolinustatin was hyfrolyzed to a simpler molecule glucose by the enzyme linustatinase to yield linamarin and lotaustralin (Cressey & Saunders 2012; FAO/WHO. World Health Organization Food Additive Series: 65 2012). Humans and animals can suffer from toxicity due to cyanides in the dose range of 0.5 mg

TABLE 6.2   Occurrence of Various Cyanogenic Glycosides in Human Food

S. no.	Source	Glycoside
1	Almonds, stone fruit, pome fruit	Amygdalin
2	Sorghum	Dhurrin
3	Cassava, lima beans, linseed/flaxseed, spinach	Linamarin
4	Cassava, linseed/flaxseed	Linustatin
5	Cassava, lima beans	Lotaustralin
6	Linseed	Neolinustatin
7	Stone fruit, pome fruit, pip fruit	Prunasin
8	Elderberries	Sambunigrin
9	Bamboo shoots	Taxiphyllin

Infrared Spectroscopy 103

to 3.5 mg hydrogen cyanide per kg body weight. Acute cyanide poisoning shows some clinical symptoms that include increased respiration, sudden fall in blood pressure, increased pulse, headache, mental confusion, bluish skin reflecting decreased oxygen, twitching, and convulsions (WHO Food Additive Series 30. Geneva. 1993, FSANZ 2004; Speijers 1993; Kwok 2008). Figure 6.2 describes the chemical structures of some important cyanogenic glycosides.

The isolation method of amygdalin from *Pruni domesticae* was optimized by using central composite design (CCD) and multilayer perception (MLP) models. In this experiment for isolation of amygdalin, the effect of time, ethanol concentration, solid to solvent ratio, and temperature were optimized using mathematical models. The study outcomes suggested for optimal conditions for better extraction of amygdalin were: time of 120 minutes, ethanol 100%, solid to solvent ratio of 1:25 (m/v), and 34.4°C of temperature. The isolated amygdalin was characterized by using FT-IR, UV, and mass data. The content of amygdalin in extract was determined by HPLC method, and the experimental value (25.30 g/100 g dried extract) was very close to the predicted value by the MLP model (25.42 g/100 g of dried extract) (Savic et al. 2015).

Sheikh et al. (2022) studied the effect of microwave heating (450 W for 6 minutes), hydrothermal treatment (6, 9, and 12 hours at 45°C), and their combination on

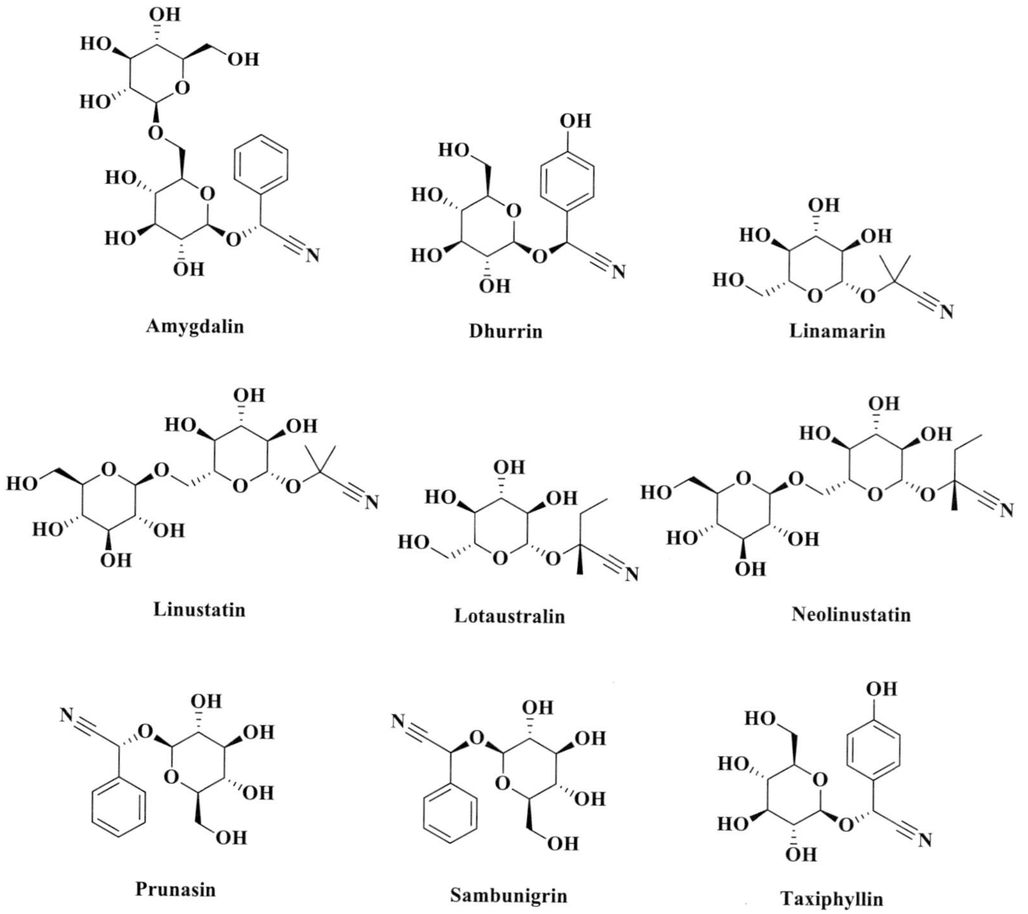

FIGURE 6.2 Chemical structures of important cyanogenic glycosides.

cyanogenic glycoside content in plum kernels (*Prunus domestica* L.). The study shows a significant reduction of cyanogenic glycoside content in plum kernels (37.81, 72.17, 84.41, 91.24, and 98.02%). The FT-IR spectra near 1,157 cm^{-1} provided valuable insights on the reduction of cyanogenic glycoside content during combined effects of microwave heating and hydrothermal treatment. The method was found to be an effective tool for neutralizing toxic effects of cyanogenic glycoside in plums.

6.5.3 Pyrrolizidine Alkaloids

Pyrrolizidine alkaloids (PAs) occur naturally in the plant kingdom (more than 6,000 species) as toxins. They are distributed widely and are produced by plants as a self-defense mechanism but they also affect humans, wildlife, and livestock (Bates et al. 2014; Bates et al. 2016; Roberts et al. 2018; EFSA 2005; EFSA 2011). Pyrrolizidine alkaloids (PAs) comprise a saturated or a 1,2-unsaturated necine base. They are categorized into platynecine, retronecine, heliotridine, and otonecine on the basis of necine bases present. All alkaloids except otonecine can undergo N-oxide formation. About 660 pyrrolizidine alkaloids and their N-oxides were recognized from over 6,000 plant species belonging to families Leguminoceae, Compositae, Orchidaceae, and Boraginaceae (Fu et al. 2004). The content of the alkaloid is solely dependent on the part of plant as well as plant species, which was further influenced by climatic conditions and nature of the soil (Chen et al. 2019; Dusemund et al. 2018; Ebmeyer et al. 2019; EFSA 2007). Pyrrolizidine alkaloids and their N-oxides are toxic when plants containing pyrrolizidine alkaloids or food intoxicated with these alkaloids are consumed (Fu et al. 2004). They are oxidized to DHP (pyrrolic ester) in the liver by cytochrome P450 enzymes, which act as electrophiles to form adducts with proteins and deoxyribose nucleic acids (Chen et al. 2010; EFSA 2007, 2011; Wang et al. 2005). High doses may cause liver poisoning. As per Margin of Exposure Methodology by European Food Safety Authority, doses of pyrrolizidine alkaloids > 0.007 µg /kg body weight/day are undoubtful in relation to risk of causing cancer. A health-based guidance value of 0.1 µg PA/kg body weight/day was laid down by the German Federal Institute for Risk Assessment in regards to non-cancerous effects (Willmot et al. 1920). Figure 6.3 displays major pyrrolizidine alkaloids.

De Jesus Inacio et al. (2020) analyzed pyrrolizidine alkaloids (PAs) and their N-oxides in bee pollen by near-infrared spectroscopy. The study results show the presence of 17 pyrrolizidine alkaloids (PAs) and their N-oxides ranging from 0.4 to 400 µg/kg. The main pyrrolizidine alkaloids (PAs) and their N-oxides were quantified are echimidine, echimidine N-oxide, indicine N-oxide, intermidine, lycopsamine, lycopsamine N-oxide, retrorsine, retrorsine N-oxide, senecionine, senecionine N-oxide, seneciphylline, seneciphylline N-oxide, heliotrine, and heliotrine N-oxide.

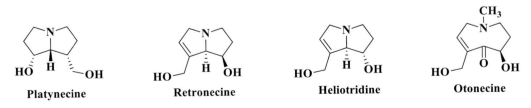

FIGURE 6.3 Chemical structures of major pyrrolizidine alkaloids.

Farsam et al. (2000) isolated and characterized four pyrrolizidine alkaloids viz. europine, europine N-oxide, ilamine, and ilamine N-oxide from *Heliotropium crossifolium*. The chemical structures of isolated pyrrolizidine alkaloids were characterized by IR, NMR, and mass data. For europine FTIR ($CHCl_3$): v = 3400 (OH), 1736 (C=O), 1617 (C=C), 1280 cm^{-1} (C-O).

Oliveira et al. (2017) characterized he monocrotaline pyrrolizidine alkaloid by IR and Raman spectroscopies. The authors proved monocrotaline chemical structures on the basis of experimental and theoretical IR and Raman spectroscopic data.

6.5.4 Furanocoumarins

Coumarins are phenolic in nature which are distributed extensively in the Kingdom Plantae. Coumarin was first isolated by Vogel in the year 1820 (Küpeli Akkol et al. 2020; Wink 1988). The word *coumarin* originates from the word *coumarou* (*Dipteryx odorata* Willd, Leguminosae), which reflects distinguishing fragrance of these compounds and once was known as *Coumarouna odorata* Aubl. The basic structure of coumarin is composed of 1,2-benzopyrone with isoprenoid chains that are attached to carbon/oxygen/both. This basic structure forms the basis for the categorization of coumarins as simple, furano, and pyrano coumarins along with their alkoxylated, alkylated, and hydroxylated derivatives alongside their glycosides (Murray et al. 1982).

Furanocoumarins are naturally occurring compounds in plants in reaction to stress and defense to counter attacks from bacteria, fungi, and insects (Fracarolli et al. 2016). Furanocoumarins are categorized into linear type (psoralens), where the furan ring shows attachment with carbons 6 and 7 position, and angular type, generically called as angelicins, where the ring shows attachment with coumarin at carbon 7 and 8. Generally, dihydrofuranocoumarins such as nodakenetin (linear type) and archangelicin (angular type) have a reduced furan ring. 7-Hydroxycoumarin, commonly known as umbelliferone, is often regarded as the parent, both structurally and biogenetically, of the more complex coumarins (Roberts & Wink 1998). 5-Methoxypsoralen was the first furanocoumarin that was isolated by Kalbrunner from bergamot oil in the year 1838. Basically, furanocoumarins are available in roots but their concentration is higher in fruits and leaves as resins. The utmost exceptional feature of furanocoumarins is their pronounced capability to cause sensitization of cells to visible light, sunlight, and specifically, near UV light. Its well-established mechanism reveals that after intercalation into deoxyribonucleic acid and molecular complexing, the stimulated furanocoumarins interact with thymine. Furanocoumarins, because of their phototoxicity, reflect their defensive mechanism against herbivores and phytopathogenic microbes (Del Río et al. 2014).

The maximum furanocoumarins level (Järvenpää et al. 1997) are observed in plants belonging to the Apiaceae family on fresh-weight basis; for example, celery 45 $\mu g\,g^{-1}$ (Diawara et al. 1995), parsnip 145 $\mu g\,g^{-1}$ (Lombeart et al. 2001), and parsley 112 $\mu g\,g^{-1}$ (Beier et al. 1994). The results were quantified on a fresh-weight basis. The lowest content of furanocoumarins is observed in fruits of the Rutaceae family, for example, bergamottin in grapefruit juices is 6 $\mu g\,g^{-1}$ (Fukuda et al. 2000). The presence of various furanocoumarins was also proved in many other plants, such as Fabaceae, Pittosporaceae, Solanaceae, Amaranthaceae, Rosaceae, Cyperaceae, and Moraceae (Murray et al. 1982) and the fruit of some of these, e.g., figs. This last family is also used for human consumption. Figure 6.4 summarizes the chemical structures of some important furanocoumarins.

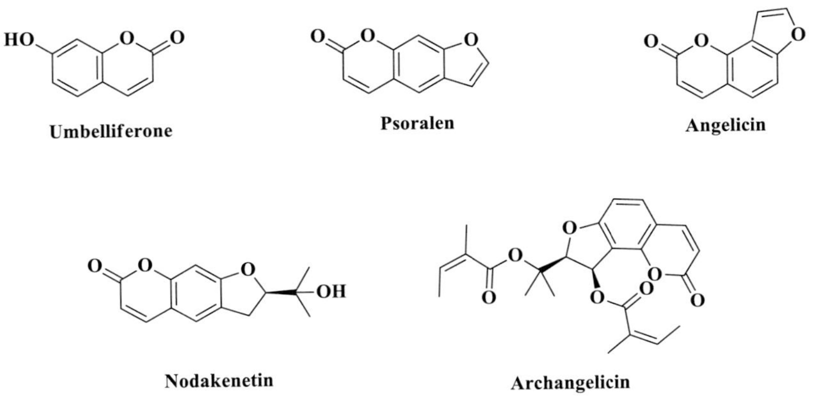

FIGURE 6.4 Chemical structures of some important furanocoumarins.

Moreno-Ley et al. (2019) applied fourier transform mid-infrared (MID-FTIR) spectroscopy coupled with chemometry for detection of coumarin adulteration in vanilla extracts. The vanilla extracts were usually adulterated with *Dipteryx odorata,* and this plant species contains toxic coumarin and 1,2-benzopyrone. The IR regions between 3,175 and 2,800 cm^{-1} and 1,800 and 850 cm^{-1} were selected for the construction of the model. The coumarins in vanilla extracts ranged from 0.25 to 10 ppm. The study finding suggested that MID-FTIR can be effectively employed for detection and quantification of toxic coumarins in vanilla extracts. In another study by Sharp et al. (2012), FTIR was utilized to analyze vanilla beans. The FTIR coupled with chemometric was applied for successful identification and characterization of different chemicals, and the major chemical compounds identified in vanilla beans were vanillin, anise alcohol, 4-methylguaiacol, *p*-hydroxybenzaldehyde/trymethylpyrazine, *p*-cresol/ anisole, guaiacol, isovaleric acid, and acetic acid. The IR bands associated with vanillin and its derivatives were observed at 1,523, 1,573, 1,516, 1,292, 1,774, 1,670, 1,608, and 1,431 cm^{-1}. In another study, Kus et al. (2007) applied FTIR for characterization of 2H-1-benzopyran-2-one isolated from the argon matrix. FTIR coupled with UV, NMR, and mass successfully characterized coumarin in the argon matrix.

6.5.5 Lectins

Lectins are proteins that are detected in plant extracts and originally termed *phytohemagglutinins.* They possess a non-catalytic-binding domain that binds to a carbohydrate recognition/binding domain in a reversible fashion (Peumans et al. 1995). The distribution of lectins ranges from viruses to vertebrates and the biological recognition of cells and proteins is due to its carbohydrate specificity. In plants, lectins act as a defense as well a recognition molecule (Barkai-Golan et al. 1978; Bohlool & Schmidt 1974). In animals, lectins act as cell adhesion and recognition of protein and synthesis of glycoprotein as well. In microbial realm, lectins are known by different names like hemagglutinins and adhesins (Gouget et al. 2006; Sharon & Lis 2004). When used in the diet, they are known to cause allergies, autoimmune disorders, and interference in absorption of nutrients thus they are called anti-nutrition molecules (Cordain et al. 2000; Pusztai et al. 1993). Transgenic plants with insecticidal lectins are an important application of lectins as a defense mechanism in plants like tobacco, potato, rice, sugarcane, and cotton (Afolabi-Balogun et al. 2011;

Vandenborre et al. 2011). Because of enormous functionalities, lectins hold the potential for the biomedical industry despite their anti-nutrition effects. Recent progress in computational analysis as well as bioinformatics have led the pharmaceutical industry to identify efficiently lectins for future drug discovery processes.

FTIR spectroscopy was used by Bonnin et al. (1999) to investigate wheat germ agglutinin (WGA). Deconvolution of the FTIR spectra of WGA revealed the presence of certain K-helices and L-sheets, in contrast to many other lectins. These results back up previous WGA crystal data. The conformational changes in WGA produced by GlcNAc-bearing liposomes or GlcNAc oligomers were studied using FTIR spectroscopy. When GlcNAc binds to WGA, the number of twists and K-helices decreases, and L-sheets appear, inducing peptidic N-H deuteration to different degrees.

Jiang et al. (2019) devised a two-step method for isolating and analyzing lectins from Zihua snap bean seeds (*Phaseolus vulgaris*). To isolate and purify lectins, the researchers used an aqueous double-phase technique and sephadex G-75, SDS-PAGE, Nano LC-ESI-MS/MS, and FTIR.

6.6 CONCLUSIONS

IR spectroscopy with its ease of handling and low cost, has received widespread recognition for analyzing food toxins in agricultural yields, foods, and plants. The fingerprint application of infrared spectroscopy along with chemometrics can be employed to elucidate important specific principles that are difficult to detect by means of traditional analytical techniques. In order to successfully execute the analysis, calibration is an important and critical criterion that needs to be done to make infrared spectroscopy a successful and reliant application. Furthermore, lack of technical knowledge and training is an important aspect that is needed to be focused upon.

REFERENCES

Achamlale, S., Rezzonico, B. and Grignon-Dubois, M. (2009). Rosmarinic acid from beach waste: Isolation and HPLC quantification in *Zostera detritus* from *Arcachon lagoon. Food Chem*, 113, 878–883.

Afolabi-Balogun, N. B., Inuwa, H. M., Sani, I., Ishiyaku, M. F., Bakare-Odunola, M. T., Nok, A. J. and van Emmenes, L. (2011). Expression of mannose-binding insecticidal lectin gene in transgenic cotton (Gossypium) plant. *Cotton Genomics and Genetics*, 22, 2.

Ahamad, J., Kamran, J. N., Ali, M. and Mir, S. R. (2013). New glycoside esters from the aerial parts of *Artemisia absinthium* Linn. *The Natural Products Journal*, 3(4), 260–267.

Ahamad, J., Kamran, J. N., Ali, M. and Mir, S. R. (2014a). Isoflavone glycosides from aerial parts of *Artemisia absinthium* Linn. *Chem Natural Comp*, 49(6), 696–700.

Ahamad, J., Kamran, J. N., Ali, M. and Mir, S. R. (2014b). New isoquinoline alkaloids from the stem bark of *Berberis aristata. Indian Journal of Chemistry-B*, 53, 1237–1241.

Alonso García, A., Cancho Grande, B. and Simal Gandara, J. (2004). Development of a rapid method based on solid-phase extraction and liquid chromatography with ultraviolet absorbance detection for the determination of polyphenols in alcohol-free beers. *J Chromatogr, A*, 1054, 175–180.

Barkai-Golan, R., Mirelman, D. and Sharon, N. (1978). Studies on growth inhibition by lectins of penicillia and aspergilli. *Archives of microbiology*, 116(2), 119–124.

Bates, B., Cox, L., Nicholson, S., Page, P., Prentice, A., Steer, T. and Swan, G. (September 2016). *National Diet and Nutrition Survey: Results from years 5 and 6 (combined) of the rolling programme (2012/2013–2013/2014)*. London: Public Health England.

Bates, B., Lennox, A., Prentice, A., Bates, C. J., Page, P., Nicholson, S., and Swan, G. editors., (2014). *National Diet and Nutrition Survey: Results from years 1, 2, 3 and 4 (combined) of the rolling programme (2008/2009–2011/2012): A survey carried out on behalf of public health England and the food standards agency*. Public Health England.

Beier, R. C., Ivie, G. W. and Oertli, E. H. (1994). Linear furanocoumarins and graveolone from the common herb parsley. *Phytochemistry*, 36(4), 869–872.

Berthod, A., Billardello, B. and Geoffroy, S. (1999). Polyphenols in counter current chromatography. *An example of large scale separation. Analysis*, 27, 750–757.

Essential Biochemistry (1982). 5th Edition, Charlotte W. Pratt, Kathleen Cornely, ISBN: 978-1-119-71285-5, April 2021, 816 Pages, Johns Wiley & Sons, Chichester.

Bittner, L. K., Schoenbichler, S. A., Bonn, G. K. and Huck, C. W. (2013). Near infrared spectroscopy (NIRS) as a tool to analyze phenolic compounds in plants. *Curr Anal Chem*, 9, 417–423.

Bohlool, B. B. and Schmidt, E. L. (1974). Lectins: A possible basis for specificity in the *Rhizobium*-legume root nodule symbiosis. *Science*, 185(4147), 269–271.

Bolarinwa, I. F., Oke, M. O., Olaniyan, S. A. and Ajala, A. S. (2016). A review of cyanogenic glycosides in edible plants. *Toxicology-New Aspects to This Scientific Conundrum*, 179–191.

Bonnin, S., Besson, F., Gelhausen, M., Chierici, S. and Roux, B. A. (1999). FTIR spectroscopy evidence of the interactions between wheat germ agglutinin and N-acetylglucosamine residues. *FEBS Letters*, 456, 361–364.

Bucić-Kojić, A., Planinić, M., Tomas, S., Bilić, M. and Velić, D. (2007). Study of solid-liquid extraction kinetics of total polyphenols from grape seeds. *J Food Eng*, 81, 236–242.

Bunaciu, A. A., Aboul-Enein, H. Y. and Fleschin, S. (2012). FTIR spectrophotometric methods used for antioxidant activity assay in medicinal plants. *Appl Spectrosc Rev*, 47, 245–255.

Cao, X., Wang, C., Pei, H. and Sun, B. (2009). Separation and identification of polyphenols in apple pomace by high-speed counter-current chromatography and high-performance liquid chromatography coupled with mass spectrometry. *J Chromat A*, 1216, 4268–4274.

Caridi, D., Trenerry, V. C., Rochfort, S., Duong, S., Laugher, D. and Jones, R. (2007). Profiling and quantifying quercetin glucosides in onion (*Allium cepa* L.) varieties using capillary zone electrophoresis and high performance liquid chromatography. *Food Chem*, 105, 691–699.

Chen, Z. and Huo, J. R. (2010). Hepatic veno-occlusive disease associated with toxicity of pyrrolizidine alkaloids in herbal preparations. *Neth J Med*, 68(6), 252–260.

Chen, L., Peijnenburg, A., de Haan, L. and Rietjens, I. M. (2019). Prediction of *in vivo* genotoxicity of lasiocarpine and riddelliine in rat liver using a combined *in vitro*-physiologically based kinetic modelling-facilitated reverse dosimetry approach. *Archives of Toxicology*, 93(8), 2385–2395.

Chen, G., Zhang, H. and Ye, J. (2000). Determination of rutin and quercetin in plants by capillary electrophoresis with electrochemical detection. *Anal Chim Acta*, 423, 69–76.

Cimpoiu, C. (2006). Analysis of some natural antioxidants by thin-layer chromatography and high performance thin-layer chromatography. *J Liq Chromatogr Relat Technol*, 29, 1125–1142.

Codex Committee on Contaminants in Foods (2008). *Discussion Paper on Cyanogenic Glycosides*. Rome: FAO/WHO.

Conn, E. E. (1981). Cyanogenic glycosides. *The Biochemistry of Plants: A Comprehensive Treatise*, 7, 479–500.

Cordain, L., Toohey, L., Smith, M. J. and Hickey, M. S. (2000). Modulation of immune function by dietary lectins in rheumatoid arthritis. *British J Nutrition*, 83(3), 207–217.

Cozzolino, D. (2015a). Infrared spectroscopy as a versatile analytical tool for the quantitative determination of antioxidants in agricultural products, foods and plants. *Antioxidants*, 4(3), 482–497.

Cozzolino, D. (2015b). The role of visible and infrared spectroscopy combined with chemometrics to measure phenolic compounds in grape and wine samples. *Molecules*, 20, 726–735.

Cressey, P. and Reeve, J. (2019). Metabolism of cyanogenic glycosides: A review. *Food Chem Toxicol*, 125, 225–232.

Cressey, P. and Saunders, D. (2012). Determination of presence of cyanogenic residues in apple juices in Australia and New Zealand. Unpublished. Report FW12032 June 2012. Institute of Environmental Science & Research Limited contracted by the New Zealand Ministry for Primary Industries.

De Jesus Inacio, L., Lanza, I., Merlanti, R., Contiero, B., Lucatello, L., Serva, L., et al. (2020). Discriminant analysis of pyrrolizidine alkaloid contamination in bee pollen based on near-infrared data from lab-stationary and portable spectrometers. *European Food Research and Technology*, 246, 2471–2483.

Del Río, J. A., Díaz, L., García-Bernal, D., Blanquer, M., Ortuno, A., Correal, E. and Moraleda, J. M. (2014). Furanocoumarins: Biomolecules of therapeutic interest. *In Studies in Natural Products Chemistry*, 43, 145–195.

Diawara, M. M., Trumble, J. T., Quiros, C. F. and Hansen, R. (1995). Implications of distribution of linear furanocoumarins within celery. *J Agri Food Chem*, 43(3), 723–727.

Dinan, L., Harmatha, J. and Lafont, R. (2001). Chromatographic procedures for the isolation of plant steroids. *J Chromatogr A*, 935(1–2), 105–123.

Dolan, L. C., Matulka, R. A. and Burdock, G. A. (2010). Naturally occurring food toxins. *Toxins*, 2(9), 2289–2332.

Dusemund, B., Nowak, N., Sommerfeld, C., Lindtner, O., Schäfer, B., Lampen, A. (2018). Risk assessment of pyrrolizidine alkaloids in food of plant and animal origin. *Food Chem Toxicol*, 115, 63–72.

Ebmeyer, J., Behrend, J., Lorenz, M., Günther, G., Reif, R., Hengstler, J. G., Braeuning, A., Lampen, A. and Hessel-Pras, S. (2019). Pyrrolizidine alkaloid-induced alterations of prostanoid synthesis in human endothelial cells. *Chemico-Biological Interactions*, 298, 104–111.

EFSA Panel on Contaminants in the Food Chain (CONTAM) (2011). Scientific opinion on pyrrolizidine alkaloids in food and feed. *EFSA Journal*, 9(11), 2406.

European Food Safety Authority (EFSA) (2005). Opinion of the Scientific Committee on a request from EFSA related to a harmonised approach for risk assessment of substances which are both genotoxic and carcinogenic. *EFSA Journal*, 3(10), 282.

European Food Safety Authority (EFSA) (2007). Opinion of the Panel on contaminants in the food chain [CONTAM] related to pyrrolizidine alkaloids as undesirable substances in animal feed. *EFSA Journal*, 5(5), 447.

FAO/WHO. WHO Food Additive Series: 65. Safety evaluation of certain food additives and contaminants. Prepared by the 74th Meeting of the Joint FAO/WHO Expert Committee on Food Additives (JECFA). 2012. World Health Organization, Geneva.

Farsam, H., Yassa, N., Sarkhail, P. and Shafiee, A. (2000). New pyrrolizidine alkaloids from *Heliotropium crossifolium*. *Planta Med*, 66(4), 389–391.

Fracarolli, L., Rodrigues, G. B., Pereira, A. C., Júnior, N. S., Silva-Junior, G. J., Bachmann, L., Wainwright, M., Bastos, J. K. and Braga, G. U. (2016). Inactivation of plant-pathogenic fungus Colletotrichum acutatum with natural plant-produced photosensitizers under solar radiation. *Journal of Photochemistry and Photobiology B: Biology*, 162, 402–411.

FSANZ. Final Assessment Report Proposal P257 (2004). *Advice on the Preparation of Cassava and Bamboo Shoots*. Canberra: FSANZ, pp. 1–59.

Fu, P. P., Xia, Q., Lin, G. and Chou, M. W. (2004). Pyrrolizidine alkaloids—Genotoxicity, metabolism enzymes, metabolic activation, and mechanisms. *Drug Metabolism Reviews*, 36(1), 1–55.

Fukuda, K., Guo, L., Ohashi, N., Yoshikawa, M. and Yamazoe, Y. (2000). Amounts and variation in grapefruit juice of the main components causing grapefruit-drug inter-action. *Journal of Chromatography B: Biomedical Sciences and Applications*, 741(2), 195–203.

García-Cañas, V., Simó, C., León, C. and Cifuentes, A. (2010). Advances in Nutrigenomics research: Novel and future analytical approaches to investigate the biological activity of natural compounds and food functions. *Journal of Pharmaceutical and Biomedical Analysis*, 51(2), 290–304.

Gouget, A., Senchou, V., Govers, F., Sanson, A., Barre, A., Rouge, P., Lont-Lexica, R. and Canut, H. (2006). Lectin Receptor Kinases Participate in Protein-Protein Interactions to Mediate Plasma Membrane-Cell Wall Adhesions in Arabidopsis. *Plant Physiol*, 140, 81–90.

Haque, M. R. and Bradbury, J. H. (2002). Total cyanide determination of plants and foods using the picrate and acid hydrolysis methods. *Food Chem*, 77, 107–114.

Harborne J. B. (2014). *Introduction to Ecological Biochemistry*. Academic press: Cambridge.

Ignat, I. V., Popa, I. and Valentin, I. (2001). A critical review of methods for char-acterisation of polyphenolic compounds in fruits and vegetables. *Food Chem*, 126, 1821–1835.

Järvenpää, E. P., Jestoi, M. N. and Huopalahti, R. (1997). Quantitative determination of phototoxic furocoumarins in celeriac (*Apium graveolens* L. var. Rapeceum) using supercritical fluid extraction and high performance liquid chromatography. *Phytochemical Analysis*, 8(5), 250–256.

Jiang, B., Wang, X., Wang, L., Lv, X., Li, D., Liu, C. and Feng, Z. (2019). Two-step isolation, purification, and characterization of lectin from Zihua snap bean (*Phaseolus vulgaris*) seeds. *Polymers*, 11, 785. 10.3390/polym11050785

Kaiser, N., Douches, D., Dhingra, A., Glenn, K. C., Herzig, P. R., Stowe, E. C. and Swarup, S. (2020). The role of conventional plant breeding in ensuring safe levels of naturally occurring toxins in food crops. *Trends in Food Science & Technology*, 100, 51–66.

Kennedy, D. O. and Wightman, E. L. (2011). Herbal extracts and phytochemicals: Plant secondary metabolites and the enhancement of human brain function. *Advances in Nutrition*, 2(1), 32–50.

Krits, P., Fogelman, E. and Ginzberg, I. (2007). Potato steroidal glycoalkaloid levels and the expression of key isoprenoid metabolic genes. *Planta*, 227(1), 143–150.

Küpeli Akkol, E., Genç, Y., Karpuz, B., Sobarzo-Sánchez, E. and Capasso, R. (2020). Coumarins and coumarin-related compounds in pharmacotherapy of cancer. *Cancers*, 12(7), 1959.

Kus, N., Breda, S., Reva, I., Tasal, E., Ogretir, C. and Fausto, R. (2007). FTIR spectroscopic and theoretical study of the photochemistry of matrix-isolated coumarin. *Photochem Biol*, 83, 1237–1253.

Kwok, J. (2008). Cyanide poisoning and cassava. *Food Safety Focus*. (19th issue February, 2008). Incident focus. http://www.cfs.gov.hk/english/multimedia

Levasseur-Garcia, C. (2018). Updated overview of infrared spectroscopy methods for detecting mycotoxins on cereals (corn, wheat, and barley). *Toxins*, 10(1), 38.

Lombeart, G. A., Sieements, K. H., Pellars, P., Mankotia, M. and Ng, W. (2001). Furanocoumarins in celery and parsnip. *J AOAC International*, 84, 1135–1143.

Lu, X. and Rasco, B. A. (2012). Determination of antioxidant content and antioxidant activity in foods using infrared spectroscopy and chemometrics: A review. *Crit Rev Food Sci Nutr*, 52, 853–875.

Makkar, H. P., Norvsambuu, T., Lkhagvatseren, S. and Becker, K. (2009). Plant secondary metabolites in some medicinal plants of Mongolia used for enhancing animal health and production. *Tropicultura*, 27(3), 159–167.

Mazid, M., Khan, T. A. and Mohammad, F. (2011). Role of secondary metabolites in defense mechanisms of plants. *Biology and medicine*, 3(2), 232–249.

McGoverin, C. M., Weeranantanaphan, J., Downey, G. and Manley, M. (2010). The application of near infrared spectroscopy to the measurement of bioactive compounds in food commodities. *Journal of Near Infrared Spectroscopy*, 18(2), 87–111.

McMullin, D., Mizaikoff, B. and Krska, R. (2015). Advancements in IR spectroscopic approaches for the determination of fungal derived contaminations in food crops. *Analytical and Bioanalytical Chemistry*, 407(3), 653–660.

Milner, S. E., Brunton, N. P., Jones, P. W., O'Brien, N. M., Collins, S. G. and Maguire, A. R. (2011). Bioactivities of glycoalkaloids and their aglycones from Solanum species. *Journal of Agricultural and Food Chemistry*, 59(8), 3454–3484.

Moreno-Ley, C. M., Hernández-Martínez, D. M., Osorio-Revilla, G., Tapia-Ochoategui, A. P., Dávila-Ortiz, G. and Gallardo-Velázquez, T. (2019). Prediction of coumarin and ethyl vanillin in pure vanilla extracts using MIDFTIR spectroscopy and chemometrics. *Talanta*, 197, 264–269.

Muñoz-González, C., Moreno-Arribas, M. V., Rodríguez-Bencomo, J. J., Cueva, C., Martín-Álvarez, P. J. and Pozo-Bayón, M. A. (2012). Feasibility and application of liquid-liquid extraction combined with gas chromatography-mass spectrometry for the analysis of phenolic acids from grape polyphenols degraded by human faecal microbiota. *Food Chem*, 133, 526–535.

Murray R. D. H., Mendez J. and Brown S. A. (1982). *The Natural Coumarins*. Chichester, UK: Wiley.

Nawrocka, A. and Lamorska, J. (2013). Determination of food quality by using spectroscopic methods. In *Advances in Agrophysical Research*. Stanisław Grundas, Andrzej Stępniewski, IntechOpen, 1–410.

Oliveira, R. P., Demuner, A. J., Alvarenga, E. S., Parma, M. C., de Almeida Barbosa, L. C., de Moura Guimarães, L. and Aguiar, A. R. (2017). Experimental and theoretical studies on the characterization of monocrotaline by infrared and Raman spectroscopies. *Journal of Molecular Structure*. 10.1016/j.molstruc.2017.01.050

Pedreschi, F., Segtnan, V. H. and Knutsen, S. H. (2010). On-line monitoring of fat, dry matter and acrylamide contents in potato chips using near infrared interactance and visual reflectance imaging. *Food Chem*, 121, 616–620.

Peumans, W. J. and Van Damme, E. J. (1995). Lectins as plant defense proteins. *Plant Physiology*, 109(2), 347.

Pusztai, A., Ewen, S. W., Grant, G., Brown, D. S., Stewart, J. C., Peumans, W. J., Van Damme, E. J. and Bardocz, S. (1993). Antinutritive effects of wheat-germ agglutinin and other N-acetylglucosamine-specific lectins. *British Journal of Nutrition*, 70(1), 313–321.

Reese, R. A. and Xu, B. (2019). Single-molecule detection of proteins and toxins in food using atomic force microscopy. *Trends in Food Science & Technology*, 87, 26–34.

Roberts, C., Steer, T., Maplethorpe, N., Cox, L., Meadows, S., Nicholson, S., Page, P. and Swan, G. (2018). National diet and nutrition survey: Results from years 7 and 8 (combined) of the rolling programme (2014/2015–2015/2016).

Roberts, M. F. and Wink, M. (1998). *Alkaloids-Biochemistry, Ecological Functions and Medical Applications*. New York: Plenum, p. 486.

Roddick, J. G. (1996). Steroidal glycoalkaloids: Nature and consequences of bioactivity. *Adv Exp Med Biol*, 404, 277–295.

Sánchez-Patán, F., Monagas, M., Moreno-Arribas, M. V. and Bartolomé, B. (2011). Determination of microbial phenolic acids in human faeces by UPLC-ESI-TQ MS. *J Agric Food Chem*, 59, 2241–2247.

Savic, I. M., Nikolic, V. D., Savic-Gajic, I. M., Nikolic, L. B., Ibric, S. R. and Gajic, D. G. (2015). Optimization of technological procedure for amygdalin isolation from plum seeds (*Pruni domesticae* semen). *Front Plant Sci*, 6, 276.

Schilter, B., Constable, A. and Perrin, I. (2014). Naturally occurring toxicants of plant origin, Editor(s): Yasmine Motarjemi, Huub Lelieveld. In *Food Safety Management*, Academic Press, 45–57. ISBN 9780123815040, https://doi.org/10.1016/B978-0-12-381504-0.00003-2

Segtnan, V. H., Kita, A., Mielnik, M., Jorgensen, K. and Knutsen, S. H. (2006). Screening of acrylamide contents in potato crisps using process variable settings and near-infrared spectroscopy. *Mol Nutr Food Res*, 50, 811–817.

Seigler D. S. (1991). Cyanide and cyanogenic glycosides. Rosenthal, G. S. and Berenbaum, M. R., Herbivores: Their Interactions with Secondary Plant Metabolites: The Chemical Participants, Academic Press Cambridge, 1–452.

Shahidi, F. and Wanasundara, P.K.J.P.D. (1997). Cyanogenic glycosides of flaxseeds. *ACS (Am. Chem. Soc.) Symp. Ser.*, 662, 171–185.

Sharon, N. and Lis, H. (2004). History of lectins: From hemagglutinins to biological recognition molecules. *Glycobiology*, 14(11), 53R–62R.

Sharp, M. D., Kocaoglu-Vurma, N. A., Langford, V., Rodriguez-Saona, L. E. and Harper, W. J. (2012). Rapid discrimination and characterization of vanilla bean extracts by attenuated total reflection infrared spectroscopy and selected ion flow tube mass spectrometry. *J Food Sci*, 77(3), C284–C292.

Sheikh, M. A., Saini, C. S. and Sharma, H. K. (2022). Synergistic effect of microwave heating and hydrothermal treatment on cyanogenic glycosides and bioactive compounds of plum (*Prunus domestica* L.) kernels: An analytical approach. *Current Research in Food Science*, 5, 65–72.

Speijers G. (1993). *Cyanogenic Glycosides, WHO Food Additives Series:30. Toxicological Evaluation of Certain Food Additives and Naturally Occurring Toxicants*. Geneva: JECFA, pp. 299–337.

Tamasi, G., Pardini, A., Croce, R., Consumi, M., Leone, G., Bonechi, C., Rossi, C. and Magnani, A. (2021). Combined experimental and multivariate model approaches for glycoalkaloid quantification in tomatoes. *Molecules*, 26, 3068. 10.3390/molecules26113068

Tek N. (2006). *Chromatographic Determination of Glycoalkaloids in Eggplant.* Izmir Institute of Technology.

Tilahun, S., Sung An, H., Hwang, I. G., Choi, J. H., Baek, M. W., Choi, H. R., Park, D. S. and Jeong, C. S. (2020). Prediction of α-solanine and α-chaconine in potato tubers from hunter color values and VIS/NIR spectra. *Journal of Food Quality*, 8884219. 10.1155/2020/8884219

Vaananen T. (2007). *Glycoalkaloid Content and Starch Structure in Solanum Species and Interspecific Somatic Potato Hybrids.* University of Helsinki.

Vandenborre, G., Smagghe, G. and Van Damme, E. J. (Sep 1 2011). Plant lectins as defense proteins against phytophagous insects. *Phytochemistry*, 72(13), 1538–1550.

Vetter, J. (2000). Plant cyanogenic glycosides. *Toxicon*, 38(1), 11–36.

Wang, Y. P., Yan, J., Fu, P. P. and Chou, M. W. (2005). Human liver microsomal reduction of pyrrolizidine alkaloid N-oxides to form the corresponding carcinogenic parent alkaloid. *Toxicology Letters*, 155(3), 411–420.

WHO: World Health Organization. Cyanogenic glycosides. Toxicological evaluation of certain food additives and naturally occurring toxicants. WHO Food Additive Series 30. Geneva. 1993. http://www.inchem.org/documents/jecfa/jecmono/v30je18.htm.

Willmot, F. and Robertson, G. (1920). Senecio disease, or cirrhosis of the liver due to Senecio poisoning. *The Lancet*, 196(5069), 848–849.

Wink, M. (1988). Plant breeding: Importance of plant secondary metabolites for protection against pathogens and herbivores. *Theoretical and Applied Genetics*, 75(2), 225–233.

Yazaki, K. (2005). Transporters of secondary metabolites. *Current Opinion in Plant Biol*, 8, (3), 301–307.

CHAPTER 7

Mass Spectrometry in Analysis of Food Toxins

Mohd. Javed Naim[1] and Javed Ahamad[2]

[1]Department of Pharmaceutical Chemistry, Faculty of Pharmacy, Tishk International University, Erbil, Kurdistan Region, Iraq

[2]Department of Pharmacognosy, Faculty of Pharmacy, Tishk International University, Erbil, Kurdistan Region, Iraq

CONTENTS

7.1 Introduction	115
7.2 Current Analytical Methods	116
7.3 Mass Spectrometry	116
7.4 Mass Spectrometry in Food Toxin Analysis	117
7.4.1 Glycoalkaloids	117
7.4.2 Cyanogenic Glycosides	119
7.4.3 Pyrrolizidine Alkaloids	121
7.4.4 Furanocoumarins	123
7.5 Conclusion	124
References	125

7.1 INTRODUCTION

Food toxins are naturally occurring chemical compounds that are sought to serve some selective and defined role in plants and developed as chemical defense mechanisms against microorganisms and insects as well as herbivorous animals. They are found widely in great molecules and are basically produced by plant kingdom or are products of bacterial metabolism (Reese et al. 2019). They show varied chemical structures which possess different levels of toxicity and individual vulnerability and they can be harmful to humans only if consumed in appropriate quantities (Kaiser et al. 2020). Cyanogenic glycosides, glycoalkaloids, pyrrolizidine alkaloids, and furanocoumarins are some important naturally occurring toxins in plants (Fracarolli et al. 2016; Milner et al. 2011; Peumans and Van Damme 1995; Roberts et al. 2018; Vetter 2000). Because of their toxicity, an uprising demand is there to analyze them efficiently in order to protect humans as well as animals from their acute as well as chronic effects. So we need to identify them as well as quantify them among the plant species in order to keep track on their acceptable and harmful levels. For this, we need to develop and validate analytical techniques in order to properly screen and analyze

DOI: 10.1201/9781003222194-9

them. Chromatographic methods offer an ease of advantage in analyzing multiple samples at a time but they are time consuming and involve many steps to carry out before getting desired results. In order to update them, nowadays these are coupled with spectroscopic techniques to provide greater selectivity and sensitivity to yield precise results/accurate toxin content. Mass spectrometry is an important technique that comes into the limelight as a boon to analyze the toxin molecules based on their mass to charge ratio as well as their fragments accurately (Tevell Åberg et al. 2013; Josephs et al. 2010). This chapter highlights the advancement and potential of mass spectrometry as a substitute to prevailing approaches for the identification and quantification of food toxin.

7.2 CURRENT ANALYTICAL METHODS

There are various analytical methods available that can be employed in the analysis of food toxins, for example, electrophoresis, DSC (differential scanning colorimetry), genomics, proteomics, several chromatographic techniques, vibrational spectroscopy, elemental analysis, IR spectroscopy, NMR spectroscopy, immunological methods along with chemometrics and bioinformatics (Danezis et al. 2016). However, almost all of these techniques still necessitate considerable preparation of sample for analysis and may yield unclear data in absence of not following correct procedures. When methods are available, they are usually dedicated to a specific (group of) substance(s) in a certain commodity. With these above-mentioned analytical discrepancies, the accessibility of a generic technique appropriate for an extensive diversity of plant toxins would be exceedingly advantageous. Several mass spectroscopic techniques have been reported, that can prove to be a boon in the analysis of food toxins (Mane et al. 2012; Vlachos et al. 2016).

7.3 MASS SPECTROMETRY

Mass spectrometry is an analytical technique that is now being widely used throughout the world. This technique measures the m/e ratio so as it calculates the possible molecular masses of various food toxins/chemical components. It can be used for the identification of toxins like a fingerprint by its fragmentation technique that can measure the masses of various fragments that a molecule has undergone during the process of analysis. Its analysis varies from smaller molecules to larger macromolecules (Tevell Åberg et al. 2013; Domon and Aebersold 2006; Giessing and Kirpekar 2012; Ahamad et al. 2013, 2014a,b).

The excellent capability and potential of liquid chromatography and mass spectrometry (LC-MS) was recognized a few decades back and until now many techniques based on LC-MS are available (Verpoorte and Niessen 1994; Josephs et al. 2010; Zhou et al. 2010). TOF-MS is a high-resolution technique to evaluate toxins or chemical compounds that lack references or standards (Li et al. 2010; Yan et al. 2010). It also helps the analyst to reevaluate data after the arrival of new toxins. The high resolving capability of LC with full scan MS can resolve co-eluting compounds that shows similar masses exactly. The appearance of Orbitrap mass spectrometer in the year 2008 has unlocked a new option for routine analysis (Bateman et al. 2009).

7.4 MASS SPECTROMETRY IN FOOD TOXIN ANALYSIS

7.4.1 Glycoalkaloids

Glycoalkaloids are abundantly found among the members of the Solanaceae family. Potato (*Solanum tuberosum* L.), tomato (*S. lycopersicum* L.), and eggplant (*S. melongena* L.) are the most common sources of glycoalkaloids (Friedman et al. 1997; Friedman 2002; Friedman 2015). The most predominant glycoalkaloids present in potatoes are α-solanine and α-chaconine, and several other glycoalkaloids such as *β*-chaconine, *γ*-chaconine, *β*1-solanine, *β*2-solanine, and *γ*-solanine are also present in small quantities. Tomato contains α-tomatine and dehydrotomatine, and eggplant contains solasonine and solamargine as their main glycoalkaloids (Friedman et al. 1997). Glycoalkaloids are also reported in *Veratrum* and *Fritillaria* genera from the Liliaceae family (Wang et al. 2019; Fang et al. 2019; Cardenas et al. 2015). Glycoalkaloids, especially from potatoes e.g., α-solanine and α-chaconine, are known to cause gastrointestinal problems such as gastritis, gastrointestinal disturbance, nausea, vomiting, diarrhea, fever, low blood pressure, and in high doses, it causes fast pulse rate along with neurological and occasional death in humans and farm animals. In recent years, an increasing number of toxicological events were reported by food contaminations with glycoalkaloids. Excessive accumulation of GAs during growth, harvesting, and postharvest procedures are the major factors that lead to accumulation of GAs in these food items. Ingestion of food items with excess GAs could cause human health problems because of their dose-dependent toxicity (Morris and Petermann 1985). Hence, cultivation, postharvesting, and storage conditions and technologies should be developed to reduce production and accumulation of GAs in food crops (Mondy and Ponnawpalam 1985). GAs, specially α-solanine and α-chaconine, are known to cause gastrointestinal problems such as gastritis, gastrointestinal disturbance, nausea, vomiting, diarrhea, fever, low blood pressure, and in high doses, it causes fast pulse rate along with neurological and occasional death in human and livestock (Langkilde et al. 2009). These, GAs are also reported as anti-nutritional compounds (Itkin et al. 2011; Itkin et al. 2013). In recent years, they have been extensively investigated for their occurrence, bio-synthesis, and diverse biological functions in tomatoes and potatoes (Schwahn et al. 2014; Iijima et al. 2013; Heinig et al. 2014). Lycotetraose, solatriose, chocotriose, and commertetraose are most frequently detected glycoside moieties decorating the SGA (steroidal glycoalkaloids) structure (Figure 7.1).

Shakya et al. (2008) proposed the LC-ESI-MS method for identification of solanidane type glycoalkaloids in four potato species. The authors used solvent A (10 mM formic acid pH 3.5 with NH_4OH) and solvent B (100% methanol with 5 mM ammonium formate) on a C18 column (Zorbax Eclipse XDB). The experiment was carried out with an ESI source in positive as well as negative ion mode and the most intense peak for solanine was observed as m/z 868, (M+H)⁺. The study reported 50 solanidane type glycoalkaloids in four species of potatoes. Dall'Asta et al. (2009) proposed a multi-residual LC-DAD-MS/MS method for simultaneous detection and quantification of glycoalkaloids in seven tomato varieties found in Italy. The authors used a methanol-water (80:20 *v/v*, 0.1% formic acid) mixture as extraction solvent and RP-LC (C18 column) for separation of glycoalkaloids using water (from 90%) to acetonitrile (upto 90%) in gradient elution. The detection of glycoalkaloids was performed at 205 nm. The tomatine content varied from 1.0±0.1 to 201.9±10.1 mg/kg of dried weight basis in different potato samples. The proposed method was found to be quick and to be a single step for detection and quantification glycoalkaloids in potato samples.

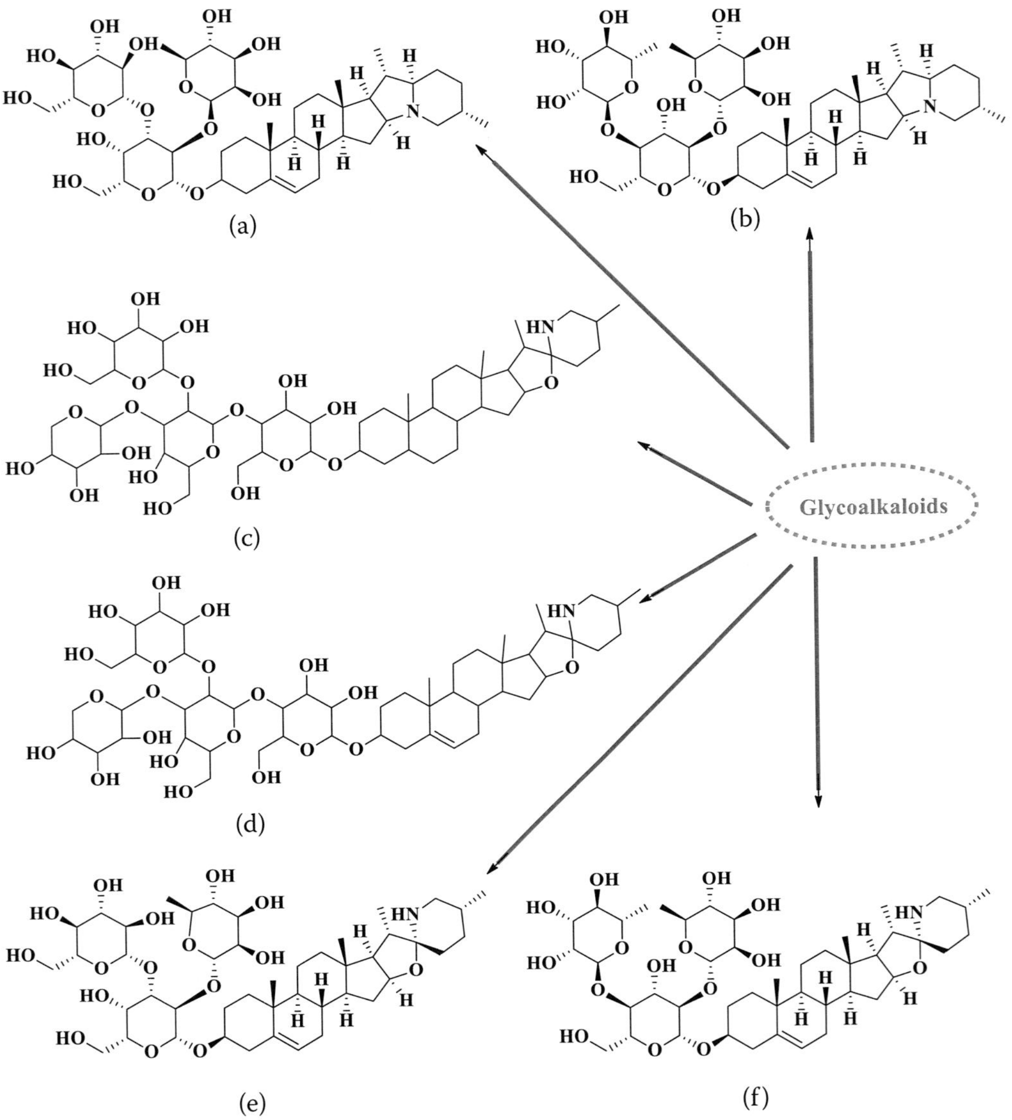

FIGURE 7.1 Chemical structure of (a) α-solanine, (b) α-chaconine, (c) α-tomatine, (d) dehydrotomatine, (e) solasonine, and (f) solamargine.

Nielsen et al. (2020) proposed LC-ESI/MS for quantification of glycoalkaloids in potato protein isolates. The authors utilized 0.1% formic acid (solvent A) and acetonitrile and 0.1% formic acid (solvent B) in a gradient elution on a Kinetex C18 column. The study reported total glycoalkaloid content ranging from 2,061–3,842 µg/g to 29–316 µg/g. Tamasi et al. (2021) determined tomatine content in eight varieties of tomatoes at different maturity stages by HPLC-ESI-QqQ-MS/MS and ATR-FT-MIR methods coupled with chemometry. The tomatine content varies in green (664 to 1874 mg/kg), turning (325 to 621 mg/kg), and pink (144 to 280 mg/kg) on the dry weight basis of tomato. Tomatine shows main IR bands in the region between 1,100 and 950 cm^{-1}. The authors compared the

IR outcomes with HPLC-ESI-QqQ-MS/MS, and the study reported that mass and IR methods were found to be fast, suitable and effective in quantification of glycoalkaloids (R^2 = 0.998 and 0.840, respectively).

7.4.2 Cyanogenic Glycosides

Cyanogenic glycosides are abundantly found in several plants, and about 60 cyanogenic glycosides are distributed in 2,600 plant species of more than 130 plant families (Bolarinwa et al. 2015; Mazza & Cottrell 2008; Bolarinwa et al. 2014; Buhrmester et al. 2000; Ganjewala 2010; Vetter 2000). Cyanogenic glycosides are mainly reported in plant species such as almonds, sorghum, spinach, elderberries, and linseed, etc. (Table 7.1). These glycosides are stored in the vacuoles of plant cells and are non-toxic to the plants, and are seperated from the plant by hydrolysing endogenous enzymes such as β-1,6-glycosidases and hydroxynitrile lyases (Abraham et al. 2016). These glycosides after reaction with endogenous enzymes release hydrogen cyanide which leads to potential toxicity in human and animals (Lee et al. 2013; Senica et al. 2016). Most of the cyanogenic glycoside containing plants produce endogenous enzymes that cause the release of HCN when their tissues are ruptured by crushing, disease, or herbivorous animals (Sánchez-Pérez et al. 2010). Cyanogenic glycosides in plants work as a chemical defense against herbivores and pathogens (Zagrobelny et al. 2004). The basic chemical structure of cyanogenic glycosides is presented in Figures 7.2 and 7.3 and in Table 7.1.

The O-glycosides of cyanohydrins are the most common source of hydrogen cyanide (HCN). Because HCN is a powerful respiratory inhibitor, the presence of cyanogenic

TABLE 7.1 Occurrence of Various Cyanogenic Glycosides in Human Food

S. no.	Source	Glycoside	R₁	R₂
1	Almonds, stone fruit, pome fruit	Amygdalin	Phenyl	H
2	Sorghum	Dhurrin	p-hydroxyphenyl	H
3	Cassava, lima beans, linseed/flaxseed, spinach	Linamarin	Methyl	Methyl
4	Cassava, linseed/flaxseed	Linustatin	Methyl	Methyl
5	Cassava, lima beans	Lotaustralin	Methyl	Ethyl
6	Linseed	Neolinustatin	Methyl	Ethyl
7	Stone fruit, pome fruit, pip fruit	Prunasin	Phenyl	H
8	Elderberries	Sambunigrin	Phenyl	H
9	Bamboo shoots	Taxiphyllin	p-hydroxyphenyl	H

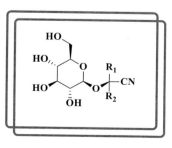

FIGURE 7.2 Generalized chemical structure of cyanogenic glycoside.

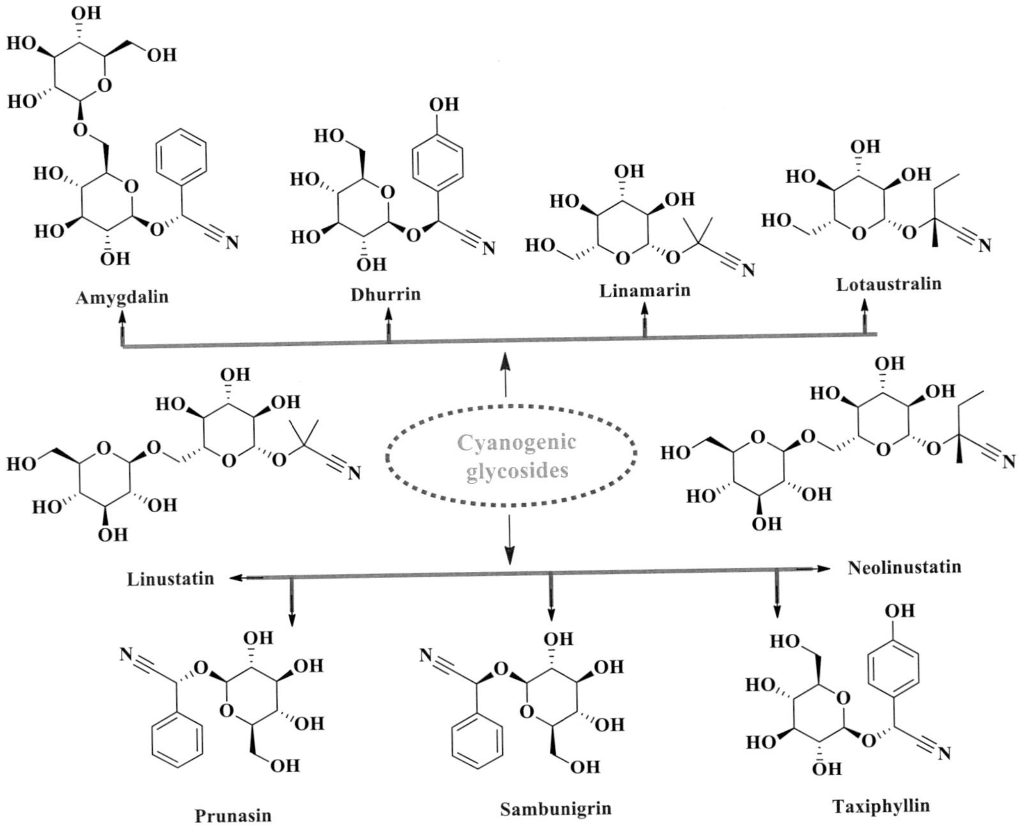

FIGURE 7.3 Chemical structures of important cyanogenic glycosides.

glycosides in human foods can cause acute or chronic cyanide poisoning, with symptoms such as anxiety, headache, vomiting, nausea, abdominal cramps, diarrhea, dizziness, weakness, and mental confusion (depending on the dose). In humans, acute cyanide toxicity (0.5–3.5 mg/kg b.w.t.) can cause loss of consciousness, hypotension, paralysis, coma, and death (Conn 1973; Conn 1969; Seigler 1975; Burns et al. 2012; Geller et al. 2006).

Consumption of bitter almonds, apricot kernels, cassava, bamboo shoots, lima beans, yams, maize, flax, sorghum, pulses, and chick peas have been reported to cause acute cyanide poisoning (Sahin 2011; Sanchez-Verlaan et al. 2011; Akintonwa & Tunwashe 1992).

Several researchers have used electron impact (EI) and chemical ionization mass spectrometry (CIMSI) method for identification, characterization, and analysis of cyanogenic glycosides in food plants (Cairns et al. 1978). These mass spectrometric approaches are useful, but they usually necessitate derivatization, which can raise the molecular weight above the mass spectrometer's mass range. Field desorption (FDMS) should be well suited for the analysis of cyanogenic glycosides due to its specific relevance to the study of these compounds (Beckey & Schulten, 1975). When analyzed by FDMS, glycosides generally yield more structural information than when analyzed by EIMS. Derivatization of the sample is unnecessary, and earlier research has shown that the presence of a mixture of components should not impede analysis (Sphon et al. 1977; Watts et al. 1975). The molecular weight of two new cyanogenic glycosides was determined using FDMS, and the rest of the structural study was completed using NMR (Seigler et al. 1975; Seigler et al. 1978; Butterfield et al. 1975).

Appenteng et al. (2021) developed a UHPLC-MS/MS analytical method for identification and quantification of cyanogenic glycosides in two genotypes (ozone and ozark in juice, skin, stem, and seeds) of American Elderberry with C18 column using UHPLC coupled to Xevo TQ-S triple-quadrupole mass spectrometer. The mobile phase used was 0.1% HCOOH in water (A) and 0.1% formic acid in acetonitrile. The identified cyanogenic glycosides were amygdalin (1.57–6.38 µg/g in ozone and 0.36–3.48 µg/g in ozark), dhurrin (0.12–0.94 µg/g in ozone and 0.22–1.91 µg/g in ozark), prunasin (0.58–2.84 µg/g in ozone and 0.36–3.07 µg/g in ozark), and linamarin (0.12–0.75 µg/g in ozone and 0.13–0.90 µg/g in ozark). Zhao et al. (2019) reported an analytical technique using UPLC/ESI-HRMS for assessment of cyanogenic glycosides in flaxseed with C18 column using waters Acquity UPLC coupled to high-resolution Bruker micrOTOF-Q II system and electrospray ionization. The mobile phase used were A: Milli-Q water and formic acid (99.9/0.1 v/v) and B: acetonitrile and formic acid (99.9/0.1 v/v) where both mobile phase contains 50 µM NaCl to promote Na^+ adduct formation. They identified linustatin (coated seeds: 1.41±0.04 mg/g; non-coated seeds 1.37±0.06 mg/g) and neolinustatin (coated seeds: 1.54±0.04 mg/g; non-coated seeds: 1.55±0.01 mg/g) in flaxseeds and the fragments of whom were observed at m/z 405.1353 for linustatin and m/z 419.1518 for neolinustatin.

7.4.3 Pyrrolizidine Alkaloids

Pyrrolizidine alkaloids (PAs) are food toxins found in several plant families such as Apocynaceae, Asteraceae, Boraginaceae, Compositae, Fabaceae, Leguminosae, Ranunculaceae, and Scrophulariaceae. Pyrrolizidine alkaloids (PAs) and their N-oxides are composed of necine and its derivative. About 660 PAs and their N-oxides have been identified in over 6,000 plant species (Roeder 2000; Boppre 2011). These PAs are a group of phytochemicals that contain two fused, five-membered rings which share a bridgehead nitrogen atom, forming a tertiary alkaloid. Several medicinal plants contain pyrrolizidine alkaloids such as comfrey (*Symphytum officinale*), coltsfoot (*Tussilago farfara*), and petasites (*Petasites japonicus*), etc. Pyrrolizidine alkaloids and their N-oxides are also present in milk (cows and goats), honey, staple foods, herbal teas, and herbal medicines (Chen et al. 2017). Pyrrolizidine alkaloids are present in plants as free necines and as mixtures of bases with their N-oxides (Sixto et al. 2019). Pyrrolizidine alkaloids are known for their hepatotoxic and carcinogenic properties (Mudge et al. 2015; Merz and Schrenk 2016; Edgar et al. 2015).

Despite the fact that PAs are known to be harmful and that incidents of human and animal poisoning have been reported, no maximum residual limits in food and feed have been defined (Valese et al. 2016). Acceptable tolerance levels for herbal preparations and extracts have been established in some European nations, including Germany, Austria, and the Netherlands. In Germany, the daily PA intake limit has been set at 1 µg if eaten for up to 6 weeks; however, it has been reduced to 0.1 µg if used for longer periods (Dübecke et al. 2011). The Federal Institute of Risk Assessment (Bundesinstitut fuer Riskobewertung, BfR, Germany) and the UK Committee on Toxicity have concluded that PA exposure from food should be kept to a minimum, recommending an exposure limit of 0.007 µg of PAs kg^{-1} (b.w./day) from a variety of foods (Merz and Schrenk 2016).

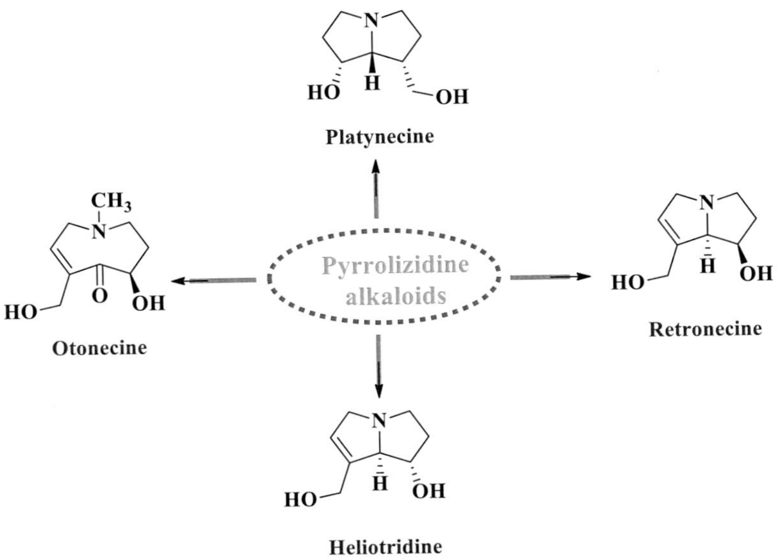

FIGURE 7.4 Chemical structures of major pyrrolizidine alkaloids.

The increased number of reports of PAs and its N-oxides contamination in foods such as grain, milk, meat, eggs, and honey highlights the significance of doing a risk assessment on these alkaloids. The lack of appropriate analytical tools to determine them in different food products is the biggest problem in gathering enough data for their investigation. Selective and extremely sensitive testing methods must be employed to ensure that herbs are free from toxic PAs. Liquid chromatography with tandem quadrupole mass spectrometry is the most frequently used method for PAs analysis (Martinello et al. 2017; Schulz et al. 2015) (Figure 7.4).

Beales et al. (2004) developed solid-phase extraction of PAs and their N-oxides from various honey samples and subsequent characterization by LC-MS method. The study outcomes showed that honey samples contain PAs and their N-oxide upto 2000 ppb that are originated from *Echium plantagineum*. The study also highlighted that about 30% honey samples do not contain PAs and their N-oxide residues. Kast et al. (2014) proposed an HPLC-MS/MS method for quantification of PAs in honey collected from various places. The authors analyzed 18 PAs and their N-oxide in *Echium Eupatorium* and *Senecio* genera. They reported that 54% of honeys contain PAs ranging from 18 to 55 µg/kg. Bolechová et al. (2015) proposed the UPLC-MS/MS method for detection and quantification of pyrrolizidine alkaloids using formic acid (0.1%, eluent A) and methanol (containg 0.1% formic acid and 1 mM ammonium formate, eluent B) in a gradient elution on a C18 column. The proposed method was applied for detection of 5 PAs in 41 animal feed samples such as milk feed, complete feed, and alfalfa. The identified PAs are senecionine, retrorsine, seneciphylline, monocrotaline, and senkirkine and their content ranges from 6.3 µg/kg to 424 µg/kg.

De Jesus Inacio et al. (2020) developed a LC-MS/MS method for analysis of PA contamination in bee pollen. The authors used a gradient elution using water with 0.1% formic acid (Solvent A) and methanol with 0.1% formic acid (solvent B). The 17 PAs/ its N-oxide were analyzed by MS with the ESI source operating in positive-ion mode. The study shows that the bee pollen contain PAs and their N-oxide in the range of 2 to 3,356 µg/kg. Seventy-seven percent of the samples contain PAs and their N-oxide. Prada et al. (2020)

studied PAs and their *N*-oxide distribution in six species of *Crotalaria* by LC-ESI-MS method. The authors reported 45 PAs in different parts of *Crotalaria* species. The authors utilized formic acid (0.05% to 0.35%) and water (solvent A) and formic acid and acetonitrile (solvent B) in gradient elution on an Hypersil GOLD aQ column. The retrorsine-type alkaloids were quantified in *Crotalaria* as product ions appear at *m/z* 138.09134, 120.08078, *m/z* 94.06513, and m/z 136.07569 and *m/z* 118.06513 for their *N*-oxides; and for platynecine-type alkaloids ion appears at *m/z* 140.10699, *m/z* 122.09643, and *m/z* 96.08078. The seeds and flowers had higher content of PAs ranging from 2 to 21 mg/g, and leaves and roots contain 0.02 to 9 mg/g of PAs.

Lu et al. (2021) developed a single-step detection and quantification method of PAs in five species of *Senecio* by a LC-MS method. The authors used a C12 column and gradient elution with 1% formic acid in water (Eluent A) and acetonitrile (Eluent B) as a mobile phase. The main PAs were identified as retronecine type and their *N*-oxide. The authors identified 85 PAs in five species of *Senecio*; 75 of these were of the retronecine-type and 10 PAs were of the otonecine-type. In another study by Williams et al. (2011), PAs and their *N*-oxides were detected in a high concentration in *Cryptantha crassipes* by HPLC-ESI/MS method. The authors reported PAs and their *N*-oxides ranging from 3% to 5% on the dry weight basis of leaves, and major alkaloids were: lycopsamine-*N*-oxide, leptanthine, echimiplatine, amabiline, echiumine, and dihydroechiumine. Pyrrolizidine alkaloids were also analyzed by several researchers in deferent plant species, and the researchers reported that these plants contain a high quantity of PAs and their *N*-oxides (Zhang W. et al. 2017; Zhou et al. 2010).

7.4.4 Furanocoumarins

Coumarins are basically phenolic compounds/derivatives that are widely dispersed throughout the plant kingdom. They are classified as simple coumarins, furanocoumarins, pyranocoumarins, or pyran-ring substituted coumarins and their hydroxylates (with OH group), alkoxylated derivatives (with alkoxy group), and alkylated derivatives (contains alkyl group), along with their glycosides (Murray et al. 1982). Furanocoumarins are naturally occurring compounds released/produced as a defense mechanism against bacteria, fungi, and insects as well as stress (Fracarolli et al. 2016). Furanocoumarins are subdivided into psoralens and angelicin whereas dihydrofuranocoumarins are subdivided into nodakenetin and archangelicin. Umbelliferone is a complex 7-hydroxycoumarin that is considered as the parent compound structurally and biogenetically (Kozawa et al. 1978; Roberts & Wink 1998). Sensitization of cells by furanocoumarins (visible and UV) results in toxicity, mutagenicity, and carcinogenicity. The highest concentration is found in plants from the Apiaceae family like celery (45 µg/g) (Diawara et al. 1995), parsnip (145 µg/g) (Lombeart et al. 2001), and parsley (112 µg/g) (Beier et al. 1994), whereas it is slightly lower in citrus fruits, which can be easily attributed to the bargamottin concentration in grapefruit juice (6 µg/g) (Fukuda et al. 2000). Furanocoumarins were also reported to be present in the plants of Solanaceae, Cyperaceae, Fabaceae, Moraceae, Pittosporaceae, Rosaceae, and Amaranthaceae (Murray et al. 1982) (Figure 7.5).

Using ultra-fast liquid chromatography combined with triple-quadrupole mass spectrometry, Zhang L. et al. (2017) identified nine andafocoumarins (1–9) in four cultivars of *Angelicae dahuricae* Radix. The authors utilized a C18 column and ACN (A) and H_2O (B) as mobile phase, with a gradient elution of 30–40% A at 0–2 minutes, 40–85% A at 2–2.01 minutes, 85–905 A at 2.01–6 minutes, 90% A at 6–10 minutes, and 90–95% A at 10–12 minutes. Andafocoumarins 1 and 2 were key constituents in the overall levels

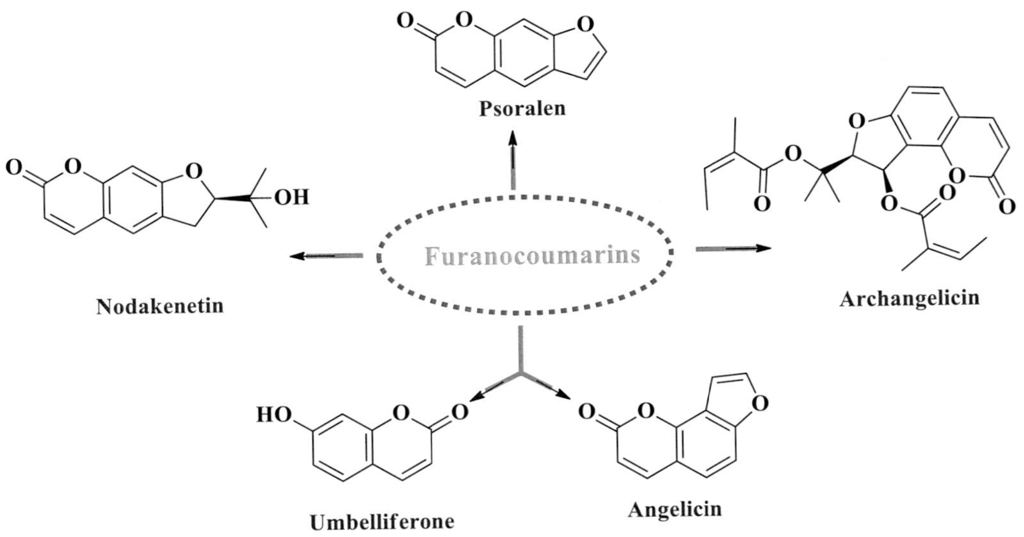

FIGURE 7.5 Chemical structures of some important furanocoumarins.

of these nine coumarins, which ranged from 89.41 to 1022.57 µg/g. Dugrand et al. (2013) developed a rapid and fast UPLC-MS method that used a C18 reversed phase column to identify and quantitate coumarins and furanocoumarins in citrus peel extracts from six different varieties. The solvents consisted of 0.1% formic acid in water (A) and 0.1% formic acid in methanol (B). The important furanocoumarins identified were umbelliferone (found in traces in lemon, grapefruit and bergamot varieties and and not be quantified) and psoralen (1.95±0.98 mg/kg in bergamot variety). Heinke et al. (2012) analyzed several furanocoumarins from Yemenite *Dorstenia* species (*Dorstenia gigas* and *D. foetida*) through liquid chromatography/electrospray tandem mass spectrometry using RP-18 column. The gradient system consists of $H_2O:CH_3CN$ 90:10 (each contains 0.2% HOAc) to 5:95 within 15 minutes, followed by a 15-minute isocratic period (gradient system 1) or followed by a 30-minute isocratic period (gradient system 2) and flow rate 70 mL/minute. A total of 34 coumarins were identified among which psoralen and angelicin appeared as major furanocoumarins.

7.5 CONCLUSION

Pyrrolizidine alkaloids (PAs) and their *N*-oxides are known to cause severe toxicities in humans and grazing animals such as genotoxicity, carcinogenicity, hepatotoxicity, and mutagenicity. Half of the known PAs have been reported to be toxic. Hence, it became an important issue to identify and analyze these toxic plant compounds in food items. The present chapter was designed to compile up-to-date information regarding the role of mass spectrometer in the identification and quantification of PAs and their *N*-oxides and in the identification of glycoalkaloids, cyanogenic glycosides, and furanocoumarins. Mass spectrometry nowadays is the most widespread analytical technique that provides accurate results due to its high selectivity and sensitivity. Further advancements in the present analytical technique may contribute towards insights in the incidence of predictable and unpredictable food toxins, their risk valuation, and defining their acceptable limits.

REFERENCES

Abraham, K., Buhrke, T. and Lampen, A. (2016). Bioavailability of cyanide after consumption of a single meal of foods containing high levels of cyanogenic glycosides: A crossover study in humans. *Arch. Toxicol.*, 90, 559–574.

Ahamad, J., Kamran, J. N., Ali, M. and Mir, S. R. (2013). New Glycoside Esters from the Aerial Parts of *Artemisia absinthium* Linn. *The Natural Products Journal*, 3(4), 260–267.

Ahamad, J., Kamran, J. N., Ali, M. and Mir, S. R. (2014a). Isoflavone glycosides from aerial parts of *Artemisia absinthium* Linn. *Chem Natural Comp*, 49(6), 696–700.

Ahamad, J., Kamran, J. N., Ali, M. and Mir, S. R. (2014b). New isoquinoline alkaloids from the stem bark of *Berberis aristata*. *Indian Journal of Chemistry-B*, 53, 1237–1241.

Akintonwa, A. and Tunwashe, O. L. (1992). Fatal cyanide poisoning from Cassava-based meal. *Hum. Exp. Toxicol.*, 11, 47–49.

Appenteng, M. K., Krueger, R., Johnson, M. C., Ingold, H., Bell, R., Thomas, A. L. and Greenlief, C. M. (2021). Cyanogenic glycoside analysis in american elderberry. *Molecules.*, 26, 1384.

Bateman, K., Kellmann, M., Muenster, H., Papp, R. and Taylor, L. (2009). Quantitative-qualitative data acquisition using a benchtop orbitrap mass spectrometer. *J. Am. Soc. Mass. Spectrom.*, 20, 1441–1450.

Beales, K. A., Betteridge, K., Colegate, S. M. and Edgar, J. A. (2004). Solid-phase extraction and LC-MS analysis of pyrrolizidine alkaloids in honeys. *J. Agricultural Food Chem*, 52(21), 6664–6672.

Beckey, H. D. and Schulten, H.-R. (1975). Field desorption mass spectrometry, *Angewandte Chemie International Edition in English*, 14(6), 403–415.

Beier, R. C., Ivie, G. W. and Oertli, E. H. (1994). Linear furanocoumarins and graveolone from the common herb parsley. *Phytochemistry*, 36(4), 869–872.

Bolarinwa, I. F., Orfila, C. and Morgan, M. (2014). Amygdalin content of seeds, kernels and food products commercially-available in the UK. *Food Chem*, 152, 133–139.

Bolarinwa, I. F., Orfila, C. and Morgan, M. R. (2015). Determination of amygdalin in apple seeds, fresh apples and processed apple juices. *Food Chem*, 170, 437–442.

Bolechová, M., Cáslavský, J., Pospíchalová, M. and Kosubová, P. (2015). UPLC-MS/MS method for determination of selected pyrrolizidine alkaloids in feed. *Food Chem*, 170, 265–270.

Boppre, M. (2011). The ecological context of pyrrolizidine alkaloids in food, feed and forage: An overview. *Food Addit. Contam. A.*, 28, 260–281.

Buhrmester, R. A., Ebinger, J. E. and Seigler, D. S. (2000). Sambunigrin and cyanogenic variability in populations of *Sambucus canadensis* L.(Caprifoliaceae). *Biochem. Syst. Ecol.*, 28, 689–695.

Burns, A. E., Bradbury, J. H., Cavagnaro, T. R. and Gleadow, R. M. (2012). Total cyanide content of cassava food products in Australia. *J. Food Compos. Anal.*, 25, 79–82.

Butterfield, C. S., Conn, E. E. and Seigler, D. S. (1975). Elucidation of structure and biosynthesis of acacipetalin. *Phytochem*, 14, 993–997.

Cairns, T., Froberg, J. E., Gonzales, S., Langham, W. S., Stamp, J. J., Howie, J. K. and Sawyer, D. T. (1978). Analytical chemistry of amygdalin. *Anal. Chem.*, 50, 317–322.

Cardenas, P. D., Sonawane, P. D., Heinig, U., Bocobza, S. E., Burdman, S. and Aharoni, A. (2015). The bitter side of the nightshades: Genomics drives discovery in Solanaceae steroidal alkaloid metabolism. *Phytochem.*, 113, 24–32.

Chen, L., Mulder, P. P. J., Louisse, J., Peijnenburg, A., Wesseling, S. and Rietjens, I. M. C. M. (2017). Risk assessment for pyrrolizidine alkaloids detected in (herbal) teas and plant food supplements. *Regul. Toxicol. Pharmacol*, 86, 292–302.

Chiara Dall'Asta, C., Falavigna, C., Galaverna, G., Sforza, S., Dossena, A. and Marchelli, R. (2009). A multiresidual method for the simultaneous determination of the main glycoalkaloids and flavonoids in fresh and processed tomato (*Solanum lycopersicum* L.) by LC-DAD-MS/MS. *J. Sep. Sci*, 32, 3664–3671.

Conn, E. E. (1969). Effect of local processing techniques on the nutrients and antinutrients content of bitter cassava (Manihot Esculenta Crantz). *J. Agric. Food Chem.*, 17, 519.

Conn E. E. (1973). *Toxicants Occurring Naturally in Foods*. Washington, DC: National Academy of Sciences, p. 299.

Danezis, G. P., Tsagkaris, A. S., Camin, F., Brusic, V. and Georgiou, C. A. (2016). Food authentication: Techniques, trends & emerging approaches. *Trends in Anal. Chem.*, 85, 123–132.

De Jesus Inacio, L., Lanza, I., Merlanti, R., Contiero, B., Lucatello, L., Serva, L., Bisutti, V., Mirisola, M., Tenti, S., Segato, S. and Capolongo, F. (2020). Discriminant analysis of pyrrolizidine alkaloid contamination in bee pollen based on near-infrared data from lab-stationary and portable spectrometers. *European Food Research and Technology*, 246(12), 2471–2483.

Diawara, M. M., Trumble, J. T., Quiros, C. F. and Hansen, R. (1995). Implications of distribution of linear furanocoumarins within celery. *J. Agri Food Chem*, 43(3), 723–727.

Domon, B. and Aebersold, R. (2006). Mass spectrometry and protein analysis. *Science*, 312, 212–217.

Dübecke, A., Beckh, G. and Lüllmann, C. (2011). Pyrrolizidine alkaloids in honey and bee pollen. *Food Addit. Contam. Part A.*, 28, 348–358.

Dugrand, A., Olry, A., Duval, T., Hehn, A., Froelicher, Y. and Bourgaud, F. (2013). Coumarin and furanocoumarin quantitation in citrus peel via ultraperformance liquid chromatography coupled with mass spectrometry (UPLC-MS). *Journal of Agricultural and Food Chemistry*, 61(45), 10677–10684.

Edgar, J. A., Molyneux, R. J. and Colegate, S. M. (2015). Pyrrolizidine alkaloids: Potential role in the etiology of cancers, pulmonary hypertension, congenital anomalies, and liver disease. *Chem. Res. Toxicol.*, 28(1), 4–20.

Fang, C., Fernie, A. R. and Luo, J. (2019). Exploring the diversity of plant metabolism. *Trends Plant Sci*, 1, 83–98.

Fracarolli, L., Rodrigues, G. B., Pereira, A. C., Júnior, N. S., Silva-Junior, G. J., Bachmann, L., Wainwright, M., Bastos, J. K. and Braga, G. U. (2016). Inactivation of plant-pathogenic fungus *Colletotrichum acutatum* with natural plant-produced photosensitizers under solar radiation. *Journal of Photochemistry and Photobiology B: Biology*, 162, 402–411.

Friedman, M. (2002). Tomato glycoalkaloids: Role in the plant and in the diet. *J. Agric. Food Chem*, 50, 5751–5780.

Friedman, M. (2015). Chemistry and anticarcinogenic mechanisms of glycoalkaloids produced by eggplants, potatoes, and tomatoes. *J. Agric. Food. Chem.*, 63, 3323–3337.

Friedman, M., McDonald, G. and Filadelfi-keszi, M. (1997). Potato glycoalkaloids: Chemistry analysis safety and plant physiology. *Crit. Rev. Plant Sci.*, 16, 55–132.

Fukuda, K., Guo, L., Ohashi, N., Yoshikawa, M. and Yamazoe, Y. (2000). Amounts and variation in grapefruit juice of the main components causing grapefruit-drug interaction. *J. Chromatography B: Biomedical Sciences and Applications*, 741(2), 195–203.

Ganjewala, D. (2010). Advances in cyanogenic glycosides biosynthesis and analyses in plants: A review. *Acta Biol. Szeged.*, 54, 1–14.

Geller, R. J., Barthold, C., Saiers, J. A. and Hall, A. H. (2006). Pediatric cyanide poisoning: Causes, manifestations, management, and unmet needs. *Pediatrics.*, 118, 2146–2158.

Giessing, A. M. and Kirpekar, F. (2012). Mass spectrometry in the biology of RNA and its modifications. *J Proteomics*, 75, 3434–3449.

Heinig, U. and Aharoni, A. 2014). Analysis of steroidal alkaloids and saponins in Solanaceae plant extracts using UPLC-qTOF mass spectrometry. In Rodríguez-Concepció n, M. (ed.). *Plant Isoprenoids. Methods in Molecular Biology (Methods and Protocols)*. New York: Humana Press, vol. 1153, pp. 171–185.

Heinke, R., Franke, K., Michels, K., Wessjohann, L., Ali, N. A. and Schmidt, J. (2012). Analysis of furanocoumarins from Yemenite Dorstenia species by liquid chromatography/electrospray tandem mass spectrometry. *J. Mass Spectrometry.*, 47(1), 7–22.

Iijima, Y., Watanabe, B., Sasaki, R., Takenaka, M., Ono, H., Sakurai, N., Umemoto, N. and Suzuki, H. (2013). Steroidal glycoalkaloid profiling and structures of glycoalkaloids in wild tomato fruit. *Phytochem.*, 95, 145–157.

Itkin, M., Heinig, U., Tzfadia, O., Bhide, A. J., Shinde, B., Cardenas, P. D. and Bocobza, S. E. (2013). Biosynthesis of antinutritional alkaloids in solanaceous crops is mediated by clustered genes. *Science.*, 341, 175–179.

Itkin, M., Rogachev, I., Alkan, N., Rosenberg, T., Malitsky, S., Masini, L. and Meir, S. (2011). Glycoalkaloid Metabolism is required for steroidal alkaloid glycosylation and prevention of phytotoxicity in tomato. *Plant Cell*, 23, 4507–4525.

Josephs, R. D., Daireaux, A., Westwood, S. and Wielgosz, R. I. (2010). Simultaneous determination of various cardiac glycosides by liquid chromatography–hybrid mass spectrometry for the purity assessment of the therapeutic monitored drug digoxin. *J. Chromatography A.*, 1217, 4535–4543.

Kaiser, N., Douches, D., Dhingra, A., Glenn, K. C., Herzig, P. R., Stowe, E. C. and Swarup, S. (2020). The role of conventional plant breeding in ensuring safe levels of naturally occurring toxins in food crops. *Trends in Food Science & Technology*, 100, 51–66.

Kast, C., Dübecke, A., Kilchenmann, V., Bieri, K., Böhlen, M., Zoller, O., Beckh, G. and Lüllmann, C. (2014). Analysis of Swiss honeys for pyrrolizidine alkaloids. *J. Apicultural Research*, 53(1), 75–83.

Kozawa, M., Morita, N., Baba, K. and Hata, K. (1978). Chemical components of the roots of Angelica keiskei Koidzumi. II. The structure of the chalcone derivatives (author's transl). *Yakugaku Zasshi: Journal of the Pharmaceutical Society of Japan*, 98(2), 210–214.

Langkilde, S., Mandimika, T., Schrøder, M., Meyer, O., Slob, W., Peijnenburg, A. and Poulsen, M. (2009). A 28-day repeat dose toxicity study of steroidal glycoalkaloids, α-solanine and α-chaconine in the Syrian Golden hamster. *Food Chem Toxicol*, 47(6), 1099–1108.

Lee, J., Zhang, G., Wood, E., Rogel Castillo, C. and Mitchell, A. E. (2013). Quantification of amygdalin in nonbitter, semibitter, and bitter almonds (*Prunus dulcis*) by UHPLC-(ESI)QqQ MS/MS. *J. Agric. Food Chem.*, 61, 7754–7759.

Li, S. L., Song, J. Z., Qiao, C. F., Zhou, Y. and Xu, H. X. (2010). UPLC–PDA–TOFMS based chemical profiling approach to rapidly evaluate chemical consistency between traditional and dispensing granule decoctions of traditional medicine combinatorial formulae. *J. Pharm. Biomed. Anal.*, 52, 468–478.

Lombeart, G. A., Sieements, K. H., Pellars, P., Mankotia, M. and Ng, W. (2001). Furanocoumarins in celery and parsnip. *J AOAC International*, 84, 1135–1143.

Lu, A. J., Lu, Y. L., Tan, D. P., Qin, L., Ling, H., Wang, C. H. and He, Y. Q. (2021). Identification of pyrrolizidine alkaloids in senecio plants by liquid chromatography-mass spectrometry. *Journal of Analytical Methods in Chemistry*. 10.1155/2021/1957863

Mane, B. G., Mendiratta, S. K., Tiwari, A. K. and Bhilegaokar, K. N. (2012). Development and evaluation of polymerase chain reaction assay for identification of buffalo meat. *Food Anal. Methods.*, 5, 296–300.

Martinello, M., Borin, A., Stella, R., Bovo, D., Biancotto, G., Gallina, A. and Mutinelli, F. (2017). Development and validation of a QuEChERS method coupled to liquid chromatography and high resolution mass spectrometry to determine pyrrolizidine and tropane alkaloids in honey. *Food Chem*, 234, 295–302.

Mazza, G. and Cottrell, T. (2008). Carotenoids and cyanogenic glucosides in saskatoon berries (Amelanchier alnifolia Nutt). *J. Food Compos. Anal.*, 21, 249–254.

Merz, K. H. and Schrenk, D. (2016). Interim relative potency factors for the toxicological risk assessment of pyrrolizidine alkaloids in food and herbal medicines. *Toxicol. Lett.*, 263, 44–57.

Milner, S. E., Brunton, N. P., Jones, P. W., O'Brien, N. M., Collins, S. G. and Maguire, A. R. (2011). Bioactivities of glycoalkaloids and their aglycones from *Solanum* species. *Journal of Agricultural and Food Chemistry*, 59(8), 3454–3484.

Mondy, N. I. and Ponnawpalam, R. (1985). Effect of magnesium fertilizers on total glycoalkaloids and nitrate-N in Katahdin tubers. *J Food Sci*, 50(2), 535–536.

Morris, S. C. and Petermann, J. B. (1985). Genetic and environmental effects on levels of glycoalkaloids in cultivars of potato (*Solanum tuberosum* L.). *Food Chem*, 18(4), 271–282.

Mudge, E. M., Jones, A. M. P. and Brown, P. N. (2015). Quantification of pyrrolizidine alkaloids in North American plants and honey by LC-MS: Single laboratory validation. *Food Addit. Contam. Part A.*, 32, 1–7.

Murray R. D. H., Mendez J. and Brown S. A. (1982). *The Natural Coumarins*. Chichester, UK: Wiley.

Nielsen, S. D., Schmidt, J. M., Kristiansen, G. H., Dalsgaard, T. K. and Larsen, L. B. (2020). Liquid chromatography mass spectrometry quantification of α-solanine, α-chaconine, and solanidine in potato protein isolates. *Foods.*, 9(4), 416.

Peumans, W. J. and Van Damme, E. J. (1995). Lectins as plant defense proteins. *Plant Physiology*. 109(2), 347.

Prada, F., Stashenko, E. E. and Martínez, J. R. (2020). LC/MS study of the diversity and distribution of pyrrolizidine alkaloids in Crotalaria species growing in Colombia. *J. Separation Science*, 43(23), 4322–4337.

Reese, R. A. and Xu, B. (2019). Single-molecule detection of proteins and toxins in food using atomic force microscopy. *Trends in Food Science & Technology*, 87, 26–34.

Roberts, C., Steer, T., Maplethorpe, N., Cox, L., Meadows, S., Nicholson, S., Page, P. and Swan, G. (2018). National Diet and nutrition survey: Results from years 7 and 8 (combined) of the rolling Programme (2014/2015–2015/2016).

Roberts M. F. and Wink M. (1998). *Alkaloids-Biochemistry, Ecological Functions and Medical Applications*. New York: Plenum, p. 486.

Roeder, E. (2000). Medicinal plants in China containing pyrrolizidine alkaloids. *Pharmazie*, 55, 711–726.

Sahin, S. (2011). Cyanide poisoning in a children caused by apricot seeds. *J. Health Med. Informat.*, 2, 1–2.

Sánchez-Pérez, R., Howad, W., Garcia-Mas, J., Arús, P., Martínez-Gómez, P. and Dicenta, F. (2010). Molecular markers for kernel bitterness in almond. *Tree Genet. Genomes*, 6, 237–245.

Sanchez-Verlaan, P., Geeraerts, T., Buys, S., Riu-Poulenc, B., Cabot, C., Fourcade, O., Megarbane, B. and Genestal, M. (2011). An unusual cause of severe lactic acidosis: Cyanide poisoning after bitter almond ingestion. *Intensive Care Med*, 37, 168–169.

Schulz, M., Meins, J., Diemert, S., Zagermann-Muncke, P., Goebel, R., Schrenk, D., Schubert-Zsilavecz, M. and Abdel-Tawab, M. (2015). Detection of pyrrolizidine alkaloids in German licensed herbal medicinal teas. *Phytomedicine.*, 22, 648–656.

Schwahn, K., de Souza, L. P., Fernie, A. R. and Tohge, T. (2014). Metabolomics assisted refinement of the pathways of steroidal glycoalkaloids biosynthesis in the tomato clade. *J. Integr. Plant Biol.*, 56, 864–875.

Seigler, D. S. (1975). Isolation and characterization of naturally occurring cyanogenic compounds. *Phytochemistry.*, 14, 9–29.

Seigler, D. S., Butterfield, C. S., Dunn, J. E. and Conn, E. E. (1975). Dihydroacacipetalin-a new cyanogenic glucoside from *Acacia sieberiana* var. Woodii. *Phytochem.*, 14, 1419–1420.

Seigler, D. S., Dunn, J. E., Conn, E. E. and Holstein, G. L. (1978). Acacipetalin from six species of Acacia of Mexico and Texas. *Phytochemistry.*, 17(3), 445–446.

Senica, M., Stampar, F., Veberic, R. and Mikulic-Petkovsek, M. (2016). Transition of phenolics and cyanogenic glycosides from apricot and cherry fruit kernels into liqueur. *Food Chem*, 203, 483–490.

Shakya, R. and Navarre, D. A. (2008). LC-MS analysis of solanidane glycoalkaloid diversity among tubers of four wild potato species and three cultivars (Solanum tuberosum). *Journal of Agricultural and Food Chemistry*, 56(16), 6949–6958.

Sixto, A., Pérez-Parada, A., Niell, S. and Heinzen, H. (2019). GC-MS and LC-MS/MS workflows for the identification and quantitation of pyrrolizidine alkaloids in plant extracts, a case study: *Echium plantagineum*. *Rev. Bras. Farmacogn.*, 29, 500–503.

Sphon, J. A., Dreifuss, P. A. and Schulten, H. R. (1977). Field desorption mass spectrometry of mycotoxins and mycotoxin mixtures, and its application as a screening technique for foodstuffs. *J. Assoc. Official Anal. Chemists.*, 60, 73–82.

Tamasi, G., Pardini, A., Croce, R., Consumi, M., Leone, G., Bonechi, C., Rossi, C. and Magnani, A. (2021). Combined experimental and multivariate model approaches for glycoalkaloid quantification in tomatoes. *Molecules.*, 26, 3068.

Tevell Åberg, A., Björnstad, K. and Hedeland, M. (2013). Mass spectrometric detection of protein-based toxins. *Biosecurity and Bioterrorism: Biodefense Strategy, Practice, and Science*, 11, S215–S226.

Valese, A. C., Molognoni, L., de SáPloêncio, L. A., de Lima, F. G., Gonzaga, L. V., Górniak, S. L., et al. (2016). A fast and simple LC-ESI-MS/MS method for detecting pyrrolizidine alkaloids in honey with full validation and measurement uncertainty. *Food Control*, 67, 183–191.

Verpoorte, R. and Niessen, W. M. A. (1994). Liquid chromatography coupled in the analysis of alkaloids. *Phytochem. Anal.*, 5, 217–232.

Vetter, J. (2000). Plant cyanogenic glycosides. *Toxicon.*, 38(1), 11–36.

Vlachos, A., Arvanitoyannis, I. S. and Tserkezou, P. (2016). An updated review of meat authenticity methods and applications. *Critical Rev. Food Sci. Nutrition*, 56, 1061–1096.

Wang, S., Alseekh, S., Fernie, A. R. and Luo, J. (2019). The structure and function of major plant metabolite modifications. *Mol. Plant.*, 12, 899–919.

Watts, C. D., Maxwell, J. R., Games, D. E. and Rossiter, M. (1975). Analysis of carotenoids by field desorption mass spectrometry. *Organic Mass Spectrometry*, 10, 1102–1110.

Williams, M. T., Warnock, B. J., Betz, J. M., Beck, J. J., Gardner, D. R., Lee, S. T., Molyneux, R. J. and Colegate, S. M. (2011). Detection of high levels of pyrrolizidine-N-oxides in the endangered plant *Cryptantha crassipes* (Terlingua Creek Cat's-eye) using HPLC-ESI-MS. *Phytochemical Analysis*, 22(6), 532–540.

Yan, G., Sun, H., Sun, W., Zhao, L., Meng, X. and Wang, X. (2010). Rapid and global detection and characterization of aconitum alkaloids in Yin Chen Si Ni Tang, a traditional Chinese medical formula, by ultra-performance liquid chromatography-high resolution mass spectrometry and automated data analysis. *J. Pharm. Biomed. Anal.*, 53, 421–431.

Zagrobelny, M., Bak, S., Rasmussen, A. V., Jorgensen, B., Naumann, C. M., Lindberg and Moller, B. (2004). Cyanogenic glucosides and plant-insect interactions. *Phytochem*, 65, 293–306.

Zhang, L., Wei, W. and Yang, X. W. (2017). Simultaneous quantification of nine new furanocoumarins in Angelicae dahuricae radix using ultra-fast liquid chromatography with tandem mass spectrometry. *Molecules.*, 22(2), 322.

Zhang, W., Huai, W., Zhang, Y., Shen, J., Tang, X., Xie, X., Wang, K. and Fan, H. (2017). Ultra-performance liquid chromatography hyphenated with quadrupole-Orbitrap mass spectrometry for simultaneous determination of necine-core-structure pyrrolizidine alkaloids in Crotalaria sessiliflora L. without all corresponding standards. *Phytochemical Analysis*, 28(5), 365–373.

Zhao, M., Bergaentzlé, M., Flieller, A. and Marchioni, E. (2019). Development and validation of an ultra-high performance liquid chromatography-high resolution mass spectrometry method for simultaneous quantification of cyanogenic glycosides and secoisolariciresinol diglucoside in flaxseed (*Linum usitatissimum* L.). *J. Chromatography A.*, 1601, 214–223.

Zhou, Y., Li, N., Choi, F. F., Qiao, C. F., Song, J. Z., Li, S. L., Liu, X., Cai, Z. W., Fu, P. P., Lin, G. and Xu, H. X. (2010). A new approach for simultaneous screening and quantification of toxic pyrrolizidine alkaloids in some potential pyrrolizidine alkaloid-containing plants by using ultra performance liquid chromatography–tandem quadrupole mass spectrometry. *Analytica Chimica Acta*, 681(1–2), 33–40.

CHAPTER 8

NMR in Analysis of Food Toxins

Naila H. Alkefai[1], Showkat R. Mir[2,], Saima Amin[2],*
Javed Ahamad[3], Kavita Munjal[4], and Vinod Kumar Gauttam[5]

[1]Department of Pharmaceutical Chemistry, Faculty of Pharmacy, University of Hafer Albatin, Hafer Albatin, Saudi Arabia
[2]Phytopharmaceuticals Research Lab., School of Pharmaceutical Education and Research, Jamia Hamdard, New Delhi, India
[3]Department of Pharmacognosy, Faculty of Pharmacy, Tishk International University, Erbil, Kurdistan Region, Iraq
[4]Department of Pharmacognosy, M. M. College of Pharmacy, M. M. (Deemed to be University), Mullana, Ambala, Haryana, India
[5]IES Institute of Pharmacy, Bhopal, India
[*]School of Pharmaceutical Education and Research, Jamia Hamdard, New Delhi, India

CONTENTS

8.1 Introduction to Food Toxins	131
8.2 Fundamentals of NMR Spectroscopy	132
8.3 NMR and Metabolomics Studies	133
8.4 NMR in Analysis of Food Composition and Authentication	133
8.5 NMR in Detection of Food Toxins and Adulterants	135
8.6 Conclusions and Future Perspective	136
References	136

8.1 INTRODUCTION TO FOOD TOXINS

Food is the main source of nutrition and energy in our body. Food that we consume needs to be palatable, tasty and nutritious. Plants, animals, and their products constitute the main sources of food worldwide. Few mineral sources also contribute to our food bowl. Different populations have developed different food habits wherein some products make the main course of their diet (staple diet) while the other commodities are used occasionally. Researchers seem to be fairly convinced that certain foods contain anti-nutritional substances and that their long term or over-consumption can have serious implications. Besides the chemicals that act as manure, pesticides, weedicides, veterinary drugs, or environmental pollutants that get into our food supply, toxins can be present due to their natural occurrence in food. The toxins found inherently in foods of plant and animal origin can be harmful when consumed in sufficient quantities. Natural toxins are poisonous secondary metabolites produced by living organisms. Food toxins can be defined as the natural substances covering a large variety of molecules, generated by fungi, algae, bacteria, plants, or animal metabolism

DOI: 10.1201/9781003222194-10

with harmful effects on humans or other vertebrates in small concentrations. They can be harmful to humans or animals when consumed [1]. The most prevalent cause of food poisoning is its infestation with organisms such as bacteria, viruses, and parasites, or their toxins, usually classified as biological sources of food toxins. These toxins can cause diverse clinical symptoms. Acute poisoning may result allergic reactions, stomachache, diarrhea, fever, liver impairment, respiratory paralysis, and even death. They are also associated with effects on the immune, reproductive, and nervous systems and can lead to different types of cancers as long-term implications [2]. Common examples of food toxins in plants include glycol-alkaloids in Solanaceous plants such as potatoes, cyanide-generating compounds in bitter apricot seeds and bitter almonds, enzyme inhibitors and lectins in soya beans, green beans, and other legumes. Natural toxins of animal origin may be a product of metabolism or a chemical that is passed along the food chain. While poisoning after eating terrestrial animals is relatively uncommon, poisoning due to marine toxins is very common. Toxins in shellfish, crustacean, and finfish have been reported to cause adverse effects following their consumption. Environmental pollutants are the other causes of food toxicity. They include factors emanating from the application of fertilizers and pesticides, to a variety of chemical wastes emitted by industries, all of which may find their way into food via a variety of routes. Thus, direct pathogenic infestations, irrigation with contaminated water, pre-harvest application of fertilizers and pesticides, exposure to chemical and industrial wastes, and the use of veterinary drugs are the major factors resulting in toxicity with food [3].

The Joint FAO/WHO Expert Committee on Food Additives (JECFA) is an international expert scientific committee, administered jointly by the Food and Agriculture Organization of the United Nations (FAO) and World Health Organization (WHO) that looks after the evaluation of naturally occurring food toxins. The organization works on the safety evaluation and risk assessment of intentionally added food additives, and flavoring agents. They have developed a series of toxicological monographs that contain detailed descriptions of the biological and toxicological data to be considered in the evaluation and the intake assessment of food ingredients. Assessment of toxins and contaminants in food commodities has always been a challenge to analytical chemists. Such assessments are also helpful in regulating their permissible limits in food products. Implementation of these regulations necessitate the development of high-throughput analytical methods that can rapidly analyze mixtures at molecular levels. The routine analytical methods are of limited use in food analysis given the complexity of the food matrices. Faced with this challenge, the analytical chemists and food scientists are looking towards the use of nuclear magnetic resonance (NMR) techniques for the analysis of food products. NMR techniques have been effectively employed for quality control, compositional studies, and characterization of food products. Protein structure, amino acid composition, carotenoids, organic acids, lipid fractions, and the mobility of water in foods have been successfully determined using NMR spectroscopy. Food metabolites are also identified and quantified using NMR methods. Among the most recent advancements is the use of NMR in the analysis of vegetable oils, fish oils, fish and meat products, milk, cheese, wheat, fruit juices, coffee, green tea, wine, and beer. Furthermore, NMR spectroscopy is being increasingly used in foodomics, a novel field that combines food science and nutritional research [4].

8.2 FUNDAMENTALS OF NMR SPECTROSCOPY

The magnetic characteristics of certain atomic nuclei are the foundation of nuclear magnetic resonance spectroscopy. The following is a simple explanation for atomic nuclear magnetism.

The atomic nuclei can be thought of as a spherical body rotating around its axis. As protons are present in the nucleus, it is positively charged. When a positively charge moiety moves around its axis, it generates current. Electric current is the flow of this charge along a certain path. Each electric current generates a magnetic field in its immediate surroundings. On the application of an external magnetic field around the nucleus of an atom that is rotating around the axis, the atomic nucleus acts like a magnet. The nucleus generates a magnetic field that has a magnetic moment. NMR is a technique for analysing the magnetic characteristics of nuclei with odd mass numbers or even mass numbers but odd atomic numbers. These nuclei have a nuclear spin S, which is a type of angular momentum and is defined by a nuclear spin quantum number, I. For nuclei with an odd mass number, I takes half-integer values, whereas for those with an even mass number but odd atomic number, I takes integer values. All nuclei with $I \neq 0$ are associated with a nuclear magnetic moment, and thus generate a tiny magnetic field. Its magnitude is a product of S and gyromagnetic ratio. The nuclear magnetic moment, interacts with the NMR instrument's external field B_0, making it possible to investigate resonating nuclei. Nuclei with spin $I = 1/2$ result in distinct and interpretable NMR signals. Fortunately, the most commonly used nuclei in food science NMR applications are ^1H, ^{13}C, and ^{31}P have $I = 1/2$ value (magnetic dipolar). ^2H, which has $I = 1$ and is a magnetic quadrupole, is also used in a number of food applications. The most common features of NMR signals that are used in food science and nutritional studies are their positions (chemical shifts), shape (coupling constant), and area (integrals) [5,6].

8.3 NMR AND METABOLOMICS STUDIES

Metabolomics is described as the thorough analysis of metabolites in a biological system, including their identification and quantification. Metabolomics is the study of molecules with a molecular size of less than or equal to 1500 Daltons. Sugars, polymers, and amino acids are not quantified in metabolomics research since their molecular weight is greater than 1500 Daltons. The advancement of metabolomic research has enabled it to be used in a variety of sectors, including nutrition, the environment, chemotaxonomy, pharmacology, and the discovery of biomarkers for specific diseases [7]. For more than 40 years, LC-MS has been utilized to undertake metabolic research, metabolic profiling, and metabolomics in biofluids and tissues. This strong relationship between metabolic measures and LC-MS has thrived because of its many unique strengths for determining the chemical composition of complicated mixtures. However, in recent years, NMR has emerged to challenge the supremacy of LC-MS in metabolomics and metabolic investigations [8]. NMR-based metabolomics studies combine the sensitivity of NMR spectroscopy and statistical analysis to investigate a food product as a whole and to identify. Besides compositional and toxin profiles, properties such as taste, aroma, and flavor can be correlated to the spectral patterns produced by NMR using chemometrics.

8.4 NMR IN ANALYSIS OF FOOD COMPOSITION AND AUTHENTICATION

Food represents a highly complex system made up of a variety of chemical components and physical structures at various levels of organisation. When it comes to food analysis and characterization, it is not just the chemical composition that determines nutritional content, but also its physical distribution in various compartments and structures, which determines the physical qualities of foods. The palatability and texture of a food product and its acceptance by consumers are determined by these physical parameters. Nuclear magnetic resonance (NMR) spectroscopy refers to a group of techniques that look at how electromagnetic radiation interacts with matter. NMR spectroscopy helps to study the physical, chemical, and biological aspects of matter at all scales, from microscopic to macroscopic. For the examination of food products, NMR spectroscopy is a very successful and multifunctional technology that works well when paired with chemometrics.

The use of NMR techniques in the field of authentication and quality control of food has grown manifolds over the years. Because of its low instrumental variability, it is possible to collect vast data sets of legitimate spectra [9]. The quantity of metabolites in foods can be influenced by a number of circumstances such as soil composition, altitude, and latitude besides manufacturing processes. The territorial characteristics effect the unique organoleptic and nutritional attributes of foods. It is necessary to integrate the analytical data to these diverse factors in order to achieve a correct interpretation of the differences associated to growing locations. As a result, it is advantageous to employ a versatile all-around technique such as NMR spectroscopy in combination with multivariate statistical analysis. It is regarded as a potent tool for determining food authenticity by investigating the metabolic profile of organic samples [10,11]. This technology is appropriate as it can yield metabolite fingerprints for a wide range of food stuffs. NMR spectroscopy is a reliable approach for analyzing mixtures at the molecular level without the need for separation or purification, making it excellent for food science applications [5,12]. To determine the association between all characteristics and peak intensities of the NMR spectra, many types of multivariate statistical analysis primarily principle component analysis (PCA) are applied [13].

There are many applications of NMR in the authentication of plant products, spices, beverages, animal products, and protein-based foods [14–15]. To detect adulteration and for quality control purposes, NMR-based metabolomics studies detect changes in the amounts of specific chemicals such as fatty acids, terpenes, sterols, and oxidation products. Recently studies on authentication of ten *Garcinia* food supplements containing *Garcinia* fruits that mainly contains hydroxycitric acid as the anti-obesity marker has been reported using NMR techniques [16]. Similarly, authentication of turmeric accessions was possible with ^{1}H-NMR spectroscopy-based metabolite fingerprinting using multivariate analysis [17]. It is also useful in establishing the botanical origin of honey and rule out sugar adulteration. Moreover, differences between major beer types and determination of the geographical origin of beer have been possible with NMR-based analysis. Wine authentication is critical in detecting malpractice and fraud, and it has been done using a variety of physical and chemical analytical approaches in the fats. Recently, ^{1}H NMR spectroscopy, a simpler and faster (in terms of sample preparation) approach in wine analysis, has seen a lot of attention. This technique allows both targeted quantitative analysis of wine constituents and non-targeted detection of a wine metabolomics fingerprint [18]. NMR can also detect crude drug adulterations through the quantification of marker compounds. Analysis of spice is another emerging area where NMR can be helpful for the profiling of constituents and sensory markers [19]. The reader is referred to a comprehensive review detailing the applications of NMR in food sciences and food safety compiled by Hatzakis [20] and the references therein for further readings.

8.5 NMR IN DETECTION OF FOOD TOXINS AND ADULTERANTS

Cultivation and collection, processing, transport, storage, and delivery are the key operations involved in the supply chain of foods. Because of the involvement of different stakeholders in the supply chain, from producers to consumers, food adulteration has reportedly increased in the recent years. Adulteration of food has become a big hazard, posing health risks and lowering the food quality. The annual economic loss of food due to adulteration is estimated to be $30–40 billion [21–22]. Despite the fact that fraud and adulteration have been done since ancient times, food authenticity issues have only recently been more visible, and public attention has been focused on the scope of the problem and the significant implications of food fraud. Furthermore, with the increasing awareness related to the food quality and nutrition, consumers have now become more concerned than ever about food safety, accessibility, cost, and origin [23]. The reports of health risks associated with the consumption of food toxins are increasing day by day. Asthma and cancer have been linked to formalin in preserved foods such as vegetables, fruits, seafood, and meat. Chemical pesticides have been linked to serious health issues like nerve damage and cancers. Adulterants such as chalk powder, brick powder, and hazardous chemicals are used in order to increase profits and cut costs [24]. These chemicals are frequently hazardous to one's health and may even cause cancer. It is the need of the hour that the food products must adhere to stringent quality standards in order to protect the public. To prevent food from deteriorating due to contaminated enzymes and microbes, a number of additives are utilised. Chemical preservatives such as sodium benzoate, benzoic acid, nitrites, sulfites, sodium sorbate, and potassium sorbate are commonly employed in food preparation [25,26]. The safety and efficacy of preservatives again becomes an important parameter. Antimicrobial preservatives such as sorbates/sorbic acid are added to foods. Sorbite reactions are uncommon; however, they have been reported to cause urticarial manifestations and contact dermatitis. Benzoates are also antimicrobial preservatives used in foods, and they have been linked to allergies, asthma, and skin rashes [27]. Many foods also contain synthetic colors to give them a fresh appearance and to compensate for natural color fluctuations. These colors are frequently hazardous to one's health and may even cause cancer. Artificial colors are used to enhance the color of the turmeric powder. Non-food-grade colours like copper sulphate, malachite green, and Sudan red are commonly used on vegetables to make them look fresh and colorful. The presence and amount of dye-based adulteration in vegetables has been assessed using ^{1}H-nuclear magnetic resonance (NMR) relaxometry [28].

The recent melamine scandal in China has highlighted a significant flaw in current food safety systems. For a long time, the presence of melamine in milk products was overlooked due to this targeted approach [29]. To prevent such incidents in the future, non-targeted screening procedures should be developed and implemented with time. In this regard, the tremendous avenues offered by NMR is now being recognized and exploited. Recent advancements have enabled NMR spectroscopy to be utilized as a screening method for determining the composition of food, both broadly and precisely [30]. Proton-NMR has a vital role in food because of the wide variety of components can be detected and discriminated without being separated or purified. The technological advancements in NMR such as cryo-probes, magnet design, and polarization techniques have resulted in substantial improvements in NMR instrument sensitivity and resolution, thereby broadening its scope in diverse fields including food sciences [31]. NMR spectroscopy is quick and can resolve almost any combination of chemicals. Moreover, it can detect a chemical in complex matrices such as foods. NMR is now a high-throughput, very sensitive method that is being increasingly used for the detection of food adulterants and toxins [32]. Ahammer et al. (2017) have identified the structure of an allergen from the apple samples using extensive 1D and 2D NMR data [33]. Bioactivity guided fractionation of *Petroselinum crispum* L. leaves resulted in the isolation of phytotoxic compounds, the structures of which were identified using a NMR technique [34]. NMR studies on *Matricaria chamomilla* have helped to identify a compound with potential risk of inducing dermal sensitization [35]. HR-MAS NMR has been used in quantitative analysis of fish, cheese, meat, vegetables, and herbs [36–39]. Samples of apple juice were successfully checked for adulteration using PCA of their NMR data [40]. Another study comparing apple juices from frozen and fresh pulp was carried out using NMR-based metabolomics analysis [41]. Fadzillah et al. (2017) carried authentication of butter using HR-NMR and identified the presence of lard in butter samples [42]. Accessions of coffee have been evaluated for adulterants using targeted statistical analysis of their NMR spectra [43]. NMR-based authentication and quantitative analysis of sucrose-based sugar products [44], sunflower lecithins [45], wine and alcoholic beverages [46,47] and other food products [48] have also been reported.

8.6 CONCLUSIONS AND FUTURE PERSPECTIVE

Appropriate methods of food processing and thorough cooking can be employed to destroy or reduce the level of toxin in food. In cases where the toxin cannot be reduced or removed, intake should be limited. Thorough cooking destroys most of the food toxins. Soaking in water and boiling can help to get rid of some cyanide-generating compounds in the foods. Removal of parts containing toxins is another method of reducing the adverse effects. In general, conventional ways of food processing and cooking should be followed. A thorough understanding of types of food toxins, their origin, and acceptable limits in the food and detection is very important. Technological advancements in the field of NMR spectroscopy have made this technique particularly suitable for food analysis. Its applications in the field of food sciences is bound to increase in the days to come.

REFERENCES

1. Dolan, L.C., Matulka, R.A. and Burdock, G.A. (2010). Naturally occurring food toxins. *Toxins (Basel)*, 2(9), 2289–2332.
2. Rather, I.A., et al. (2017). The sources of chemical contaminants in food and their health implications. *Frontiers in Pharmacology*, 8(2017), 830.
3. Alengebawy, A., et al. (2021). Heavy metals and pesticides toxicity in agricultural soil and plants: Ecological risks and human health implications. *Toxics*, 9(3), 42.
4. Parlak, Y. and Güzeler, N. (2016). Nuclear magnetic resonance spectroscopy applications in foods. *Curr Res Nutr Food Sci*, 4, 161–168.
5. Levitt, M. (2005). *Spin dynamics*. Chichester: John Wiley & Sons.
6. Tampieri, A. , (2021). A brief introduction to the basics of NMR spectroscopy and selected examples of its applications to materials characterization. *J Physical Sciences Reviews*, 6(1).
7. Tyagi, S., et al. (2010). Applications of metabolomics – A systematic study of the unique chemical fingerprints: An overview. *Int. J. Pharm. Sci. Rev. Res*, 3(1), 83–86.

8. Wishart, D.S. (2019). NMR metabolomics: A look ahead. *Journal of Magnetic Resonance*, 306, 155–161.

9. Le Gall, G. and Colquhoun, I.J. (2003). NMR spectroscopy in food authentication. 131–155.

10. Masetti, O., Sorbo, A. and Nisini, L. (2021). NMR tracing of food geographical origin: The impact of seasonality, cultivar and production year on data analysis. *Separations*. 8(12), 230.

11. Dimitrakopoulou, M.-E., et al. (2021). Nontargeted 1H NMR fingerprinting and multivariate statistical analysis for traceability of Greek PDO Vostizza currants. 86(10), 4417–4429.

12. Munjal, K., et al. (2020). Polyphenol-enriched fraction and the compounds isolated from Garcinia indica fruits ameliorate obesity through suppression of digestive enzymes and oxidative stress. 16(70), 236–245.

13. Mannina, L., Sobolev, A.P. and Viel, S. (2012). Liquid state 1H high field NMR in food analysis. *Progress in Nuclear Magnetic Resonance Spectroscopy*, 66, 1–39.

14. Pacholczyk-Sienicka, B., Ciepielowski, G. and Albrecht, Ł. (2021). The application of NMR spectroscopy and chemometrics in authentication of spices. 26(2), 382.

15. Ichim, M.C. and Booker, A. (2021). Chemical authentication of botanical ingredients: A review of commercial herbal products, *Frontiers in pharmacology*, 12, 666850.

16. Seethapathy, G.S., et al. (2018). Authentication of garcinia fruits and food supplements using DNA barcoding and NMR spectroscopy. *Scientific Reports*, 8(1), 10561.

17. Windarsih, A., Rohman, A. and Swasono, R.T. (2018). Authentication of turmeric using proton-nuclear magnetic resonance spectroscopy and multivariate analysis. *International Journal of Applied Pharmaceutics*, 10(6), 174–180.

18. Solovyev, P.A., et al. (2021). NMR spectroscopy in wine authentication: An official control perspective. *Compr Rev Food Sci Food Saf*, 20(2), 2040–2062.

19. Kuballa, T., et al. (2018). Application of NMR for authentication of honey, beer and spices. *Current Opinion in Food Science*, 19, 57–62.

20. Hatzakis, E. (2019). Nuclear Magnetic Resonance (NMR) spectroscopy in food sciences: A comprehensive review. *Comprehensive Reviews in Food Science and Food Safety*, 18, 189–220.

21. Negi, A., et al. (2021). Detection of food adulterants in different foodstuff. in *Food Chemistry: Food Chemistry: The Role of Additives, Preservatives and Adulteration*, pp. 117–164.

22. Yadav, A., Kumar, R. and Pandey, A.K. (2015). Some common food toxins and their effect on human health. *Journal of Kalash Science*, 2015, 1–6.

23. Sharma, A., et al. 2017). Food Adulteration: A Review. *International Journal for Research in Applied Science & Engineering Technology (IJRASET)*, 5, 686–689.

24. Hassoun, A., et al. (2020). Fraud in Animal Origin Food Products: Advances in Emerging Spectroscopic Detection Methods over the Past Five Years. *Foods*, 9(8), 1069.

25. Pongsavee, M. (2015). Effect of sodium benzoate preservative on micronucleus induction, chromosome break, and Ala40Thr superoxide dismutase gene mutation in lymphocytes. *BioMed Research International*, 2015, 103512–103512.

26. Yadav, R. and Gupta, R. (2021). Impact of chemical food preservatives through local product on human health – A review. *High Technology Letters*, 27, 767–773.

27. Hannuksela, M. and Haahtela, T. (1987). Hypersensitivity reactions to food additives. *Allergy*, 42(8), 561–575.
28. Shomaji, S., et al. (2021). Detecting dye-contaminated vegetables using low-field NMR relaxometry. *Foods*, 10(9), 2232.
29. Lachenmeier, D.W., et al. (2009). NMR-spectroscopy for nontargeted screening and simultaneous quantification of health-relevant compounds in foods: The example of melamine. *Journal of Agricultural and Food Chemistry*, 57(16), 7194–7199.
30. Schmitt, C., et al. (2020). Detection of peanut adulteration in food samples by nuclear magnetic resonance spectroscopy. *Journal of Agricultural and Food Chemistry*, 68(49), 14364–14373.
31. Okuno, Y., et al. (2019). Laser- and cryogenic probe-assisted NMR enables hypersensitive analysis of biomolecules at submicromolar concentration. *Proceedings of the National Academy of Sciences*, 116(24), 11602–11611.
32. Hachem, R., et al. (2016). Proton NMR for detection, identification and quantification of adulterants in 160 herbal food supplements marketed for weight loss. *J Pharm Biomed Anal*, 124, 34–47.
33. Ahammer, L., Grutsch, S., Kamenik, A.S., Liedl, K.R. and Tollinger, M. (2017). Structure of the major apple allergen mal d 1. *Journal of Agricultural and Food Chemistry*, 65(8), 1606–1612.
34. Sbai, H., Saad, I., Ghezal, N., Greca, M.D. and Haouala, R. (2016). Bioactive compounds isolated from Petroselinum crispum L. leaves using bioguided fractionation. *Industrial Crops & Products*, 89, 207–214.
35. Avonto, C., Rua, D., Lasonkar, P.B., Chittiboyina, A.G. and Khan, I.A. (2017). Identification of a compound isolated from German chamomile (Matricaria chamomilla) with dermal sensitization potential. *Toxicology & Applied Pharmacology*, 318, 16–22.
36. Nestor, G., Bankefors, J., Schlechtriem, C., Br̈annas, E., Pickova, J. and Sandstr̈om, C. (2010). High-resolution ¹H magic angle spinning NMR spectroscopy of intact Arctic char (Salvelinus Alpinus) muscle. Quantitative analysis of n-3 fatty acids, EPA and DHA. *Journal of Agricultural and Food Chemistry*, 58(20), 10799–10803.
37. Mazzei, P. and Piccolo, A. (2012). 1H HRMAS-NMR metabolomic to assess quality and traceability of mozzarella cheese from Campania buffalo milk. *Food Chemistry*, 132(3), 1620–1627.
38. Ritota, M., Casciani, L., Failla, S. and Valentini, M. (2012b). HRMAS-NMR spectroscopy and multivariate analysis meat characterisation. *Meat Science*, 92(4), 754–761.
39. Petruľová-Poracḱa, V., Repčak, M., Vilkov́a, M. and Imrich, J. (2013). Coumarins of Matricaria chamomilla L.: Aglycones and glycosides. *Food Chemistry*, 141(1), 54–59.
40. Gall, G.L., Puaud, M. and Colquhoun, I.J. (2001). Discrimination between orange juice and pulp wash by 1H nuclear magnetic resonance spectroscopy: Identification of marker compounds. *Journal of Agricultural and Food Chemistry*, 49(2), 580–588.
41. Santucci, C., Brizzolara, S. and Tenori, L. (2015). Comparison of frozen andfresh apple pulp for NMR-based metabolomic analysis. *Food Analytical Methods*, 8(8), 2135–2140.

42. Fadzillah, N.A., Rohman, A., Salleh, R.A., Amin, I., Shuhaimi, M., Farahwahida, M.Y., et al. (2017). Authentication of butter from lard adulteration using high-resolution of nuclear magnetic resonance spectroscopy and high-performance liquid chromatography. *International Journal of Food Properties*, 20(9), 2147–2156.

43. Monakhova, Y.B., Ruge, W., Kuballa, T., Ilse, M., Winkelmann, O., Diehl, B., et al. (2015). Rapid approach to identify the presence of Arabica and Robusta species in coffee using 1H NMR spectroscopy. *Food Chemistry*, 182, 178–184.

44. Monakhova, Y.B. and Diehl, B.W. (2016a). Authentication of the origin of sucrose-based sugar products using quantitative natural abundance ^{13}C NMR. *Journal of the Science of Food and Agriculture*, 96(8), 2861–2866.

45. Monakhova, Y.B. and Diehl, B.W.K. (2016b). Quantitative analysis of sunflower lecithin adulteration with soy species by NMR spectroscopy and PLS regression. *Journal of the American Oil Chemists' Society*, 93(1), 27–36.

46. Monakhova, Y.B., Godelmann, R., Hermann, A., Kuballa, T., Cannet, C., Sch¨afer, H., et al. (2024). Synergistic effect of the simultaneous chemometric analysis of 1H NMR spectroscopic and stable isotope (SNIF-NMR, ^{18}O, ^{13}C) data: Application to wine analysis. *Analytica Chimica Acta*, 833, 29–39.

47. Monakhova, Y.B., Kuballa, T. and Lachenmeier, D.W. (2012). Non-targeted NMR analysis to rapidly detect hazardous substances in alcoholic beverages. *Applied Magnetic Resonance*, 42(3), 343–352.

48. Monakhova, Y.B., Kuballa, T. and Lachenmeier, D.W. (2013). Chemometric methods in NMR spectroscopic analysis of food products. *Journal of Analytical Chemistry*, 68(9), 755–766.

SECTION III

Analytical Techniques Exploited in Detection and Quantification: *Quantitative Application*

CHAPTER 9

HPLC for Detection and Quantification of Food Toxins of Plant Origin

Devarajan Thangadurai[1], D. Divya[2], Suraj S. Dabire[3],
Jeyabalan Sangeetha[4], Mahantesh Dodamani[3], Saher Islam[5],
Ravichandra Hospet[6], Simran Panigatti[1], Muniswamy David[3],
Zaira Zaman Chowdhury[7], and Vishal Ahuja[8]

[1]Department of Botany, Karnatak University, Dharwad, Karnataka, India
[2]Pinnacle Biosciences, Kanyakumari, Tamil Nadu, India
[3]Department of Zoology, Karnatak University, Dharwad, Karnataka, India
[4]Department of Environmental Science, Central University of Kerala, Kasaragod, Kerala,
India
[5]Department of Biotechnology, Lahore College for Women University, Lahore, Pakistan
[6]Department of Food Protectants and Infestation Control, CSIR-Central Food
Technological Research Institute, Mysore, Karnataka, India
[7]Nanotechnology and Catalysis Research Center (NANOCAT), Institute of Advanced
Studies (IAS), University of Malaya, Kuala Lumpur, Malaysia
[8]Department of Biotechnology, Himachal Pradesh University, Shimla, Himachal Pradesh,
India

CONTENTS

9.1	Introduction		144
9.2	Sample Preparation		145
	9.2.1	Solids	146
	9.2.2	Liquids	147
9.3	Principles and Methods		147
	9.3.1	Principles	148
		9.3.1.1 Working Principle of HPLC	148
		9.3.1.2 Four Primary Types of Isolation Mechanism in HPLC	148
		9.3.1.3 Normal Phase Chromatography	149
		9.3.1.4 Reversed-Phase Chromatography	149
	9.3.2	Methods	149
		9.3.2.1 Types of Detectors Used in HPLC	149
		9.3.2.2 Refractive Index (RI) Detector	149
		9.3.2.3 Detector of Ultraviolet-Visible Light	149
		9.3.2.4 Photodiode Array Detector (PDA)	149

DOI: 10.1201/9781003222194-12

	9.3.2.5 Fluorescent Detector	149
	9.3.2.6 Evaporation Light Scattering Detector	150
	9.3.2.7 Electrochemical Detector	150
	9.3.2.8 Mass Detector	150
9.3.3	Analysis of Plant Toxins and Alkaloids	150
9.4 Analysis of Different Food Compounds		150
9.4.1	Analysis of Major Biomolecules	151
	9.4.1.1 Carbohydrates	151
	9.4.1.2 Proteins and Lipids	151
9.4.2	Analysis of Minor Components	151
	9.4.2.1 Vitamins	151
	9.4.2.2 Phenolic Compounds/Flavonoids	152
	9.4.2.3 Food Additives/Pigments	152
9.4.3	Analysis of Unwanted Compounds/Food Toxins	152
	9.4.3.1 Mycotoxins	154
	9.4.3.2 Allergens	156
	9.4.3.3 Pesticides	156
9.5 Conclusion		156
References		157

9.1 INTRODUCTION

Plants produce a wide range of secondary metabolites with a large number of different structures, from simple organic compounds to complex molecules like proteins (Nobukazu 2016). Natural toxins are chemical molecules that are not really hazardous to the organisms that produced them, but can pose a health risk to humans if consumed through food. Food quality has steadily increased as science and technology have progressed (Natalia et al. 2020). Food analysis, on the other hand, is a difficult task due to the great complexity of these matrices, which makes extraction and detection of analytes difficult (Souza-Silva et al. 2015). As a result, sample preparation is an important step in food analysis to ensure proper analyte isolation and preconcentration, as well as adequate matrix interference cleanup prior to instrumental analysis (Colombo and Papetti 2020). Food samples were cleaned using commercial immunoaffinity columns before being tested with a high-performance liquid chromatography technique (Jarmila et al. 2013).

Food crops produce a vast array of non-nutrient secondary metabolites in contrast to nutrition (Essers et al. 1998; Gry et al. 2007; Dolan et al. 2010). Currently, chemical food safety is mainly focused on synthetic chemicals such as pesticides and antibiotic residues from agriculture production. Whereas essential plant toxicants have typically been identified based on their toxic features as well as impacts on humans, bioactivities of non-nutrient plant secondary metabolites that are compatible with desirable health effects have received considerable attention over recent decades. Although secondary plant metabolites have been associated with human outbreaks of adverse health effects (Table 9.1), they are also considered to represent epidemiological evidence explaining the health advantages of a diet rich in vegetables and fruits (Walter 2003). Proteins can also be considered naturally derived plant toxins. Lectins are proteins that are specifically attached to carbohydrates; steroidal alkaloids in potatoes; pyridine alkaloids in tobacco; monoterpenes and phenylpropanoids in a variety of herbs, such as dill, coriander, and parsley; glucosinolates in cabbages; and cyanogenic glycosides in many fruits, like apples and peaches (Guenthardt et al. 2018).

HPLC for Detection and Quantification   145

TABLE 9.1   Medically Driven Food Prohibitions of Plant Origin

Disease/syndrome	Causative food	Causes	Comments
Disaccharide intolerance	Sucrose, dextrins	Autosomal recessive trait characterized by the deficiency or absence of enzymes sucrase and isomaltase in the intestine.	Attacks characterized by bloating and diarrhea.
Favism	Broad bean (*Vicia fava*)	X-linked recessive trait resulting in low amounts of glucose-P-dehydrogenase. Several subtypes known.	Hemolytic anemia may result from consumption of offending foods.
Gluten intolerance	Wheat, barley, gluten-containing foods	Autoimmune disease	Sensitivity to storage protein (gliadin) in some grains.
Refractory sprue	Wheat, barley, and rye	Autoimmune disorder triggered by gliadin, a gluten storage protein.	Unlike common celiac sprue, adherence to a gluten-free diet may not cause symptoms to abate.

Adapted from Dolan et al. (2010) with slight modification, Creative Commons Attribution 3.0 International.

Specific methodologies for the analysis of food toxins have already been developed in the latest generations (Nikolin et al. 2004). Several studies have shown that red kidney beans possess significant amounts of phytohemaglutinin (PHA). It is a lectin that can react to mammalian red blood cells and agglutinate it. PHA is considered to be the cause of the harmful implications associated with the consumption of overcooked red kidney bean-based food products (Noah et al. 1980). The margin of exposure between actual exposure and the level recognized to cause adverse effects in humans is comparatively small for several natural plant toxicants. However, due of their potential physiological mechanisms in the plant, their disposal is frequently challenging (Schilter et al. 2014). In terms of isolation and quantitative and qualitative estimate of active compounds, HPLC provides a broad array of applications, and this will become an effective method for analyzing, detecting, and quantifying naturally occurring food toxins of plant origin in the future.

9.2   SAMPLE PREPARATION

High-performance liquid chromatography (HPLC) is a highly preferred method of food analysis and is widely employed as an optimal chromatographic methodology for both qualitative and quantitative analysis of a vast variety of food analytes that include chemical additives, contaminants, residues, and natural toxins. HPLC involves a stepwise approach for analyte detection and quantification that includes excluding the sample matrix, isolating the target analyte(s), and individually resolving them on a chromatographic column (Gratzfeld-Huesgen and Schein 2002). Differential interactions of the target analytes with both mobile and stationary phases have a major impact on the efficiency of chromatographic separation.

Sample preparation is the first stage of the HPLC procedure. It essentially involves the introduction of a medium into which the target analytes can be partitioned. The analyte-containing mobile phase is then separated from the complex sample matrix. The selection of mobile phase or the partitioning medium depends on attributes of analyte(s) and nature of sample matrix. The sample may require heating, chemical treatment, or pH adjustment to render the target analytes more accessible to the partitioning medium (Slack and Snow 2007). The most common extraction methods for solid and liquid samples have been described below.

9.2.1 Solids

Solid food samples require proper homogenization before analyte partitioning if they are non-uniform in nature. Three types of extraction methods that are most routinely used for solid analytes are liquid extraction, extraction with supercritical fluids, and steam distillation (Majors 2002).

1. Liquid extraction generally allows highly selective matching of target analytes with extraction solvents; the choice of solvent depends on the solubility of the analyte. This method is perhaps the fastest, with extraction times typically in minutes, and cost effective, which requires only small quantities of organic solvents. The used solvents can be easily disposed of. The liquid extraction method is often used in combination with ultrasonic stimulation to facilitate efficient penetration of the solvent into the sample matrix, thus ensuring reliable partitioning of the analyte molecules (Gratzfeld-Huesgen and Schein 2002). Other than ultrasonic stimulation, heating the solvent may be used to serve the same purpose if the target analytes are known to be thermally stable. On the other hand, solvent heating cannot be used for samples that contain heat-sensitive analytes, and maintaining an appropriate temperature is necessary to prevent analyte degradation.

2. Supercritical fluid is substance at temperature and pressure above its critical value but below pressure required to compress it into liquid. Distinct liquid and gas phases cannot exist at supercritical points, and the properties of such supercritical fluids are intermediate to those of gases and liquids. Such fluids are much in ability to penetrate a solid sample matrix and dissolve the target analytes (Slack and Snow 2007). Extraction times are fairly short with supercritical fluids; however, ultrapure supercritical fluids may not always be readily available for analysis, which is a major limitation of this approach. Carbon dioxide is an example of a non-toxic, non-hazardous, non-contaminating supercritical fluid with reasonable availability. It is preferred because of its relatively low supercritical temperature and pressure, and analytes dissolved in supercritical CO_2 can be easily recovered by lowering the pressure, which converts the fluid to its gaseous state. As CO_2 has a low polarity, its solvent strength is often enhanced by adding small quantities of polar solvents, such as methanol (Gratzfeld-Huesgen and Schein 2002).

3. Steam distillation is a method particularly suitable for the extraction of volatile analytical compounds. Longer extraction time is the major limitation of this method, and the method is preferred only for the extraction of specific compounds.

Soxhlet extraction is also used for extracting analytes from food samples, particularly plant material. Whatever may be the choice of the extraction method, the extracted target analytes are next dissolved in a compatible liquid for chromatographic separation (Slack and Snow 2007). This liquid may be pretreated via certain procedures, such as liquid/liquid extraction (LLE) or solid-phase extraction (SPE), to eliminate the substances that may interfere with the separation and analysis of the target compound(s).

9.2.2 Liquids

For liquid food samples and liquid extracts of solid food samples, three major methods are employed for the partitioning of target compounds. These are liquid/liquid extraction, solid-phase extraction, and gel permeation chromatography (GPC). A filtration step may be required before analyte extraction for samples with high turbidity or opacity.

1. Liquid/liquid extraction allows using a suitable solvent system that has been customized for maximum recovery of target compounds from the sample. Manual LLE procedures use separatory funnels for analyte partitioning, whereas automated LLE systems use continuous or counter-current distribution apparatus (Diniz et al. 2004). This is a widely preferred method that can be modified according to the experimental protocols that require selective pH or the use of specific salts or reagents. The major limitation of this method is that it may require the use of highly toxic or hazardous solvents, associated with potential human health risks, disposal issues, and environmental toxicity.

2. Solid-phase extraction is typically used for both partitioning and concentrating the analytes. The method isolates the target compound in adsorbent-filled cartridges from which they are recovered by elution with an appropriate solvent. This method also can be modified according to process requirements due to the availability of a wide range of adsorbent materials (Liska et al. 1989). SPE is gaining widespread popularity as a sample preparation technique, and efforts are being made to integrate it with automated chromatography systems.

3. Gel permeation chromatography is used for the partitioning of high and low molecular weight analytes, which are eluted in order of decreasing molecular weight. The technique is also known as size-exclusion chromatography.

9.3 PRINCIPLES AND METHODS

HPLC is a widely known analytical technique for separating, identifying, and quantifying each component of a mixture (Thammana 2016). It has been applied to several categories of substances in food science: carbohydrates, vitamins, lipids, contaminants that are naturally occurring, as well as amino acids and others (Macrae 1988; Bélanger et al. 1997). Plant-originated natural toxins include lectins in legumes; steroidal glycoalkaloids in potatoes (*Solanum tuberosum*); amygdalin, which is present in pear and apple seeds; and furanoterpenoids in diseased sweet potatoes. There are various methods to apply efficient analytical methods for monitoring the presence of known or suspected analysis of specific plant food associated toxins in established, genetically modified, and new foods that are being developed (Colegate et al. 2015).

9.3.1 Principles

HPLC is kind of liquid chromatography in which the mobile phase and stationary phase are separated. The capacity of the sample constituents to disperse themselves across the two phases will determine the separation. There is a stationary phase and a mobile phase; the solvent passes through the stationary phase, carrying analytes with it. Sample gets injected by injector in mobile phase flow and passes through stationary phase in HPLC process (column). Analytes present in a sample mixture interact with solid support as they travel with the flow of the mobile phase. The rate of movement is determined by the analytes' affinity for the stationary phase. Strongly interacting analytes with the stationary phase elute slowly, whereas less interacting molecules elute fast.

9.3.1.1 Working Principle of HPLC

The mobile phase changes depending on the column chemistry and the kind of analyte separated using that column. Because adsorption and partition principles are used in the separation in bonded phase silica columns, a suitable solvent mixture must be selected based on its miscibility index and viscosity. If aqueous separations are used, additional elements such as ion paring reagents and buffers to keep the pH stable can be added to allow for the fine separation of analytes of interest from complex mixtures such as blood, urine, plant, and tissue extracts. Compounds are quantified by taking either peak height or peak area into consideration. The peak area/height increases in a well-resolved chromatographic peak. It is then examined in relation to the rise in concentration. There are two approaches for sample analysis that incorporate both external and internal calibration.

9.3.1.2 Four Primary Types of Isolation Mechanism in HPLC

The separation process could take distinct phases depending on the mechanism of interaction nature of the stationary phase:

1. Adsorption chromatography is a basic kind of separation in which the stationary phase is solid and the mobile phase can be either liquid or gaseous and the separation is based on repeated adsorption-desorption stages.
2. Partition chromatography involves the partitioning of solute molecules between two immiscible liquids, with the separation dependent on the partitioning of the mobile and stationary phases.
3. Ion-exchange chromatography separates ions and polar molecules based on their ion exchange, with the stationary phase consisting of an ionic surface with the opposite charge as the sample.
4. Size exclusion chromatography is used in the separation of large molecules/macromolecular complexes such as industrial polymers and proteins according to their molecular size using a column filled with a substance with a properly controlled pore size.

The HPLC method mainly deals with two stages: mobile phase and stationary phase. In practice, adsorption chromatography is the most widely used and depends on the polarity of the two phases.

9.3.1.3 Normal Phase Chromatography

The stationary phase is polar (for example, silica or alumina) and the mobile phase is non-polar (e.g., hexane). In this phase, polar substances are held more firmly by the column, enabling non-polar chemicals to be eluted first (Bélanger et al. 1997).

9.3.1.4 Reversed-Phase Chromatography

The stationary phase (e.g., hydrocarbons) is non-polar and elution solvent (or mobile phase) is polar (e.g., water or methanol). As this is the inverse of conventional phase chromatography, non-polar substances will be retained on the column for a longer time.

9.3.2 Methods

HPLC deals with methods for determining the chemical composition of samples. A compound can often be measured by several methods. And the choice of analytical methods is based on many considerations, such as chemical properties, sample preparation, and HPLC condition. The present trend is to use mass detectors in the different HPLC types and configurations. This is because of the universal, selective, and sensitive detection they provide (Alcaide-Molina et al. 2009).

9.3.2.1 Types of Detectors Used in HPLC

To identify any substance using HPLC, a detector must be selected first. Once a detector has been chosen and optimized for detection, a separation process must be carried out. The most important part of a HPLC system is detectors connected to the outlet end of the column, and their role is to monitor the column effluent in real time; HPLC uses various detectors such as photodiode-array detector (PDA), light scattering detector, mass detector (LCMS), refractive index detector, fluorescence detector, infrared detector, electrochemical detector, conductivity detector, and an UV/VIS detector.

9.3.2.2 Refractive Index (RI) Detector

The refractive index detectors are low sensitivity detectors and are mostly used for carbohydrate analysis.

9.3.2.3 Detector of Ultraviolet-Visible Light

Analyte detection is based on UV absorption. The detection is caused by the molecule's absorption of UV rays at a certain wavelength. UV-visible detectors use a light source (wavelength ranges from 200–400 nm) and a tungsten light source (wavelength ranges from 400–800 nm). It is considered the "workhorse" of detectors for HPLC systems (Bélanger et al. 1997).

9.3.2.4 Photodiode Array Detector (PDA)

A photodiode array detector is a hybrid detector capable of constantly monitoring all visible to ultraviolet wavelengths of the incident beam. The detector uses an array of diodes instead of a photomultiplier tube or photodiode, and it can monitor wavelengths ranging from 200 nm to 800 nm.

9.3.2.5 Fluorescent Detector

In such detectors, the excitation-induced emission property of the analyte of interest is employed. It is frequently powered by a xenon light source, which can generate light in both the ultraviolet and visible ranges.

9.3.2.6 Evaporation Light Scattering Detector

The evaporation light scattering detector is based on light scattering by the solid analyte following mobile phase evaporation. The analyte must be less volatile than the mobile phase.

9.3.2.7 Electrochemical Detector

The detection is based on the electric current flowing through the electrodes as a result of an oxidizable or reducible analyte's chemical reaction.

9.3.2.8 Mass Detector

LC-MS is a strong technology with excellent sensitivity and specificity that is utilized in a variety of applications. In general, its use is focused on the particular detection and identification of substances in presence mixed combination. Reverse-phase HPLC-MS method helps in the analysis for detection and identification of toxic alkaloids in larkspur plant (Gardner and Pfister 2009). RP-HPLC for the determination of tea catachins, and widely used for separation and detection of ochratoxin A, in food and feed (Scott 2002). HPLC with UV/Vis or fluorescence detector is used in determining bacterial and fungal toxins. A new method to determine aflatoxins B1, B2, G1, and G2 in cereal foods is using solid phase microextraction coupled with HPLC and a post-column photochemical derivatization (Quinto et al 2009). Other methods used to separate biomolecules from complex are micellar electrokinetic capillary chromatography, strong anion exchange HPLC (SAX-HPLC), high-speed counter-current chromatography (HSCCC), and low-pressure size-exclusion chromatography (SEC) (Dnyaneshwar 2017).

9.3.3 Analysis of Plant Toxins and Alkaloids

Plant alkaloids, just like any other analyte, must go through a variety of extraction and preparation methods. The most frequent extraction method is sonication in a suitable solvent. The pH of the solvent is important to achieving good extraction efficiency. Apart from sonication, acid base extraction is also utilized, in which a crude extract is prepared from the plant, followed by a liquid-liquid extraction using an acid that is often diluted, which is then neutralized with an appropriate solvent. The HPLC analysis of alkaloids is often performed in an alkyl silica stationary phase, whereas the mobile phases are typically made up of methanol, acetonitrile, and water. The pH of the mobile phase is an important component in the analysis of alkaloids using HPLC (Hussain et al. 2020). These methods characterized by appropriate independent methods for accurate, sensitive, and rapid analysis of potential natural toxins.

9.4 ANALYSIS OF DIFFERENT FOOD COMPOUNDS

High-performance liquid chromatography (HPLC) has been effectively used as a tool for characterization of various food compounds for over 40 years (Hurst et al. 1979). Food compounds that are existing in a mixture solution may be separated, analyzed, and quantified using this technique. There are a variety of substances that can be examined using HPLC in the food analysis process. These includes carbohydrates, vitamins, supplements, pigments, amino acids, proteins, triglycerides, lipids, and chiral molecules. Apart from this, HPLC is preferred for detecting contamination of several food materials due to mycotoxins, phytotoxins, etc. Carbohydrates, proteins, and triglycerides are the bulk or main components; vitamins, pigments, and polyphenols are the trace components; while pesticides and mycotoxins are unwanted impurities. However, their composition may vary form one food material to another (Stefano et al. 2012).

9.4.1 Analysis of Major Biomolecules

9.4.1.1 Carbohydrates

Carbohydrates, proteins, and lipids are the major biomolecules or nutrients of almost all food materials, each possess entirely different structural characteristics, physiological roles, and nutritional properties. It was recommended by various research that, HPLC is an imperative technique that assist to analysis of individual carbohydrates those possess nutritional interest includes major and minor sugars. Moreover, this aid has been successfully applied to different fields of carbohydrate research, including enzyme investigations on polysaccharides and analysis of the resultant fraction. Nowadays, rapid HPLC has much attention from food makers to perform quick quantification of food materials while maintaining high accuracy and repeatability (Tomas and Jiri 2020). However, the sample preparation and extraction process could vary according to the food matrix used for analysis. Hence, an optimization of sample mixture/mobile phase and flow rate is essential. Selective detectors and specific temperature range is preferred to accomplish accurate measurement of sugars, where a 2.0 ml/min flow rate was more adorable. Nowadays, more advancements are made with monolithic HPLC columns and microcolumns to ensure quick analysis, good selectivity, and sensitivity of simple and complex carbohydrates (Dion et al. 2008).

9.4.1.2 Proteins and Lipids

Proteins are their building blocks amino acids are the second type of biomolecules that contribute structural and functional component of all organic materials. HPLC was promising to establish protein/amino acid content of the food stuffs. A technique highly employed for proteomic research is nano-LC, which showed better outcomes than UPLC, whereas in addition to these techniques, a two-dimensional makes HPLC an imperative exploratory tool for proteomics research (Dion et al. 2008; Tranchida et al. 2004). Lipids are found in a number of forms, including phospholipids, fatty acids, sphingomyelins, acyl glycerols, steroids, glycol-sphingo lipids, and bile acids. HPLC–MS is a strong method for analyzing lipid components in food materials. Since lipids are so complex, MS linked with HPLC is preferred to enable structural study of discrete compounds. The composition of olive oil and vegetable oils of many foodstuffs has been thoroughly investigated through HPLC and HPLC-GC. Diverse lipids such as fatty acids, triacylglycerols, and phospholipids were also established from different stuffs, mostly by HPLC–MS (Alexander et al. 2011).

9.4.2 Analysis of Minor Components

9.4.2.1 Vitamins

Vitamins are vital trace components with a variety of structural characteristics. Carotenoids and provitamins are the main classes of vitamin nutrients that contribute to the color of a variety of plants. The primary analytical method for their analysis is HPLC, which is particularly combined with MS. Electrospray is often utilized for HPLC–MS, however APCI ionization is employed on occasion. There were several illustrations of vitamins in foodstuffs were done with HPLC–MS examination. Currently, UHPLC–MS analysis has largely supplanted with HPLC–MS analysis, which provides good resolution than earlier modes of examinations. The examination of apolar carotenoids was accomplished by (U)HPLC–MS due to their huge number of structural variations, where an ionization mechanism may be chosen for accurate explications (Yuan et al. 2018).

9.4.2.2 Phenolic Compounds/Flavonoids

Secondary plant metabolites known as phenolic compounds play a vital role in the quality detection of various food materials. These are mainly from plant origin and possess a variety of positive health effects; hence, much emphasis has been paid to their examination in a variety of dietary samples. Phenolic acids, tannins, flavonoids, alkaloids, and lignans are the most common secondary metabolites found in plants. A promising separation method for these applications is high-performance liquid chromatography (HPLC). Tandem mass spectrometry is preferred along with HPLC–MS study to boost selectivity and structural information flavonoids. The electrospray ionization (ESI) method is extensively employed for testing phenolic compounds. Such procedures are often tailored for either determining the polyphenols or determining a specific amount of phenolic chemicals of a variety of sources. Several recent studies have examined the use of the RPHPLC technique for the measurement of phenolic compounds in plants, food, and beverages (Kalili and de Villiers 2011). Moreover, the efficacy is depending up on the ionization mechanism and the kind of analyzer used. Many substances are examined by HPLC, such as flavonoids (fruit juices), organic acids (apple juices, citrus juices, French ciders), phenolic pigments (black tea liquors), proline isomers and amino acids (wines), anthocyanins (jams, juices), and so on (Cordella et al. 2002).

9.4.2.3 Food Additives/Pigments

Food additives are a class of substances that possess specific structure and outstanding unique properties. Several of them are tiny molecules, such as benzoic acid, that are utilized in conservation, while others are macromolecules, such as guar gum. Certain substances possess synthetic nature, whereas others are natural extracts (variable components). Identification and measurement of such chemicals are critical for determining their beneficial or detrimental effects. There are few typical methods for real-time extraction of many types of food additives or pigments from a variety of matrixes, where HPLC–MS is the most-used method for its identification (Stefano et al. 2012).

9.4.3 Analysis of Unwanted Compounds/Food Toxins

In addition to typical compounds, other toxic substances may also be present in food materials that are generally originating from metabolically, agro-chemical treatments, or packaging materials. They are usually present in very minute quantities while causes dangerous effects in humans. Due to high specificity, (U)HPLC–MS(MS) got much attention than that of other linked techniques. Such toxins/compounds include mycotoxins, allergens, and pesticides (Table 9.2).

Some older literature is dealing with isolation and structure elucidation of cyanogenic glycosides (Nahrstedt, A. 1981; Ngamriabsakul, C. et al. 2009).

TABLE 9.2   Inherent Plant Toxicants in Food Systems

Inherent toxicants	Typical food plants	Reported effects in human	Mechanism reported	References
α-Solanine	Potato	*Gastrointestinal effects*: diarrhea, vomiting, abdominal pain *Neurological effects (at higher dose)*: drowsiness, apathy, confusion, vision disturbances, death	Cholinesterase inhibition, disruption of cell membrane	Kuiper and Nawrot (1992)
Glycyrrhizic acid	Licorice	Hypokalemia, sodium retention, cardiac arrhythmia, hypertension	Suppression of the rennin–angiotensin–aldosterone system through inhibition of 11-beta-hydrosteroid dehydrogenase in liver and kidney	Van Gelderen et al. (2000)
Linamarin	Cassava	Mediated by hydrogen cyanide effects *Acute high dose*: nausea, vomiting, giddiness, headache, hyperpnea, dyspnea, convulsion, death *Moderate dose*: neurological effects (konzo)	Cyanide binding to cytochrome oxidase resulting in reduced oxygen utilization and anoxia	Speijers (1992)
Genistein	Soybean	Various hormonal effects which may be interpreted either as adverse or beneficial	Interaction with estrogen receptor beta, various interferences with thyroid hormone system	BfR (2007)
8-Methoxypsoralen	Celery	In combination with sunlight or UVA light, acute phototoxicity and skin burns. Medium-term exposure may increase skin cancer	Intercalation between base pairs of DNA to form a non-covalent DNA complex With UVA, formation of photo adducts from this complex Modification of protein Lipid peroxidation Lysosome damage	Sabine et al. (2011)
α-Thujone	Wormwood oil, absinthe	Seizure, coma	Modulation of GABA type A receptor	Scientific Committee on Food (2003)

The preferred method to analyze pyrrolizidine alkaloids is LC-MS. For discussion methods, the reader is directed to Chapter 10 of this book.

Furocumarins are quantified by HPLC-UV, HPLC-DAD, and HPLC-MS. HPLC-UV methods are found in older literature. HPLC-MS methods are discussed in Chapter 10.

Analyses of different extracts of fruits of *Heracleum candicans* were performed in a liquid chromatograph with Waters (Milford, MA, USA) pumps (Waters 515) equipped with an online degasser, a Waters PCM (pump control module), Rheodyne 7725 injection valve furnished with a 20-lL loop, a Waters 2996 photodiode array detector, and Waters Empower software. Separation was carried out using a Merck Purospher star (RP-18e, 250 · 4.6 mm, i.d., 5 µm pore size) column with guard column of same chemistry. HPLC finger print profile was established for furocoumarins. Elution was carried out at a flow rate of 0.5 mL per min) 1 with water:phosphoric acid (99.7:0.3 v/v) as solvent A and acetonitrile:water:phosphoric acid (79.7:20:0.3 v/v) as solvent B using a gradient elution in 0–5 minutes with 30–20% A, 5–6 minutes with 20–19% of A, 6–7 minutes with 19–17% of A, 7–10 minutes with 17–15% of A, 10–15 minutes with 15–12% A, 15–20 minutes with 12–0% A and 20–25 minutes with 0–70% A (Govindarajan, R. et al. 2007).

For detection of glycoalkaloids, older methods are based on HPLC-UV, HPLC-DAD; more recent methods on HPLC-MS.

A high-performance liquid chromatography (HPLC), Shimadzu (Kyoto, Japan) instrument consisting of a UV detector, multisolvent delivery system (LC-10AD), autosampler (SIL-10ADvp), controller module (SCL-10Avp), autosampler, and Class VP 5.02 software was used for the analysis of glycoalkaloids in *Solanum lycocarpum*. A Zorbax SB-C18 analytical reverse phase column (250 × 4.6 mm i.d.; particule size 5 μm) (Agilent Technologies, USA), coupled with a guard column from the same company was used (Tiossi, R. F. J et al. 2012).

The sample analyses were carried out by employing an isocratic elution system using a mobile phase composed of acetonitrile and sodium phosphate buffer (pH 7.2; 0.01 M) in a ratio of 36.5:63.5 (v/v) at a flow rate of 1 mL/min. A 20 µL aliquot of each sample was injected and a run time of analysis of 20 min with detection at 200 nm was employed.

For the quantitation of solasonine and solamargine in plant biomass, an aliquot of 250 mg of the powdered material were extracted in three replicates in a shaker (120 rpm/ 30°C/2 hours), using 20 mL of 80% aqueous EtOH containing 3 µg mL^{-1} of veratraldehyde (IS). For both the dried crude hydroalcoholic extract (42 mg) and the dried alkaloidic extract (4 mg), the samples were directly dissolved in 20 mL of IS solution.

All the samples were filtered and analyzed by HPLC-UV using the same conditions according to the analytical method developed.

A Shimadzu Class-VP (Kyoto, Japan) single piston high-pressure liquid chromatograph with photodiode array detection was used for the separation and determination of eggplant steroidal glycoalkaloids and their aglycones. UV detection at 205 and 208 nm was chosen. Both isocratic and gradient elution methods were employed. For all work, a binary mobile phase system was set up where one delivery bottle contained the organic solvent (ACN for all work) and was designated 'B'. The second delivery bottle contained the buffer (Tris-HCl, TEAP, or ammonium dihydrogen phosphate) (Eanes, R. C. et al. 2008).

9.4.3.1 Mycotoxins

During the storage process, molds produce harmful secondary metabolites known as mycotoxins, which may quickly invade crops or foodstuffs. Even so, human health may be exposed by their metabolites, which are commonly generated by live plants. Toxins that have been conjugated or masked may also be detected and characterized using

TABLE 9.3   Sorbent-Based Microextraction for Isolating Natural Toxins in Plant Foods

Food matrix	Analytes	Sample pretreatment	Microextraction technique	Analysis	Recovery (%)	LOD	References
Cereal flours (2 g)	AF (B1, B2, G1, G2)	Extraction with 10 mL of MeOH/ phosphate buffer (80/20, v/v, pH 5.8). Evaporation to dryness and reconstitution with 4 mL of phosphate buffer. An aliquot of the extract (2 mL) subject to microextraction.	SPME Sorbent: Commercial fibers Elution: 0.1 mL MeOH	HPLC-FLD	49–59	0.035–0.2 μg/Kg	Quinto et al. (2009)
Nuts, cereals, dried fruits and spices (0.5 g)	AF (B1, B2, G1, G2)	Extraction with 1 mL of MeOH/ H_2O (80/20, v/v). An aliquot of the extract (0.1 mL) mixed with 0.1 mL of 50 mM Tris buffer and brought to a total volume of 1 mL with H_2O before microextraction.	In-tube SPME* Sorbent: SUPEL-Q PLOT capillary	HPLC-MS	81–109	0.0021–0.0028 μg /L	Nonaka et al. (2009)
Fruit juice and dried fruit (1 mL)	PAT	-	In-tube SPME* Sorbent: Carboxen-1006 PLOT capillary	HPLC-MS	> 92	0.023 μg /L	Kataoka et al. (2009)
Nut and grain samples (0.5 g)	OTA, OTB	Extraction with 1 mL of MeOH/ H_2O (80/20, v/v). Defatted with 3 mL hexane, supernatant discarded. An aliquot of the clean extract (0.1 mL) brought to a total volume of 1 mL with H_2O before microextraction.	In-tube SPME* Sorbent: Carboxen-1006 PLOT capillary	HPLC-MS	88	0.089–0.092 μg /L	Saito et al. (2012)
Coffee (10 g) and grape juice (10 mL)	OTA	Extraction of coffee samples with 100 mL of carbonate. An aliquot of the extract (10 mL) adjusted to pH 1.5 before microextraction. Grape juice samples adjusted to pH 1.5 before microextraction.	μ-SPE Sorbent: 15 mg AFFINIMIPTM OTA Elution: 0.25 mL MeOH/Acetic acid (98:2, v/v)	HPLC-FLD	91–101	0.02–0.06 μg/Kg	Lee et al. (2012)

Note

[*] Adapted from Natalia et al. (2020) with slight modification, Creative Commons Attribution 4.0 International.

LC–MS/MS procedures, which are well suited for high sample throughput multi-mycotoxin study. Mass spectrometry and electrospray, APCI, or APPI ionizations are preferred according to the sample. In MRM mode, triple-quadrupole instruments may be utilized for focused analysis, since such high-resolution scrutiny is adequate to find out unknown derivatives (Sofia et al. 2020). For reliable and exact analysis, isotope-labeled standards should be used.

9.4.3.2 Allergens

Most of the potential allergens in food materials are not well known due to the lack of accurate separation and identification techniques for such allergenic proteins and peptides. The concerns associated with allergen detection can be overcome by the excellent selectivity and sensitivity of protein samples. A wide variety of approaches are used for analyzing allergens in foods; however, the HPLC–MS technique got much attention in the scientific community. HPLC–MS analysis of allergens has been used in research on propolis, cannabidiol, and the discovery of novel lipocalin allergen variations, showed better results (Yuxin et al. 2021).

9.4.3.3 Pesticides

The widespread usage of pesticides leads various issues in the society hence investigation of various food materials is required to trace out such chemicals. HPLC–MS is used for this purpose so longer, where, HPLC-linked tandem mass spectrometry was more efficient since it simplifies sample preparation prior to analysis and enables for multi-residue analysis through a single run. The isotope-labeled standards are preferred for accurate quantification and thereby ensure exact analysis. UHPLC (ultra-high-performance liquid chromatography) with a C18 column (particle size 1.7 ppm) is often used for the purpose which enables quick (less than 15 minutes) and efficient separation. MRM channels unique to a certain molecule, which are exclusively monitored at the relevant retention time-windows in HPLC–MS applications, boosting sensitivity. Isotope dilution is another method for improving the exactness of pesticide scrutiny. Due to the high cost, the isotope dilution technique is not preferred for agri-food analysis. Pesticides and pollutants may be found in animal feed also, since current regulation restrict dangerous pollutants in animal feed in order to avoid its entry to food chain. A wide range of products are frequently evaluated through these techniques including vegetables, vegetable oil, infant food, honey, wine, fruit juice, and milk (Monika and Boguslaw 2002; Stefano et al. 2012) (Table 9.3).

9.5 CONCLUSION

Plant toxins are secondary plant metabolites that exhibit chronic toxicity or have severe anti-nutritional effects. HPLC separation of phytotoxins is generally carried out on a reversed-phase (RP) column with the mobile phase of acetonitrile or methanol and water containing a small amount of acid as the modifier. The phytoestrogens and their metabolites generally contain phenolic hydroxyl groups, which exhibit a weak acidic nature. Hence, applications of acidic modifiers, such as acetic acid, formic acid, trifluoroacetic acid, and phosphoric acid can make the analytes efficiently dissociated in a solvent system, thus improving the chromatographic separation, resolution, and peak shape. In comparison with gas chromatography, the main advantage of HPLC analysis is the ease of utilization and simple sample preparation procedure. The key limitation of

HPLC separation is the poorer chromatographic resolution. The most commonly used coupling with HPLC has been UV detection. Ultraviolet detection is, however, non-specific, and cannot achieve a good enough sensitivity. These restricting factors have limited HPLC application for analysis of food toxins of plant origin in multicomponent biological matrices.

REFERENCES

Alcaide-Molina, M., Ruiz-Jiménez, J., Mata-Granados, J. and Luque de Castro, M. (2009). High through-put aflatoxin determination in plant material by automated solidphase extraction on-line coupled to laser-induced fluorescence screening and determination by liquid chromatography-triple quadrupole mass spectrometry. *Journal of Chromatography A*, 1216(7), 1115–1125.

Alexander, F., Harald, K., Martin, T., et al. (2011). A comprehensive method for lipid profiling by liquid chromatography-ion cyclotron resonance mass spectrometry. *J Lipid Res*, 52(12), 2314–2322. doi:10.1194/jlr.D016550

Bélanger, J. M. R., Jocelyn Pare, J. R. and Sigouin, M. (1997). High performance liquid chromatography (HPLC): Principles and applications. In Bélanger, J. M. R. and Jocelyn Pare, J. R. (eds.). *Instrumental Methods in Food Analysis*. Amsterdam: Elsevier Science, pp. 37–59.

BfR. (2007). Isolated isoflavones are not without risk. BfR Expert Opinion No. 039/2007. https://www.bfr.bund.de/en/publication/bfr_opinions_2007-127800.html

Colegate, S. M., Gardner, D. R. and Lee, S. T. (2015). Plant-associated natural food toxins. In Cheung, P. and Mehta, B. (eds.). *Handbook of Food Chemistry*. Berlin: Springer, pp. 753–783. doi:10.1007/978-3-642-36605-5_9

Colombo, R. and Papetti, A. (2020). Pre-concentration and analysis of mycotoxins in food samples by capillary electrophoresis. *Molecules*, 25, 3441. doi:10.3390/molecules25153441

Cordella, C., Moussa, I., Martel, A., Sbirrazzuoli, N. and Lizzani-cuvelier, L. (2002). Recent developments in food characterization and adulteration detection: Technique-oriented perspectives. *J Agric Food chem*, 50, 1751–1764.

Diniz, M. C. T., Fatibello, F. O. and Rohwedder, J. J. R. (2004). An automated system for liquid-liquid extraction based on a new micro-batch extraction chamber with on-line detection. *Analytica Chimica Acta*, 525(2), 281–287.

Dion, M. A. M. L., Ruud, J. B. P., Saskia, M. V. and Hans, B. (2008). A review of analytical methods for the identification and characterization of nano delivery systems in food. *J Agric Food Chem*, 56(18), 8231–8247. doi:10.1021/jf8013926

Dnyaneshwar, W. (2017). Safety of traditional medicines. In Patwardhan, B. and Chaguturu, R. (eds.). *Innovative Approaches in Drug Discovery: Ethnopharmacology, Systems Biology and Holistic Targeting*. Amsterdam: Elsevier, pp. 351–365.

Dolan, L. C., Matulka, R. A. and Burdock, G. A. (2010). Naturally occurring food toxins. *Toxins*, 2(9), 2289–2332. doi:10.3390/toxins2092289

Eanes, R. C., Tek, N., Kirsoy, O., Frary, A., Doganlar, S. and Almeida, A. E. (2008). Development of practical HPLC methods for the separation and determination of eggplant steroidal glycoalkaloids and their aglycones. *Journal of liquid chromatography & related technologies*, 31(7), 984–1000. doi:10.1080/10826070801924741

Essers, A. J. A., Alink, G. M., Gerrit J. A., et al. (1998). Food plant toxicants and safety: Risk assessment and regulation of inherent toxicants in plant foods. *Environ. Toxicol. Pharmacol*, 5(3), 155–172. doi:10.1016/s1382-6689(98)00003-9

Gardner, D. R. and Pfister, J. A. (2009). HPLC-MS analysis of toxic norditerpenoid alkaloids: Refinement of toxicity assessment of low larkspurs (*Delphinium* spp.). *Phytochemical Analysis*, 20(2), 104–113. doi:10.1002/pca.1104

Govindarajan, R., Singh, D. P., Singh, A. P., Pandey, M. M., and Rawat, A. K. S. (2007). A validated HPLC method for quantification and optimization of furocoumarins in different extracts of fruits of Heracleum candicans. *Chromatographia*, 66(5), 401–405.

Gratzfeld-Huesgen, A. and Schein, A. (2002). The application of HPLC in food analysis. *Food Safety Magazine*. https://www.food-safety.com/articles/2098-the-application-of-hplc-in-food-analysis. doi:10.1365/s10337-007-0316-5

Gry, J., Black, L., Eriksen, F., et al. (2007). EuroFIR-BASIS – a combined composition and biological activity database for bioactive compounds in plant-based foods. *Trends in Food Science and Technology*, 18(8), 444. doi:10.1016/j.tifs.2007.05.008

Guenthardt, B. F., Hollender, J., Hungerbuehler, K., et al. (2018). A comprehensive toxic plant-phytotoxin (TPPT) database and its application to assess their aquatic micropollution potential. *Journal of Agricultural and Food Chemistry*, 66, 7577–7588. doi:10.1021/acs.jafc.8b01639

Hurst, W. J., Martin Jr, R. A. and Zoumas, B. L. (1979). Application of HPLC to characterization of carbohydrates in foods. *Journal of Food Science*, 44, 895–904.

Hussain, C. M., Deepak, R., Gaurav, P. and Maithri, T. (2020). *Handbook of Analytical Techniques for Forensic Samples: Current and Emerging Developments.* Amsterdam: Elsevier.

Jarmila, S., Vladimir, O., Frantisek, M. and Tomas, R. (2013). Determination of ochratoxin A in food by high performance liquid chromatography. *Analytical Letters*, 46(10), 1495–1504. doi:10.1080/00032719.2013.771266

Kalili, K. M. and de Villiers, A. (2011). Recent developments in the HPLC separation of phenolic compounds. *Journal of Separation Science*, 34(8), 854–876. doi:10.1002/jssc.201000811

Kataoka, H., Itano, M., Ishizaki, A. and Saito, K. (2009). Determination of patulin in fruit juice and dried fruit samples by in-tube solid-phase microextraction coupled with liquid chromatography–mass spectrometry. *Journal of Chromatography A*, 1216, 3746–3750. doi:10.1016/j.chroma.2009.03.017

Kuiper, G. T. and Nawrot, P. S. (1992). Solanine and chaconine. In: *WHO Food Additives Series 30*. Geneva. https://inchem.org/documents/jecfa/jecmono/v30je19.htm

Lee, T. P., Saad, B., Khayoon, W. S. and Salleh, B. (2012). Molecularly imprinted polymer as sorbent in micro-solid phase extraction of ochratoxin A in coffee, grape juice and urine. *Talanta*, 88, 129–135. doi:10.1016/j.talanta.2011.10.021

Liska, I., Krupcik, J. and Leclercq, P. A. (1989). The use of solid sorbents for direct accumulation of organic compounds from water matrixes – A review of solid-phase extraction techniques. *Journal of High Resolution Chromatography*, 12(9), 577–590.

Macrae, R. (1988). *HPLC in Food Analysis*. London: Academic Press.

Majors, R. E. (2002). Trends in sample preparation. *LCGC North America*, 20(12), 1098–1113.

Monika, M. and Boguslaw, B. (2002). HPLC determination of pesticide residue isolated from food matrices. *Journal of Liquid Chromatography and Related Technologies*, 25(13–15), 2293–2306. doi:10.1081/JLC-120014004

Nahrstedt, A. D. O. L. F. (1981). Isolation and structure elucidation of cyanogenic glycosides. In Vennesland, B. (ed.). *Cyanide in Biology*. New York: Academic Press, p. 145.

Natalia, C., Judith, G., Zarcero, M. S., et al. (2020). New advanced materials and sorbent-based microextraction techniques as strategies in sample preparation to improve the determination of natural toxins in food samples. *Molecules*, 25(3), 702. doi:10.3390/molecules25030702

Ngamriabsakul, C., & Kommen, H. (2009). The preliminary detection of cyanogenic glycosides in pra (Elateriospermum tapos Blume) by HPLC. *Walailak Journal of Science and Technology (WJST)*, 6(1), 141–147.

Nikolin, B., Imamović, B., Medanhodzić-Vuk, S. and Sober, M. (2004). High perfomance liquid chromatography in pharmaceutical analyses. *Bosn J Basic Med Sci*, 4(2), 5–9. doi:10.17305/bjbms.2004.3405

Noah, N. D., Bender, A. E., Reaidi, G. B. and Gilbert, R. J. (1980). Food poisoning from raw red kidney beans. *Br Med J*, 281(6234), 236–237.

Nobukazu, S. (2016). Secondary metabolites in plants: Transport and self-tolerance mechanisms. *Bioscience, Biotechnology and Biochemistry*, 80(7), 1283–1293. doi:10.1080/09168451.2016.1151344

Nonaka, Y., Saito, K., Hanioka, N., Narimatsu, S. and Kataoka, H. (2009). Determination of aflatoxins in food samples by automated on-line in-tube solid-phase microextraction coupled with liquid chromatography–mass spectrometry. *Journal of Chromatography A*, 1216, 4416–4422. doi:10.1016/j.chroma.2009.03.035

Quinto, M., Spadaccino, G., Palermo, C. and Centonze, D. (2009). Determination of aflatoxins in cereal flours by solidphase microextraction coupled with liquid chromatography and post-column photochemical derivatization-fluorescence detection. *Journal of Chromatography A*, 1216, 8636–8641. doi:10.1016/j.chroma.2009.10.031

Sabine, G., Michael, H., Dieter, S. and Gerhard. E. (2011). Update of the toxicological assessment of furanocoumarins in foodstuffs (Update of the SKLM Statement of 23/24 September 2004). *Opinion of the Senate Commission on Food Safety (SKLM) of the German Research Foundation (DFG)*, 55(5), 807–810. doi:10.1002/mnfr.201100011

Saito, K., Ikeuchi, R. and Kataoka, H. (2012). Determination of ochratoxins in nuts and grain samples by in-tube solidphase microextraction coupled with liquid chromatography – Mass spectrometry. *Journal of Chromatography A*, 1220, 1–6. doi:10.1016/j.chroma.2011.11.008

Schilter, B., Constable, A. and Perrin, I. (2014). Naturally occurring toxicants of plant origin. In Motarjemi, Y. and Lelieveld, H. (eds.). *Food Safety Management: A Practical Guide for the Food Industry*. Amsterdam: Elsevier, pp. 45–57. doi:10.1016/B978-0-12-381504-0.00003-2

Scientific Committee on Food. (2003). Opinion of the SCF on thujone. https://ec.europa.eu/food/system/files/2016-10/fs_food-improvement-agents_flavourings-out162.pdf

Scott, P. M. (2002). Methods of analysis for ochratoxin A. *Adv Exp Med Biol*, 504, 117–134. doi:10.1007/978-1-4615-0629-4_12

Slack, G. C. and Snow, N. H. (2007). HPLC sample preparation. In Ahuja, S. and Rasmussen, H. (eds.). *HPLC Method Development for Pharmaceuticals*. Amsterdam: Elsevier, pp. 237–268.

Sofia, A., Eygenia, S. and Theodoros, V. (2020). Advances in analysis and detection of major mycotoxins in foods. *Foods*, 9(4), 518. doi:10.3390/foods9040518

Souza-Silva, É. A., Gionfriddo, E. and Pawliszyn, J. (2015). A critical review of the state of the art of solid-phase microextraction of complex matrices II. Food analysis. *Trends in Analytical Chemistry*, 71, 236–248. doi:10.1016/j.trac.2015.04.018

Speijers, G. (1992). Cyanogenic glycosides. In *WHO Food Additives Series 30*. World Health Organization, Geneva. https://inchem.org/documents/jecfa/jecmono/v30je18.htm

Stefano, V. D., Avellone, G., Bongiorno, D., et al. (2012). Applications of liquid chromatography–mass spectrometry for food analysis. *Journal of Chromatography A*, 1259(2012), 74–85. doi:10.1016/j.chroma.2012.04.023

Thammana, M. (2016). A review on high performance liquid chromatography (HPLC). *Research and Reviews: Journal of Pharmaceutical Analysis*, 5(2), 22–28.

Tiossi, R. F. J., Miranda, M. A., de Sousa, J. P. B., Praça, F. S. G., Bentley, M. V. L. B., McChesney, J. D. and Bastos, J. K. (2012). A validated reverse phase HPLC analytical method for quantitation of glycoalkaloids in Solanum lycocarpum and its extracts. *Journal of analytical methods in chemistry*, 2012. doi:10.1155/2012/947836

Tomas, C. and Jiri, P. (2020). Rapid HPLC method for determination of isomaltulose in the presence of glucose, sucrose, and maltodextrins in dietary supplements. *Foods*, 9(9), 1164. 10.3390/foods9091164

Tranchida, P. Q., Dugo, P., Dugo, G. and Mondello, L. (2004). Comprehensive two-dimensional chromatography in food analysis. *J Chromatogr A*, 1054, 3–16.

van Gelderen C. E., Bijlsma, J. A., van Dokkum, W. and Savelkoul, T. J. (2000). Glycyrrhizic acid: The assessment of a no effect level. *Hum Exp Toxicol*, 19(8), 434–439. doi:10.1191/096032700682694251

Walter, P. (2003). Why are fruits and vegetables are very healthy? *Int J Vit Nutr Res*, 72(3), 52–53.

Yuan, Z., Wei, Z., Jia-Qing, Y., et al. (2018). A review of the extraction and determination methods of thirteen essential vitamins to the human body: An update from 2010. *Molecules*, 23(6), 1484. doi:10.3390/molecules23061484

Yuxin, Y., Wang, Y., Zhigan, H. and Long, P. (2021). HPLC method for separation of cannabidiol hemp seed oil with skin lipids and tandem HRMS technology for characterization of a chemical marker. *Cosmetics*, 8(4), 108. doi:10.3390/cosmetics8040108

CHAPTER 10

Analytical Determination of Food Toxins of Plant Origin Using LC-MS

Jeyabalan Sangeetha[1], D. Divya[2], Pavitra Chippalakatti[3], Devarajan Thangadurai[3], Jarnain Naik[4], Saher Islam[5], Ravichandra Hospet[6], Muniswamy David[4], and Zaira Zaman Chowdhury[7]

[1]Department of Environmental Science, Central University of Kerala, Kasaragod, Kerala, India
[2]Pinnacle Biosciences, Kanyakumari, Tamil Nadu, India
[3]Department of Botany, Karnatak University, Dharwad, Karnataka, India
[4]Department of Zoology, Karnatak University, Dharwad, Karnataka, India
[5]Department of Biotechnology, Lahore College for Women University, Lahore, Pakistan
[6]Department of Food Protectants and Infestation Control, CSIR-Central Food Technological Research Institute, Mysore, Karnataka, India
[7]Nanotechnology and Catalysis Research Center (NANOCAT), Institute of Advanced Studies (IAS), University of Malaya, Kuala Lumpur, Malaysia

CONTENTS

10.1	Introduction	162
10.2	LC-MS: Instrumentation and Preparative Techniques	164
	10.2.1 LC–MS/MS Instrumentation	164
	10.2.1.1 Ionization Techniques	164
	10.2.1.2 Mass Analyzers	165
	10.2.2 Preparation Techniques	165
	10.2.3 Quantitation Process	166
10.3	Quality Analysis of Various Food Compounds	166
10.4	Analysis of Food Toxins of Plant Origin	167
10.5	LC Conditions	172
10.6	Conclusion	173
References		173

DOI: 10.1201/9781003222194-13

10.1 INTRODUCTION

A diverse and enormous assortment of organic compounds are produced in plants, the great majority of them do not contribute in growth and development directly but are known to play a major role in the adoption of plants to the surrounding environment and are often involved in protection of plants against stress, as well as exert a wide range of effects on other living organisms. They are traditionally referred to as secondary metabolites (Teoh 2016; Jain et al. 2019). Some of these secondary metabolites are formed in food as defense machinery of plants, which can cause a variety of hostile health effects and pose a serious health risk to both humans and livestock when they are consumed. These chemical compounds are named food toxins of plant origin and they have diverse structures and are different in biological function as well as toxicity; and the level of toxins present in plants vary according to species as well as geographical conditions where it is grown (Anna 2007; WHO 2018). The popular notion equating natural and healthy is obviously not realistic since natural plant toxins pose a far greater health risk than that of many man-made chemicals occurring in our foods (Hajšlová et al. 2018). The commonly eaten foods may contain natural toxins like cyanogen as in cassava and bamboo shoots, hypoglycin in ackee, α-solanine and chaconine in potatoes, glucosinolates in cabbage, α-solasonine and α-solamargine contents of brinjal, coumarins and furonocoumarins in citrus plants, and lectins in kidney beans (Bender and Reaidi 1982; Kuiper-Goodman and Nawrot 1992; Blake et al. 2006; Sánchez-Mata et al. 2010; Sarangthem and Hoikhokim 2010; Choi et al. 2014; Dugrand-Judek et al. 2015; Oluwaniyi and Oladipo 2017). van Egmond (2004) mentioned another problem in food safety which is about toxins occurring in herbs. Examples are anisatin in star anise and the large group of the pyrrolizidine alkaloids and myristicin in *dill* (Zhu et al. 2019) (Table 10.1).

Some of the antinutrients like lectins, oxalates, goitrogens, phytoestrogens, phytates, saponins, and tannins have also been found to be capable of eliciting deleterious effects on human health depending on their type and quantity (Bing et al. 2014). The compounds with potential beneficial effects too have negative effects at higher doses (Gry et al. 2007), meaning that, only 'dose' separates toxic from non-toxic. The basic concept of toxicology as abandoned with adoption by Federal Food and Drug Act in 1958 is that "the dose makes the poison" and demanding "safety per se" or "safe at any dose" for all foods and ingredients (Laurie et al. 2010). The influence of these non-nutrient components of the diet on health and well-being, as well as intake level of these compounds, and the nature and dose activity relationship of their biological effect are of considerable interest in last few years.

To ensure food safety and to meet the regulations, novel analytical methods are desired to detect and quantify these toxic molecules at trace concentrations and also there is a cumulative attention to study the nature of these biotoxins in food and water in last few decades (Peltomaa et al. 2018; Picardo et al. 2018). The development of analytical methodology for the plant toxins has been a rather neglected area, and validated methods of analysis are rather scarce compared to biological contaminants such as mycotoxins. The methods in practice to determine plant toxins are based on chromatographic techniques coupled with spectroscopic techniques, such as TLC, immunochemical procedures, LC with UV, MS detectors and tandem mass spectrometry (LC-MS/MS), GC in combination with FID, MS detectors and tandem mass spectrometry (GC-MS/MS), and NMR and high-resolution mass spectrometry (LC-HRMS) to achieve enough sensitivity (van Egmond 2004; Picardo et al. 2018). The progress in plant toxin analytical methodology is slow, which has contributed to a situation where there is a chronic lack of reliable analytical standards and reference materials. An exception is the glucosinolate, plant toxins for which

Analytical Determination Using LC-MS

TABLE 10.1 Food Toxins of Plant Origin

Plant species	Toxicants
Conium maculatum	Coniine, coniceine
Phaseolus vulgaris	Phytoheamagglutinin
Persea sp.	Persin equine colic
Lathyrus vulgaris	Amino acid ODAP neurolathyrism
Datura stromonium	Atropine, hyoscamine, scopolamine
Abrus precatorius	Abrin abric acid from thoroughly chewed seed
Argemone mexicana	Toxic alkaloids sanguinarine, berberine, protopine
Gutierrezia sarothrae	Saponin in leaves
Hippomane mancinella	Purgative oil milky sap
Manihot esculenta	Hydrocyanic acid
Nerium indicum	Cardioactive glycosides
Nicotiana tabacum	Nicotine
Passiflora quandrangularis	Passiflorine hydrocyanic acid
Philodendron spp.	Tissue contains irritant juice, calcium oxalate crystal
Ricinus communis	Ricin, cardiac glycoside
Urtica dioica	Formic acid, histamine, acetylcholine in sting
Dieffenbachia sp.	Oxalic acid, asparagine
Xanthium strumarium	Hydroquinone in seeds and seedlings
Strychnos nux-vomica	Strychnine, brucine, vomicine, protostrychnine, chlorogenic acid, *n*-oxystrychnine
Calotropis gigantea	Laurane, saccharose, □-amyrin, calotroposide, calactin, calotoxin, calotropins DI and DII, gigantin
Parthenium hysterophorus	Germacrene D, *trans*-□-ocimene, □-myrcene, □-caryophyllene
Aconitum napellus	Aconitine, mesaconitine, hypaconitine, other aconitum alkaloids
Convallaria majalis	38 different cardenolides (convallarin, cannogenol-3-*o*-□-□-rhamnoside, neoconvallocide), saponin, azetidine 2-carboxylic acid
Cerbera odollam	Cerberin, cardiac glycoside
Thevetia peruviana	Thevetin A, thevetin B, thevetoxin, neriifolin, peruviside, ruvoside
Colocasia esculenta	Insoluble and soluble oxalate

Reproduced with permission from Dubey et al. (2018), Copyright © Elsevier, 2018 and adapted from Ogori (2019), Creative Commons Attribution 4.0 International.

the analytical methodology is rather advanced, and BCR-certified reference materials have become available (van Egmond 2004). Against this background, LC coupled with UV/VIS detectors have been employed well, but identification of the toxins cannot be confirmed. So, liquid chromatography–mass spectrometry (LC–MS) is the most commonly used method that enables the simultaneous quantification and identification.

LC–MS, traditionally an important part of the medical laboratory, created a growing market in a new application field – food safety testing. In general, the approaches are based on the LC coupled to tandem mass spectrometry (MS/MS), but recently it has been reported that some approaches taking the advantages of the high-resolution MS (HRMS). Holstege et al. (2001) identified the grayanotoxins in feces and urine from goats exposed to *Rhododendron* using the LC-MS/MS method and quantification of grayanotoxins was performed by the LC-MS (Puschner et al. 2001). Particularly LC–MS is suited for the analysis of food contaminants, since a large amount of information is provided about a complex mixture and enables the screening, confirmation, and quantification of hundreds of components with one analysis (Malik et al. 2010; Picardo et al. 2018). This method does not require extensive sample cleanup, and compensates for separation problems. High specificity of mass selective detection of LC-MS avoids matrix interferences (Sproll et al. 2006). Recently it has become the most popular analytical tool for accurate and reliable determination of different residues and contaminants because of high specificity, selectivity, and sensitivity (Kaplan et al. 2014).

10.2 LC-MS: INSTRUMENTATION AND PREPARATIVE TECHNIQUES

Food contains a huge number of harmful compounds including pesticides, poisons, antibiotics, etc., that may cause adverse health issues. These chemicals are typically present at low concentrations, since highly selective techniques are required for their scrutiny. Liquid chromatography (LC) linked with mass spectrometry (MS) is a separation tool that offers good selectivity and fast identification of such types of sensitive compounds. Mass spectroscopy works on the basis of the production of ions from evaluated components, which are then filtered in the course of their mass-to-charge ratio. Recently, specific LC–tandem MS (LC–MS/MS) is on trend to analyze organic toxins in food products. Most quantitative bioanalysis applications using tandem mass spectrometers (MS/MS) operate two mass analyzers, where one mass detector is set for the precursor ion of first quadrupole and the second one allied with the product ion of third quadrupole.

10.2.1 LC–MS/MS Instrumentation

10.2.1.1 Ionization Techniques

It has been indeed a big challenge of researchers/scientists for the last few decades to establish a good interface assembly between LC that functions at atmospheric pressure and MS that works at high vacuum, which makes LC compatible to MS. The presently employed ionization methods for quantitative food-sample analysis by LC–MS include electrospray ionization (ESI) and atmospheric-pressure chemical ionization (APCI), which are referred to as atmospheric pressure ionization (API). APPI (atmospheric pressure photo-ionization) has lately emerged as an intriguing substitute source for LC-MS/MS (Núñez et al. 2005).

10.2.1.1.1 Electrospray Ionization (ESI)   ESI is the most extensive ionization method widely employed to analyze polar/ionized/toxic compounds of food. For regular food analysis, LC-UV techniques with a column size of 3–4 mm are preferred. In some cases, the setup is modified with split devices in course to lower the mobile phase flow rate when it is necessary for absolute separation. Due to the concentration dependence of electrospray, the split eluent has little effect on sensitivity. For instance, fluoroquinolone antibiotics were successfully analyzed using a reverse-phase chromatography system possess column with 4-mm and a flow rate of 1 mL/min in mobile phase. Prior to introducing the ESI source, a T-piece splitter (9:1) was employed to lower the flow rate. The instability/blocking of the simple system can be minimized by avoiding low-volatility of mobile phase and effluents with a relatively more salt content. The use of electro-sprayed solution, frequent spraying of the orifice with a small amount of solvent and usage of a certain gas decreases the system instability, but enhances separation. Even so, eluent composition, mobile-phase composition, and post-column solvent also determine the effectiveness of electrospray ionization (Vyncht et al. 2002).

10.2.1.1.2 Atmospheric-Pressure Chemical Ionization (APCI) APCI is a more efficient technique than electrospray because the effluent rapidly evaporates at high temperature (350°C–500°C) under nitrogen drift. Despite the fact that APCI uses a soft-ionization mode, which also offers structural information through ions fragmentation, it results in less sensitivity and more complicated spectra than MS/MS. Such ion fragmentation could be determined by possible variations among the sample cone and skimmer lens. Nonetheless, from the standpoint of identification and characterization, this technique is intriguing. Weakly basic compounds such as insecticides like triazines and phenylureas are extremely susceptible to APCI, since such compounds possess more proton affinity. When dealing with thermally degradable compounds, high temperatures must be a concern. For instance, LC-APCI-MS is not sufficient to study an active component in the herbicide Surflan since it cannot withstand high processing temperature, whereas ESI-MS was appropriate for its detection (West et al. 2004), where the protonated molecules [M+ H+] could be in positive ionization mode and deprotonated molecules [M– H–] could be in negative ionization mode. But in certain cases, extra signals that emerged in the spectra might be due to the excitation of cations of the contaminants in mobile phase organic solvents in LC, often which may be seen as the spectra's base peak.

10.2.1.1.3 Atmospheric Pressure Photo-Ionization (APPI) Nowadays, atmospheric pressure photo-ionization (APPI) has much attention because it can evade the suppression difficulties experienced in other sources. It has been recommended as an efficient tool for complicated sample analysis recently. APPI was successfully employed in the LC–MS of carbamate pesticides found in fruits and vegetables (Gaudin et al. 2012).

10.2.1.2 Mass Analyzers

Triple-quadrupole and ion-trap mass spectrometers are the extensively linked devises to LC for investigation of poisonous components in food items, among the mass spectrometers that enable MS/MS operations. These are more significant due to their simpler operational performance, improved resilience for regular analysis, and lower cost over other methods as TOF and FT-ICR. While choosing multiple reaction monitoring (MRM) mode, triple-quadrupole analyzers have a high sensitivity, which makes them ideal for determining hazardous components of diverse food materials. An antibiotic chloramphenicol in bovine milk was successfully detected using a LC-MS/MS-triple-quadrupole apparatus. However, at least two transitions must be recorded for confirmation while using this technique. Ion-trap devices are less sensitive in contrast to triple-quadrupole analyzers, which ensure proper product-ion scan without losing its sensitivity. Ion traps provide multiple-stage fragmentation that is useful for exact screening and characterization of harmful chemicals in complex food sample as well as distinguishing isomer compounds, although these techniques are sufficient for the identification of pollutants that are entirely unknown (Ardrey 2003).

10.2.2 Preparation Techniques

Unique preparation techniques are requited for accomplishing LC-MS/MS, where accurate sample preparation, selection of appropriate column, and exact mode of extraction is obligatory. The groundwork of sample prior to injection is an important step in this high-throughput examination. This procedure is very imperative to eliminate certain unwanted components from the matrix, which leads to ionization suppression. Nowadays, online sample extraction linked to LC-MS/MS has emerged that allows direct injection analysis and also reduces the cost by offering repeated use of extraction columns for even hundreds of biological samples. An automated 96-well plate is a widely used technology that allows automated solid-phase extraction (SPE), liquid/liquid extraction (LLE), and protein precipitation (PP). Automated sample preparation requires only half the time required of manual sample preparation. Two distinct setups are available for online extraction; these are single-column extraction and dual-column extraction. In the former case, samples are put into an extraction column, and the mobile phase encloses with a strong solvent. In this method, the components directly extract onto the analytical column in single-column extraction while the samples must be devoid of unwanted species in order to avoid interference during exposure. In the dual-column conformation, extraction occurs by the solvent that was previously held in an extraction loop since this configuration gives better separation and peak morphology when compared to the single–column approach. Disposable cartridges, monolithic, limited access media (RAM), and immunoaffinity are some of the other intriguing extraction columns that are frequently employed for online extraction allied with LC-MS/MS (Zhou et al. 2005; Xu et al. 2007).

10.2.3 Quantitation Process

MS/MS has good selectivity, while analyte validation is not constantly perfect. The majority of LC–MS/MS programs for food material analysis track just one transition point only for each molecule. Its exact confirmation needs a minimum of three to four identification points (IPs) according to current regulations. Signal suppression, reported for electrospray sources, is a major issue when studying complex (food) samples with LC–MS. Several practices may be used to mitigate matrix effects while quantifying, of which isotopic dilution is the best approach (Cervino et al. 2008; Al-Taher et al. 2017). This technique neutralizes signal suppression when just one analyte is identified. Another choice for correcting matrix suppression is to allow quantitation by standard addition. However, when a large number of samples must be evaluated, this approach is inconvenient and, as a result, it is not often employed for the identification of harmful components in food. Conversely, matrix-matched calibration is another option where availability of blank matrix for each sample is obligatory. This approach is challenging since more samples with various matrices must be assured and evaluated. External calibration is a less ideal approach for the analysis of components from fruits and vegetables (Núñez et al. 2005). However, for a mixture of analytes, this approach seems to be challenging to perform, since adequate separation is required to avoid coelution amongst standards and analytes.

10.3 QUALITY ANALYSIS OF VARIOUS FOOD COMPOUNDS

There are various types of toxins that contaminate the food, especially mycotoxins, that include Alternaria toxins (*Alternaria* fungi), flatoxins and ochratoxins (*Aspergillus* sp.), fumonisins, trichothecenes and zearalenone (*Fusarium* sp.), patulin (*Penicillium* sp.), and ergot alkaloids (*Claviceps* sp.). These can lead to serious consequences in terms of human health as well as economic conditions (Krska et al. 2008). The contamination of food may occur at various stages such as preparation, storage, and transportation. The determination of these toxins in the food commodities is a challenging task because of their low occurrence and complexity of matrix (Malachová et al. 2018). Therefore, it is necessary to have a suitable analytical method for determination of various food toxins. Presently, analytical methods that are used for the determination of food toxins include thin-layer chromatography (TLC), gas chromatography (GC) with electron capture or mass spectroscopy (MS), liquid chromatography with fluorescence detection (LC-FLD), high-performance liquid chromatography (HPLC), and liquid chromatography with tandem mass spectrometry (LC-MS/MS) (Di Mavungu et al. 2009). LC-MS/MS has become the most promising instrumental technique due to its high sensitivity, selectivity, robustness, and multi-analyte that provides broad applications by analysis both qualitative and quantitatively (Sulyok et al. 2020). Presently, LC-MS is considered as a standard practice in most laboratories for food toxin analysis. The use of high-resolution mass spectrometry time-of-flight (TOF) and orbitrap spectrometry (HRMS) analyzer has enabled the nanotarget screening and unknown identification system development (Masiá et al. 2016). The LC-MS/MS analytical method was used to determine 17 mycotoxins produced by *Alternaria* fungi that occurred in tomato sauce, sunflower seed oil, and wheat flour, of which 12 were parent compounds and 5 modified *Alternaria* toxins (Puntscher et al. 2018). LC-ESI$^+$-MS/MS was used to determine 13 mycotoxins in several food matrices such as maize, wheat, oats, cornflakes, and bread (De Boevre et al. 2012). The mycotoxins such as Aflatoxin B1, B2, G1, G2, fumonisin B1, B2, deoxynivalenol, zearalenone, ochratoxin A, T-2, and HT-2 were detected in rice, maize, wheat, and peanuts were determined using LC-MS/MS (Tang et al. 2013). The levels of acrylamide were analyzed in different varieties of instant, ground, and brewed coffees using the modified liquid chromatography-tandem mass spectrometry (LC-MS/MS) method (Andrzejewski et al. 2004). Through LC-MS, different mycotoxins were detected in the range of 10–200 µg/kg in cornflakes, figs, maize, peanuts, pistachios, raisins, and wheat (Spanjer et al. 2008).

10.4 ANALYSIS OF FOOD TOXINS OF PLANT ORIGIN

Separation and analysis of cyanogenic glycosides in American elderberry were performed with a C18 column (Acquity BEH, 1.7 µm, 50 × 2.1 mm, Waters, Milford, MA, USA) using a Waters Acquity UHPLC coupled to a Xevo TQ-S triple quadrupole mass spectrometer (UHPLC-MS/MS) (Appenteng, M. K. et al. 2021). The mobile phase included 0.1% formic acid in water (mobile phase A) and 0.1% formic acid in acetonitrile (mobile phase B). The gradient used was 95% A, 0–1 minute; 95–80% A, 1–3 minutes; 80–40% A, 3–7 minutes; 40% B, 7–8 minutes and 95% A 8.1–10 minutes re-equilibration. The flow rate was 200 µL min^{-1}.

The following conditions were used for the electrospray ionization (ESI) source: source temperature 150°C, desolvation temperature 350°C, capillary voltage 3.07 kV, cone voltage 21, and nebulizer gas 500 L h^{-1} N$_2$. Argon was used as the collision gas. The collision energies were optimized and ranged from 17 to 30 eV for individual analytes. The column and sample temperatures were 40°C and 10°C, respectively. The ESI source was operated in the positive ion mode. Instrument control and data processing were performed by using MassLynx software (version 4.1, Waters, Milford, MA, USA). Cyanogenic standard solutions were prepared with concentrations ranging from 1 ng mL^{-1} to 10 µg mL^{-1}. All analyses were done in triplicate, along with a blank. The interday and intraday precisions of the method had a CV% of less than 5%.

Twelve grapevine (*Vitis vinifera* L.) cultivars were surveyed for 'cyanide potential' (i.e. the total cyanide measured in β-glucosidase-treated crude, boiled tissue extract) in mature leaves. Two related cultivars (Carignan and Ruby Cabernet) had mean cyanide potential (equivalent to 110 mg HCN kg^{-1} fr. wt) ca. 25-fold greater than that of the other ten cultivars, and so the trait is polymorphic in the species. In boiled leaf extracts of Carignan and Ruby Cabernet, free cyanide constituted a negligible fraction of the total cyanide potential because β-glucosidase treatment was required to liberate the major cyanide fraction – which is therefore bound in glucosylated cyanogenic compound(s) (or cyanogenic glucosides). In addition, cyanide was liberated from ground leaf tissue of Ruby Cabernet but not Sultana (a cultivar with low cyanide potential). Hence, the high cyanide potential in Ruby Cabernet leaves is coupled with endogenous β-glucosidase(s) activity and this cultivar may be considered 'cyanogenic'. A method was developed by Franks et al. (2005) to detect and identify cyanogenic glucosides using liquid chromatography combined with tandem mass spectrometry (LC–MS/MS). Two putative cyanogenic glucosides were found in extracts from leaves of Carignan and Ruby Cabernet and were identified as the epimers prunasin and sambunigrin. Cyanide potential measured at three times over the growing season in young and mature leaves, petioles, tendrils, flowers, berries, seeds, and roots of Ruby Cabernet was substantially higher in the leaves compared with all other tissues. This characterization of cyanogenic glucoside accumulation in grapevines provides a basis for gauging the involvement of the trait in interactions of the species with its pests and pathogens.

Popular liqueurs made from apricot/cherry pits were evaluated in terms of their phenolic composition and occurrence of cyanogenic glycosides (CGG) (Senica et al. 2016). Analyses consisted of detailed phenolic and cyanogenic profiles of cherry and apricot seeds as well as beverages prepared from crushed kernels. Phenolic groups and cyanogenic glycosides were analyzed with the aid of high-performance liquid chromatography (HPLC) and mass spectrophotometry (MS). Lower levels of cyanogenic glycosides and phenolics have been quantified in liqueurs compared to fruit kernels. During fruit pit steeping in the alcohol, the phenolics/cyanogenic glycosides ratio increased and at the end of beverage manufacturing process higher levels of total analyzed phenolics were detected compared to cyanogenic glycosides (apricot liqueur: 38.79 µg CGG per mL and 50.57 µg phenolics per mL; cherry liqueur 16.08 µg CGG per mL and 27.73 µg phenolics per mL). Although higher levels of phenolics are characteristic for liqueurs made from apricot and cherry pits, these beverages nevertheless contain considerable amounts of cyanogenic glycosides.

The retained glycoalkaloids (GAs) were eluted directly onto an analytical Xterra MS C18 column (50 mm × 2 mm inner diameter × 3 µm) from Waters (Manchester, UK) for 2 minutes with a flow rate of 0.5 mL min^{-1} and a binary liquid chromatography (LC) gradient, consisting of 0.1% formic acid (solvent A) and 0.1% formic acid in methanol (solvent B) and the following profile: 0.0 minute, 10% solvent B; 0.1–2.05 minutes, 10–90% solvent B; 2.05–2.5 minutes 90% solvent B; 2.5–3.0 minutes 90–10% solvent B; 3.01–5.0 minutes 10% solvent B Distl, M. et al. 2009). MS data were obtained using a API 3000 mass spectrometer from MDS Sciex (Concord, Canada), equipped with a turbo ion spray interface for ionization. MS detection of positively charged ions was performed in the positive mode using multiple reaction monitoring (MRM). MRM traces were set at m/z 868.3 and 98.2 for α-solanine and at m/z 852.4 and 98.2 for α-chaconine. The run time was 5 minutes and the instrument was set to the following tuning parameters: nitrogen was used as the nebulizer gas (15 l h–1) and the curtain gas (10 l h–1), argon was the collision gas with a collision cell potential of 6 V, the ion spray voltage was set to 5 kV and the temperature of the heated transfer capillary was maintained at 400°C. The complete system was operated and data processing was carried out using Analyst™ 1.4.1 from MDS Sciex (Concord, Canada).

α-solanine and α-chaconine were analyzed by LC/MS in four potato cultivars (Chuda, Y. et al. 2007). The LC/MS system used consisted of an alliance 2690 (Waters) with a 996-diode array detector (Waters) and a ZMD mass spectrometer (Micromass, Manchester, UK). The system was controlled by MassLynx software, which runs on Windows NT (Microsoft, Redmond, WA). The analytical column was a 2.1 mm id × 150 mm SymmetryShieldTM 3.5 um RP18 (Waters) with a guard column of a 2.1 mm id × 10 mm SymmetryShieldTM 3.5 um RP18. The volume injected was 5 as follows: A, water containing 0.2% formic acid, and B, MeCN containing 0.2% formic acid. Elution was performed with a linear gradient of 20–25% solvent B in solvent A for 12 minutes at a flow rate of 0.2 mL/min. MS analysis was carried out by electrospray ionization (ESI) in the positive ion mode. The source block temperature was 150°C, and the desolvation temperature was 300°C. The capillary voltage was 3.3 kV, and the cone voltage was 50 V. Nitrogen was used as the desolvation and nebulizer gas.

The LC–MS system for the quantification of steroidal glycalkaloids (SGAs) consisted of a Waters 2767 Sample Manager and a 2525 Binary Gradient Pump coupled to a single quadrupole ZQ mass spectrometer (Micromass 4000), operating in the ESI and APCI modes (Paul, A. T. et al. 2008). The MS experiment setup and data acquisition were conducted using the MassLynx software V 4.0. The optimum values of MS analyses were as follows: ESI positive ion mode; capillary voltage, 3.5 kV; cone voltage, 40 V; dissolvation temperature, 350°C; source temperature, 100°C; extractor, 2 V; RF lens, 0.2 V; nebulizing gas, 500 L/h; cone gas 70 L/h. High-purity nitrogen was used as a nebulizer and cone gas. MS analyses were performed in a mass range of 100–1500 m/z.

The analytical method was developed for separation of SGAs using the XTerraMS C-18 column (5 m, 250 mm × 4.6 mm) and mobile phase consisting of 0.5% (v/v) formic acid in H_2O (A) and ACN:2-propanol:formic acid (94.5:5:0.5, v/v/v) (B). The gradient was: 0 minute, 20% B; 14 minutes, 30% B; 20 minutes, 30% B; 27 minutes, 60% B and the column was re-equilibrated to initial condition (20% B) for 10 minutes prior to next injection. The flow rate was 1 mL/min. The mass detector received a flow of 150 L/min from the splitter. The same analytical method was scaled up for preparative isolation of SGAs. The preparative separations were carried out by LC–MS using the XTerra MS C-18 column (10 m, 250 mm × 10 mm) and same mobile phase of analytical method. The separations were carried out using gradient elution. The flow rate was 5 mL/min.

There is a lot of literature on LC/MS analyses of pyrrolizidine alkaloids. Only a limited number are discussed in the next paragraphs.

A Waters Alliance 2965 liquid chromatographic system coupled to a Waters Micromass Quattro Micro triple-quadrupole mass spectrometer (Milford, MA) was used for the analysis of pyrrolizidine alkaloids in North American plants and honey. Chromatographic separation was achieved on a Phenomenex Synergi hydro-RP C18 2.5 μm, 100 Å, 100 × 30 mm column. The mobile phases were A: 0.1% formic acid in water and B: 0.1% formic acid in acetonitrile. The run time was 21 minutes with flow rate of 0.4 mL/min and a gradient was as follows: 0–8 minutes: 10–20% B; 8–12 minutes: 20–25% B, 12–15 minutes: 25–30% B, 15–16 minutes: 30–70% B, 16–17 minutes: 70–10% B, 17–21 minutes: 10% B. The column temperature was 25°C and the injection volume was 5 μL. The mass spectrometer was equipped with electrospray ionisation operating in positive mode. Nitrogen gas was used as the nebulizing gas. The capillary voltage was set to 3.2 kV with cone, extractor and RF voltages of 20, 3.00 and 0.3 V, respectively. The source temperature and desolvation temperature were 120°C and 500°C. The cone and desolvation gas flows were 50 and 500 L/hour. The parameters for the mass analyzers were: LM1, LM2, HM1, and HM2 resolution of 12.5, 15.0, 12.5, and 15.0, respectively. The ion energy 1 and 2 were 0.3 and 0.5 and the entrance and exit potentials were both 50 with a collision energy of 0. The mass acquisition mode used was selected ion monitoring (SIM) of the molecular ions for the pyrrolizidine alkaloids found in the four plants evaluated. Ions were selected from a full scan analysis of the four PA-containing plants, known PAs and PANOs in the four PA-containing plants and evaluation of the mass spectra.

Strong-cation-exchange and solid-phase extraction of pyrrolizidine alkaloids and their N-oxides from honey samples was followed by reduction of the N-oxides and subsequent analysis of total pyrrolizidine alkaloids using high-performance liquid chromatography–atmospheric pressure chemical ionization mass spectrometry (Beales, K. A. 2004). A limited survey of 63 preprocessing samples of honey, purposefully biased toward honeys attributed to floral sources known to produce pyrrolizidine alkaloids, demonstrated levels of pyrrolizidine alkaloids up to approximately 2,000 parts per billion (ppb) in a sample attributed to *Echium plantagineum*. Up to 800 ppb pyrrolizidine alkaloids were detected in some honeys not attributed by the collector to any pyrrolizidine alkaloid-producing floral source. No pyrrolizidine alkaloids were detected in approximately 30% of the samples in this limited study, while some honeys showed the copresence of pyrrolizidine alkaloids from multiple floral sources such as *E. plantagineum* and *Heliotropium europaeum*. In addition, retail samples of blended honeys (with no labeling to suggest that pyrrolizidine alkaloid-producing floral sources were used in the blends) have been shown to contain up to approximately 250 ppb pyrrolizidine alkaloids.

A method for analyzing honey samples was developed by Betteridge et al. (2005), which enabled the simultaneous detection and identification of pyrrolizidine alkaloids and their N-oxides. Honey samples were treated with methanol or dilute sulfuric acid and then centrifuged to remove insoluble material. Subsequent strong cation exchange, solid-phase extraction of the supernatant provided a fraction that was analyzed for the presence of pyrrolizidine alkaloids and their N-oxides using high-pressure liquid chromatography coupled to electrospray ionization mass spectrometry. The procedure was validated using extracts of *Echium plantagineum* and authenticated standards of pyrrolizidine alkaloids and their N-oxides from other plant sources. Of several variations of the solid-phase extraction method assessed in this study, the best combination for generic use involved the dilution of honey with 0.05 M sulfuric acid and the subsequent application of the centrifuged solution to solid-phase extraction columns at the rate of a maximum of 10 g of honey per solid-phase extraction column.

A liquid chromatography coupled to mass spectrometry (LC-MS/MS) method was developed according to the Decision 2002/657/EC, to evaluate the contamination of Veneto honey by the following pyrrolizidine alkaloids: echimidine, heliotrine, lycopsamine, retrorsine, senecionine, and seneciphylline, as well as their N-oxides (pyrrolizidine alkaloid N-oxides (PANOs)) (Lorena, L. et al. 2016). Before the solid-phase extraction, a reduction step with zinc dust in the presence of sulfuric acid was carried out for the reduction of PANOs to PAs in honey samples. For all of the analytes, the mean recovery ranged between 82.70 and 104.16% and the limit of quantification (LOQ) was 0.25 μg kg^{-1}.

Analysis of pyrrolizidine alkaloids in North American plants and honey was performed on a Thermo Fisher Scientific Accela LC system coupled to a Quantum Discovery Max triple-quadrupole mass spectrometer fitted with a heated electrospray ionisation

(HESI) source Mudge, E. M. et al. 2015). Instrument control and data acquisition were performed with Xcalibur 2.0.7 software (Thermo Fisher Scientific, Hemel Hempstead, UK). The analytical column used was a Kinetex PFP core-shell column (150 × 2.1 mm, 2.6 µm particle size) protected with a KrudKatcherTM in-line filter (all Phenomenex Inc.). The mobile phase consisted of (A) 95:5 v/v water/acetonitrile with 0.05% formic acid and (B) 100% acetonitrile. An isocratic elution was performed 85% (A) and 15% (B) at a flow rate of 0.2 mL min^{-1} for 6 minutes (the flow rate was optimised for efficient electrospray ionisation). The column temperature was maintained at 35°C and the samples were cooled to 4°C in the autosampler. Methanol was used as a washing solution in the autosampler and the injection was performed in a partial-loop mode. The injection volume was 10 µL. The detection of the analytes was carried out using positive electrospray ionisation (ESI) in SRM mode for all PAs and PANOs using the 'EZ' method option. The dwell time was 0.3 s with a scan width in Q1 of 0.7 u full width at half maximum (FWHM). The heated ESI had a spray voltage of 3,500 V and a vaporizing temperature of 90°C. The ion transfer tube temperature was 250°C. Nitrogen gas was used as a sheath gas and auxiliary gas at pressures of 15 and 25 arbitrary TSQ Quantum units, respectively. Argon was used as the collision gas at 1.5 mTorr.

The instrument was a Nexera X2 system coupled with a triple-quadrupole mass spectrometer LCMS-8060 (Shimadzu, Duisburg, Germany) via an APCI interface set in positive ionization mode. The chromatographic system was equipped with two LC-30AD pumps, a SIL-30AC autosampler, a DGU-20A5R degassing unit, and a CTO-20AC oven. The separation was achieved by using an Ascentis Express C18 column (50 × 4.6 mm, 2.7 µm) provided by Merck. The mobile phase was A) water/methanol/THF (85:10:5, v-v:v) and B) methanol/THF (95:5, v-v). The chromatographic run was carried out at a flow rate of 2 mL min^{-1} and a temperature of 40°C in gradient mode according to the following program: 0–4.5 minutes, 15–28% B; 4.5–7.0 minutes, 28–60% B; 7.0–11.0 minutes, 60–85% B, hold for 3 minutes. The injection volume was 2 µL. The MS system operated in both full scan and MRM acquisition mode to ensure both untargeted and targeted analyses, in full scan and MRM mode, respectively. MS parameters were as follows: interface temperature was set at 450°C; desolvation line (DL) and heat block temperatures were both 300°C; nebulizing and drying gas flow were 3 and 15 L/min, respectively; the pressure of the CID gas was 270 kPa. The mass spectral range was 150–450 m/z for the untargeted analysis, while the targeted analysis was carried out in MRM mode through a synchronized method, which set specific acquisition windows according to the retention time of each target, and by applying a dwell time of 20.0 ms. In this way it was possible to obtain ten scan per peak, as required for a correct quantification. Instead, the homologous series was analysed in single ion monitoring positive mode (SIM+). The m/z monitored for the alkyl aryl ketones was: m/z 121, m/z 135, m/z 149, m/z 163, m/z 177, and m/z 191 for acetophenone, propiophenone, butyrophenone, valerophenone, hexanophenone, and heptanophenone, respectively. Target analytes (Cs, FCs, and PMFs) were detected in MRM acquisition mode. Taking into account the sensitivity of the MRM detection, the carryover phenomenon was considered by injecting a reagent blank, represented by 2 µL of pure ethanol between two consecutive sample analyses, in order to evaluate the efficacy of the washing gradient step. 2.3. Qualitative and quantitative analysis peak identification were carried out by complement.

A method for LC-MS/MS profiling of coumarins in Zantoxylum zanthoxyloides extracts and essential oils was worked out by Tine, Y. et al. (2017).

MS^2 conditions were carried on an AB Sciex (Toronto, ON, Canada) 3200 QTRAP linear triple-quadrupole fitted with an atmospheric pressure chemical ionization (APCI) ion source operating in positive mode. High-purity nitrogen was used as both a nebulizer and turbo gas. The APCI source was operated with following settings in positive mode; curtain gas (CUR) 25 psi, nebulizer gas (GS1) 31 psi, heater gas (GS2) 65 psi, ion spray voltage (IS) 5000 V, and temperature 450°C. Standard solutions (component concentration: 0.1 mg/L) were directly infused at the flow rate of 10 µL/min in the MS/MS apparatus. Multiple EPI mass spectra of each compound were recorded in the range of m/z = 50–500 at 4,000 Da/s. IDA properties were set to select 1 to 2 peaks above 300 counts with an exclusion filter after five occurrences for 30 seconds with dynamic background subtraction. The software used for data acquisition and data analysis was Analyst 1.5.2 (AB Sciex, Framingham, MA, USA).

10.5 LC CONDITIONS

The LC system consisted of a Flexar LC Perkin-Elmer (Waltham, MA, USA) with two Flexar FX-10 LC pumps, a Flexar solvent manager, a 275-Flexar autosampler, and a Flexar LC PE200 column oven. LC analyses were performed on a 100 mm × 2.1 mm i.d. 3 µm, LUNA 3U C18 column (Phenomenex, Torrance, CA, USA) and the column temperature was set at 25°C. A volume of 10 µL of sample was injected using an injection loop of 15 µL in partial loop mode. The mobile phase consisted of MilliQ water (solvent A) and ACN (solvent B). The flow rate was set at 500 µL/min. The column was equilibrated (A:B; v/v) in 90:10 (5 minutes), and elution was carried out with the following steps; 90:10 (5 minutes), 80:20 (5 minutes), 70:30 (5 minutes), 60:40 (5 minutes), 50:50 (5 minutes), a linear gradient increasing from 50% B to 100% (5 minutes), and 100% B (7 minutes).

Dietary intake of coumarins and furocoumarins through citrus beverages was estimated by a HPLC-MS/MS (Arigò, A. et al. 2021). The instrument was a Nexera X2 system coupled with a triple-quadrupole mass spectrometer LCMS-8060 (Shimadzu, Duisburg, Germany) via an APCI interface set in positive ionization mode. The chromatographic system was equipped with two LC-30AD pumps, a SIL-30AC autosampler, a DGU-20A5R degassing unit and a CTO-20AC oven. The separation was achieved by using an Ascentis Express C18 column (50 × 4.6 mm, 2.7 µm) provided by Merck. The mobile phase was A) water/methanol/THF (85:10:5, v-v:v) and B) methanol/THF (95:5, v-v). The chromatographic run was carried out at a flow rate of 2 mL min^{-1} and a temperature of 40°C in gradient mode according to the following program: 0–4.5 minutes, 15–28% B; 4.5–7.0 minutes, 28–60% B; 7.0–11.0 minutes, 60–85% B, hold for 3 minutes. The injection volume was 2 µL. The MS system operated in both full scan and MRM acquisition mode to ensure both untargeted and targeted analyses, in full scan and MRM mode, respectively. MS parameters were as follows: interface temperature was set at 450°C; desolvation line (DL) and heat block temperatures were both 300°C; nebulizing and drying gas flow were 3 and 15 L/min, respectively; the pressure of the CID gas was 270 kPa. The mass spectral range was 150–450 m/z for the untargeted analysis, while the targeted analysis was carried out in MRM mode through a synchronized method, which set specific acquisition windows according to the retention time of each target, and by applying a dwell time of 20.0 ms. In this way it was possible to obtain ten scan per peak, as required for a correct quantification. Instead, the homologous series was analyzed in single ion monitoring positive mode (SIM+). The m/z monitored for the alkyl aryl ketones was: m/z 121, m/z 135, m/z 149, m/z 163, m/z 177, and m/z 191 for acetophenone, propiophenone, butyrophenone, valerophenone, hexanophenone, and heptanophenone, respectively.

The impact of secondary metabolites on fruit quality, plant growth and human health has led to an increased demand for analytical methods to characterize and quantify these metabolites in recent years. A versatile, sensitive, and rapid method based on UHPLC-QqQ-MS/MS was developed for simultaneous qualitation and quantitation of coumarins, furocoumarins, flavonoids, and phenolic acids (Zhao, X. J. et al. 2020). The chromatographic elution and multiple reaction monitoring mode transitions were optimized to achieve good separation and accurate quantitation of 47 analytes, including 13 groups of isomers, during a single 13 minute chromatographic run. This method was validated with good precision and recoveries, wide linear ranges, and low limits of detection and quantitation ($0.014–1.50$ µg L^{-1}). The validated method was further applied to quantify the analytes in flavedo, albedo, and pulp from two pummelo varieties, *C. grandis* 'Shatianyu' and *C. grandis* 'Guanximiyu'. This method combines high sensitivity, good selectivity, and short chromatographic run time.

10.6 CONCLUSION

Plants produce massive amounts of bioactive compounds, some of which can induce adverse effects on health. Analysis of the food toxins of plant origin is observed to be crucial to ensure the safety and standard of foods. A wide range of endogenous toxins in some food plants is generally characterized by LC-MS mediated analysis. Toxins present in the aqueous phase are compounds with medium to high polarity. Therefore, LC is a more convenient approach, offering separation without derivatization. LC coupled with detectors such as UV/VIS has been well studied. A sensitive analytical method based on LC-MS enables comprehensive identification and characterization of toxins. The use of smaller column particles (sub 2 mm) helps to improve the speed, sensitivity, and resolution. Combined liquid chromatographic and mass spectrometric techniques have been contributing in a decisive way to the progress of food toxins of plant origin analysis, particularly in complex biological matrices. A validated method for the simultaneous detection and analysis of a wide range of plant toxins in a variety of food and botanicals was discussed.

REFERENCES

Al-Taher, F., Cappozzo, J., Zweigenbaum, J., Lee, H. J., Jackson, L. and Ryu, D. (2017). Detection and quantitation of mycotoxins in infant cereals in the US market by LC-MS/MS using a stable isotope dilution assay. *Food Control*, 72, 27–35.

Andrzejewski, D., Roach, J. A., Gay, M. L. and Musser, S. M. (2004). Analysis of coffee for the presence of acrylamide by LC-MS/MS. *Journal of Agricultural and Food Chemistry*, 52(7), 1996–2002.

Anna, S. P. T. (2007). An overview of natural toxins in food. *Food Safety Focus, Food and Environmental Hygiene Department, Centre for Food Safety. Government of Hong Kong*, 11, 3–4. https://www.cfs.gov.hk/english/multimedia/multimedia_pub/multimedia_pub_fsf_11_02.html

Appenteng, M. K., Krueger, R., Johnson, M. C., Ingold, H., Bell, R., Thomas, A. L. and Greenlief, C. M. (2021). Cyanogenic glycoside analysis in american elderberry. *Molecules*, 26(5), 1384. doi:10.3390/molecules26051384

Ardrey, R. E. (2003). *Liquid Chromatography-mass Spectrometry: An Introduction*. Chichester, England: John Wiley & Sons Ltd.

Arigò, A., Rigano, F., Russo, M., Trovato, E., Dugo, P. and Mondello, L. (2021). Dietary Intake of Coumarins and Furocoumarins through Citrus Beverages: A Detailed Estimation by a HPLC-MS/MS Method Combined with the Linear Retention Index System. *Foods*, 10(7), 1533. doi:10.3390/foods10071533

Beales, K. A., Betteridge, K., Colegate, S. M. and Edgar, J. A. (2004). Solid-phase extraction and LC–MS analysis of pyrrolizidine alkaloids in honeys. *Journal of Agricultural and Food Chemistry*, 52(21), 6664–6672. doi:10.1021/jf049102p

Bender, A. E. and Reaidi, G. B. (1982). Toxicity of kidney beans (*Phaseolus vulgaris*) with particular reference to lectins. *Journal of Plant Foods*, 4(1), 15–22. doi:10.1080/0142968X.1982.11904243

Betteridge, K., Cao, Y. and Colegate, S. M. (2005). Improved method for extraction and LC-MS analysis of pyrrolizidine alkaloids and their N-oxides in honey: application to Echium vulgare honeys. *Journal of Agricultural and Food Chemistry*, 53(6), 1894–1902. doi:10.1021/jf0480952

Bing, Z., Zeyuan, D., Yao, T. and Rong, T. (2014). Toxins in foods of plant origin. In Yu, L., Wang, S. and Sun, B.-G. (eds.). *Food Safety Chemistry: Toxicant Occurrence, Analysis and Mitigation*. Boca Raton: CRC Press, pp. 305–320. doi:10.1201/b17712-16

Blake, O. A., Bennink, M. R. and Jackson, J. C. (2006). Ackee (*Blighia sapida*) hypoglycin A toxicity: Dose response assessment in laboratory rats. *Food Chem. Toxicol.*, 44(2), 207–213. doi:10.1016/j.fct.2005.07.002

Cervino, C., Asam, S., Knopp, D., Rychlik, M. and Niessner, R. (2008). Use of isotope-labelled aflatoxins for LC-MS/MS stable isotope dilution analysis of foods. *Journal of Agricultural and Food Chemistry*, 56(6), 1873–1879.

Choi, S. H., Park, S., Lim, Y. P., et al. (2014). Metabolite profiles of glucosinolates in cabbage varieties (*Brassica oleracea* var. *capitata*) by season, color, and tissue position. *Hortic. Environ. Biotechnol.*, 55, 237–247. doi:10.1007/s13580-014-0009-6

Chuda, Y., Tsuda, S., Ohara-Takada, A., Kobayashi, A., Suzuki, K., Ono, H., Yoshida M., Nagata T., Kobayashi S. and Mori, M. (2007). Quantification of light-induced glycoalkaloids, α-solanine and α-chaconine, in four potato cultivars (Solanum tuberosum L.) distributed in Japan by LC/MS. *Food Science and Technology Research*, 10(3), 341–345. doi:10.3136/fstr.10.341

De Boevre, M., Di Mavungu, J. D., Maene, P., et al. (2012). Development and validation of an LC-MS/MS method for the simultaneous determination of deoxynivalenol, zearalenone, T-2-toxin and some masked metabolites in different cereals and cereal-derived food. *Food Additives and Contaminants: Part A*, 29(5), 819–835.

Di Mavungu, J. D., Monbaliu, S., Scippo, M. L., et al. (2009). LC-MS/MS multi-analyte method for mycotoxin determination in food supplements. *Food Additives and Contaminants*, 26(6), 885–895.

Distl, M., Sibum, M. and Wink, M. (2009). Combination of on-line solid-phase extraction with LC-MS for the determination of potentially hazardous glycoalkaloids in potato products. *Potato Research*, 52(1), 39–56. doi:10.1007/s11540-008-9106-1

Dubey, N. K., Dwivedy, A. K., Chaundhari, A. K. and Das, S. (2018). Common toxic plants and their forensic significance. In Mandal, S., Mandal, V. and Konishi, T. (eds.). *Natural Products and Drug Discovery: An Integrated Approach*. Amsterdam, Netherlands: Elsevier, pp. 349–374.

Dugrand-Judek, A., Olry, A., Hehn A., et al. (2015). The distribution of coumarins and furanocoumarins in *Citrus* species closely matches *Citrus* phylogeny and reflects the organization of biosynthetic pathways. *PLoS One*, 10(11), e0142757. doi:10.1371/journal.pone.0142757.

Franks, T. K., Hayasaka, Y., Choimes, S. and Van Heeswijck, R. (2005). Cyanogenic glucosides in grapevine: polymorphism, identification and developmental patterns. *Phytochemistry*, 66(2), 165–173. doi:10.1016/j.phytochem.2004.11.017

Gaudin, M., Imbert, L., Libong, D., Chaminade, P., Brunelle, A., Touboul, D. and Laprévote, O. (2012). Atmospheric pressure photoionization as a powerful tool for large-scale lipidomic studies. *Journal of the American Society for Mass Spectrometry*, 23(5), 869–879.

Gry, J., Black, L., Eriksen, F., et al. (2007). EuroFIR-BASIS - a combined composition and biological activity database for bioactive compounds in plant-based foods. *Trends in Food Science and Technology*, 18, 434–444. doi:10.1016/j.tifs.2007.05.008.

Hajšlová, J., Schulzová, V., Botek, P. and Lojza, J. (2018). Natural toxins in food crops and their changes during processing. *Czech Journal of Food Sciences*, 22, 29–34. doi:10.17221/10606-CJFS.

Holstege, D. M., Puschner, B. and Le, T. (2001). Determination of grayanotoxins in biological samples by LC-MS/MS. *Journal of Agricultural and Food Chemistry*, 49(3), 1648–1651. doi:10.1021/jf000750s.

Jain, C., Khatana, S. and Vijayvergia, R. (2019). Bioactivity of secondary metabolites of various plants: A review. *Int. J. Pharm. Sci. Res.*, 10(2), 494–504.

Kaplan, M., Olgun, E. O. and Karaoglu, O. (2014). Determination of grayanotoxins in honey by liquid chromatography tandem mass spectrometry using dilute-and-shoot sample preparation approach. *Journal of Agricultural and Food Chemistry*, 62(24), 5485–5491. doi:10.1021/jf501560t.

Krska, R., Schubert-Ullrich, P., Molinelli, A., Sulyok, M., MacDonald, S. and Crews, C. (2008). Mycotoxin analysis: An update. *Food Additives and Contaminants*, 25(2), 152–163.

Kuiper-Goodman, T. and Nawrot, P. S. (1992). Solanine and chaconine. In *WHO Food Additives Series 30*. Geneva: World Health Organization. http://www.inchem.org/documents/jecfa/jecmono/v30je19.htm

Laurie, C. D., Ray, A. M. and George, A. B. (2010). Naturally occurring food toxins. *Toxins*, 2, 2289–2332. doi:10.3390/toxins2092289.

Lorena, L., Roberta, M., Alessandra, R., Clara, M. and Francesca, C. (2016). Evaluation of some pyrrolizidine alkaloids in honey samples from the Veneto region (Italy) by LC-MS/MS. *Food Analytical Methods*, 9(6), 1825–1836. doi:10.1007/s12161-015-0364-7

Malachová, A., Stránská, M., Václavíková, M., et al. (2018). Advanced LC–MS-based methods to study the co-occurrence and metabolization of multiple mycotoxins in cereals and cereal-based food. *Analytical and Bioanalytical Chemistry*, 410(3), 801–825.

Malik, A. K., Blasco, C. and Picó, Y. (2010). Liquid chromatography-mass spectrometry in food safety. *J. Chromatogr. A.*, 1217(25), 4018–4040. doi:10.1016/j.chroma.2010.03.015.

Masiá, A., Suarez-Varela, M. M., Llopis-Gonzalez, A. and Picó, Y. (2016). Determination of pesticides and veterinary drug residues in food by liquid chromatography-mass spectrometry: A review. *Analytica Chimica Acta*, 936, 40–61.

Mudge, E. M., Jones, A. M. P. and Brown, P. N. (2015). Quantification of pyrrolizidine alkaloids in North American plants and honey by LC-MS: single laboratory validation. *Food Additives & Contaminants: Part A*, 32(12), 2068–2074. doi:10.1080/19440049.2015.1099743

Núñez, O., Moyano, E. and Galceran, M. T. (2005). LC–MS/MS analysis of organic toxics in food. *TrAC Trends in Analytical Chemistry*, 24(7), 683–703.

Ogori, A. F. (2019). Plant toxins. *Ame. J. Biomed. Sci. Res*, 4(3), 173–175. doi:10.34297/AJBSR.2019.04.000793.

Oluwaniyi, O. and Oladipo, J. (2017). Comparative studies on the phytochemicals, nutrients and antinutrients content of cassava varieties. *Journal of the Turkish Chemical Society, Section A: Chemistry*, 4, 661. doi:10.18596/jotcsa.306496.

Paul, A. T., Vir, S. and Bhutani, K. K. (2008). Liquid chromatography–mass spectrometry-based quantification of steroidal glycoalkaloids from Solanum xanthocarpum and effect of different extraction methods on their content. *Journal of Chromatography A*, 1208(1–2), 141–146. doi:10.1016/j.chroma.2008.08.089

Peltomaa, R., Benito-Peña, E. and Moreno-Bondi, M. C. (2018). Bioinspired recognition elements for mycotoxin sensors. *Anal. Bioanal. Chem.*, 410, 747–771. doi:10.1007/s00216-017-0701-3.

Picardo, M., Filatova, D., Nuñez, O. and Farré, M. (2018). Recent advances in the detection of natural toxins in freshwater environments. *TrAC Trends in Analytical Chemistry*, 112, 75–86. doi:10.1016/j.trac.2018.12.017.

Puntscher, H., Kütt, M. L., Skrinjar, P., et al. (2018). Tracking emerging mycotoxins in food: Development of an LC-MS/MS method for free and modified *Alternaria* toxins. *Analytical and Bioanalytical Chemistry*, 410(18), 4481–4494.

Puschner, B., Holstege, D. M., Lamberski, N. and Le, T. (2001). Grayanotoxin poisoning in three goats. *Journal of the American Veterinary Medical Association*, 218(4), 573–575. doi:10.2460/javma.2001.218.573.

Sánchez-Mata, M. C., Yokoyama, W. E., Hong, Y. J. and Prohens, J. (2010). α-solasonine and α-solamargine contents of gboma (*Solanum macrocarpon* L.) and scarlet (*Solanum aethiopicum* L.) eggplants. *Journal of Agricultural and Food Chemistry*, 58(9), 5502–5508. doi:10.1021/jf100709g.

Sarangthem, K. and Hoikhokim. (2010). Cyanogen content in bamboo plants. *Asian Journal of Biological Sciences*, 5, 178–180.

Senica, M., Stampar, F., Veberic, R. and Mikulic-Petkovsek, M. (2016). Transition of phenolics and cyanogenic glycosides from apricot and cherry fruit kernels into liqueur. *Food Chemistry*, 203, 483–490. doi:10.1016/j.foodchem.2016.02.110

Spanjer, M. C., Rensen, P. M. and Scholten, J. M. (2008). LC–MS/MS multi-method for mycotoxins after single extraction, with validation data for peanut, pistachio, wheat, maize, cornflakes, raisins and figs. *Food Additives and Contaminants*, 25(4), 472–489.

Sproll, C., Perz, R. C. and Lachenmeier, D. W. (2006). Optimized LC/MS/MS analysis of morphine and codeine in poppy seed and evaluation of their fate during food processing as a basis for risk analysis. *Journal of Agricultural and Food Chemistry*, 54(15), 5292–5298. doi:10.1021/jf0608975.

Sulyok, M., Stadler, D., Steiner, D. and Krska, R. (2020). Validation of an LC-MS/MS-based dilute-and-shoot approach for the quantification of >500 mycotoxins and other secondary metabolites in food crops: Challenges and solutions. *Analytical and Bioanalytical Chemistry*, 412(11), 2607–2620.

Tang, Y. Y., Lin, H. Y., Chen, Y. C., et al. (2013). Development of a quantitative multi-mycotoxin method in rice, maize, wheat and peanut using UPLC-MS/MS. *Food Analytical Methods*, 6(3), 727–736.

Teoh, E. S. (2016). Secondary metabolites of plants. In Soon, T. E. (ed.) *Medicinal Orchids of Asia*. Cham: Springer, pp. 59–73. doi:10.1007/978-3-319-24274-3.

Tine, Y., Renucci, F., Costa, J., Wélé, A. and Paolini, J. (2017). A method for LC-MS/MS profiling of coumarins in Zanthoxylum zanthoxyloides (Lam.) B. Zepernich and Timler extracts and essential oils. *Molecules*, 22(1), 174. doi:10.3390/molecules22010174 van Egmond, H. P. (2004). Natural toxins: Risks, regulations and the analytical situation in Europe. *Anal. Bioanal. Chem.*, 378, 1152–1160. doi:10.1007/s00216-003-2373-4.

Vyncht, G. V., Jànosi, A., Bordin, G., Toussaint, B., Maghuin-Rogister, G., De Pauw, E. and Rodriguez, A. R. (2002). Multiresidue determination of (fluoro) quinolone antibiotics in swine kidney using liquid chromatography–tandem mass spectrometry. *Journal of Chromatography A*, 952(1–2), 121–129.

West, S. D., Hastings, M. J., Shackelford, D. D. and Dial, G. E. (2004). Determination of oryzalin in water, citrus fruits, and stone fruits by liquid chromatography with tandem mass spectrometry. *Journal of Agricultural and Food Chemistry*, 52(19), 5781–5786.

WHO. (2018). *Natural Toxins in Food*. Geneva: World Health Organization. https://www.who.int/news-room/fact-sheets/detail/natural-toxins-in-food#

Xu, R. N., Fan, L., Rieser, M. J. and El-Shourbagy, T. A. (2007). Recent advances in high-throughput quantitative bioanalysis by LC–MS/MS. *Journal of Pharmaceutical and Biomedical Analysis*, 44(2), 342–355.

Zhao, X. J., Guo, P. M., Pang, W. H., Zhang, Y. H., Zhao, Q. Y., Jiao, B. N. and Kilmartin, P. A. (2020). A rapid UHPLC-QqQ-MS/MS method for the simultaneous qualitation and quantitation of coumarins, furocoumarins, flavonoids, phenolic acids in pummelo fruits. *Food Chemistry*, 325, 126835. doi:10.1016/j.foodchem. 2020.126835

Zhou, S., Song, Q., Tang, Y. and Naidong, W. (2005). Critical review of development, validation, and transfer for high throughput bioanalytical LC-MS/MS methods. *Current Pharmaceutical Analysis*, 1(1), 3–14.

Zhu, X., Wang, Y. K. and Yang, X. N. (2019). Metabolic activation of myristicin and its role in cellular toxicity. *J. Agric. Food Chem.*, 67(15), 4328–4336. doi:10.1021/acs.jafc.9b00893.

CHAPTER 11

Quantitative Determination of Food Toxins of Plant Origin by GC-MS

Devarajan Thangadurai[1], D. Divya[2], Poojashree Nagappa Kunnur[1], Saher Islam[3], P. Lokeshkumar[4], Jeyabalan Sangeetha[5], Pavitra Chippalakatti[1], Ravichandra Hospet[6], Muniswamy David[4], Zaira Zaman Chowdhury[7], Vishal Ahuja[8], and Vedavyas Shivanand Chavan[1]

[1]Department of Botany, Karnatak University, Dharwad, Karnataka, India
[2]Pinnacle Biosciences, Kanyakumari, Tamil Nadu, India
[3]Department of Biotechnology, Lahore College for Women University, Lahore, Pakistan
[4]Department of Zoology, Karnatak University, Dharwad, Karnataka, India
[5]Department of Environmental Science, Central University of Kerala, Kasaragod, Kerala, India
[6]Department of Food Protectants and Infestation Control, CSIR-Central Food Technological Research Institute, Mysore, Karnataka, India
[7]Nanotechnology and Catalysis Research Center (NANOCAT), Institute of Advanced Studies (IAS), University of Malaya, Kuala Lumpur, Malaysia
[8]Department of Biotechnology, Himachal Pradesh University, Shimla, Himachal Pradesh, India

CONTENTS

11.1	Introduction	180
11.2	Fast GC-MS	180
11.3	GC-TOF-MS	181
11.4	Low-Pressure (LP) GC-MS	182
	11.4.1 Working Principle and Instrumentation	183
	11.4.2 Significances of LPGC-MS	184
	11.4.3 Applications of LPGC-MS	184
11.5	GC/Supersonic Molecular Beam (SMB)-MS	184
11.6	GC-MS of Food Toxins of Plant Origin	188
11.7	Conclusion	188
References		189

DOI: 10.1201/9781003222194-14

11.1 INTRODUCTION

Plants have evolved various toxins to combat with environmental conditions. Several cases of food poisoning with plants species such as *Scopolia* and *Datura* have been narrated when humans get exposed to such toxins by consumption of dietary supplements, herbal teas, herbal medicine, or food containing phytotoxins (Mikolich et al. 1975; Urich et al. 1982; Hayman 1985; Guharoy and Barajas 1991; Nogue et al. 1995; Chang et al. 1999; Thabet et al. 1999). *Scopolia* and *Datura* plants release tropane alkaloids (phytotoxin) in leaves, stems, flowers, and seeds. Another phytotoxin, pyrrolizidine alkaloids are produced by plant families including Asteraceae, Boraginaceae, Apocynaceae and Fabaceae (Hartmann and Witte 1995; Kempf et al. 2008). Pyrrolizidine alkaloids serve as constitutive defense compounds for plant and they are potential feeding disincentive for most of the insects, while livestock animals are mostly dissuade by their savour. Spices and various herbs can also contain toxic contaminants like endogenous compounds (alkenylbenzenes) that may have both carcinogenic and genotoxic effects in mammals (Alajlouni et al. 2016). Lopez et al. (2015) determined alkenylbenzenes in various food products including fish, fruits, and meat samples by GC-MS. Phenol and other trivial phenols like cresols, catechol, orcinols, pyrogallol, and guaiacol are usually considered as industrial chemicals. Phenols occur extensively in common food plants that cause minimal oral toxicity (Singleton 1981).

In consideration of several reports about severe intoxications of phytotoxins to livestock and humans through food chain has tugged more attention (Stegelmeier et al. 1999; Edgar et al. 2002; Fu et al. 2004). To ensure the correct level of users' protection, sensitive analytical approaches need to be established. Gas chromatography-mass spectrometry (GC-MS) is one of those sensitive analytical tools that can be routinely practiced in a laboratory for the quantitative detection of food toxins of plant origin (Kowalczyk et al. 2018).

GC is used for separation analysis of multi constituent mixtures like hydrocarbons, essential oils, and solvents (Kadhim et al. 2016). Different temperature ranges can be set to make result readings more significant; for instance, separation between such materials that behave almost similarly during GC procedure (Pierangeli et al. 2009; Altameme et al. 2015). GC-MS is a highly recommended approach to monitor and track organic solvents of plant origin. This technique is exclusively useful to analyse esters, alcohols, fatty acids, terpenes, aldehydes, tropane alkaloids, and pyrrolizidine alkaloids. Being a powerful and unique technology, GC-MS provides a rare opportunity for performing the exploration of new complexes for the identification and characterization of synthesized compounds or derivatives (Hameed et al. 2015; Jasim et al. 2015).

This chapter discusses different food toxins not strictly of plant origin which are deterrent to humans as well as animals and gives detailed view about GC-MS application to accurately determine these toxins before these toxins become a part of food supply chain. Toxins in this chapter are not limited to toxins of plant origin. Here, the concept of toxins is seen in a broader perspective.

11.2 FAST GC-MS

GC-MS is an advanced technique that cannot be compared with other modern analytical tools. It has a broad range of applications that caters to academic research, industrial as well as quality control applications (Ashish et al. 2014). However, GC analysis requires a long time, characteristically 20–60 minutes. And many analyses need high throughput. In

Quantitative Determination by GC-MS 181

TABLE 11.1   Column Type and Future Approaches to Fast GC-MS

Column type	Characteristics (compared with standard)
Packed	Particles sizes smaller
Open tubular	ID smaller (0.1–0.15 mm)
Multi-capillaries	919 capillaries of 40 mm in 1 m parallel
Vacuum outlet	Larger ID column is connected to vacuum outlet
Supersonic GC-MS	–
Pressure tunable GC/GC-MS	–

GC-MS: gas chromatography with mass spectrometry, ID: internal diameter.
Reproduced with permission from Moeder (2014) with slight modification, Copyright © Springer-Verlag (2014).

view of the long time associated with standard GC analysis, several fast GC systems have been developed that incorporate low thermal mass devices that provide a fast heating temperature program and cooling rates for the GC separation columns and many replicate analyses are performed in the same time that it would take to perform a single conventional run (Michal et al. 2005b; Alexander et al. 2011). Faster GC-MS by increasing the speed of analysis has always been an important goal for GC separations without decreasing chromatographic resolution (Steven and Jana 2002; Kawana et al. 2008) and the mass spectrometer introduces additional important considerations which change several central aspects of the fast GC design (Alexander et al. 2011). Especially fast GC techniques satisfy the present-day demands on faster and cost-effective analysis (Matisová and Hrouzková 2012). Fast GC uses a concept of resistive heating (Stuart et al. 1980). GC peak widths may be reduced in fast GC, consequently MS requires combination of fast scan speed and ion source response time (Alexander et al. 2011). It produces very clean mass spectra due to the high inertness and low bleed of columns (Mathieu et al. 2013). The Fast-GC-MS peaks are extremely narrow (Kawana et al. 2008). Faster GC-MS increases the speed of analysis in different ways: (1) with reduced column-film thickness, (2) short column, (3) increased column diameter, (4) reduced carrier-gas viscosity, (5) heating the column more quickly, and/or (6) increased carrier-gas flow (Steven and Jana 2002; Peter et al. 2020). For proper operation of any "fast" column with MS, the peak broadening caused by extra column effects must be small enough to preserve the column efficiency (Michal et al. 2005a). Fast GC-MS requires high quadrupole scan rate and low inter-scan dead time on order to have a sufficiently high sampling rate (Baier 2014; Luigi and Hans-Ulrich 2022). There are many approaches to fast GC-MS: (1) using micro-bore columns with time-of-flight (TOF)-MS, (2) using low-pressure (LP)-GC-MS to aid separations at increased flow rate, and (3) using supersonic molecular beam (SMB)-MS, otherwise known as supersonic GC-MS having increased flow rates and short columns (Steven and Jana 2002). Recently, a Fast-GC-MS technique using hydrogen gas rather than helium is receiving much attention due to the superior chromatographic resolution that can be achieved, along with a shorter analysis time and lower price per unit volume compared to helium (Kateřina and Steven 2003; Kawana et al. 2008) (Table 11.1).

11.3 GC-TOF-MS

Quantitative metabolomics is a tool which helps in understanding the biochemistry and metabolism with absolute measurement preceding the statistical and bioinformatics analysis (Tiago et al. 2016). Being unique and versatile, gas chromatography is used in measuring trace amounts of organic contaminants and toxic substances in complex food and environmental matrices through vaporization without disintegration (Wedad et al. 2017). For determination and identification of wide range of target and non-target organic components occurring in various biotic matrices and also for food analysis applications, single quadrupole or ion trap mass analysers are used (Thomas and Hajšlová 2007). GC-TOF-MS technology offers high mass accuracy and fast acquisition time; thus, it allows accurate deconvolution of overlapping peaks such as those typically found in plant extracts (Tiago et al. 2016).

The most appropriate derivatization procedure used in GC-MS studies is methoxyamination followed by sialylation (José and Sonia 2016). The importance of food analysis arises from nutrition and health concerns and also it includes many reasons for food analysis such as checking for food adulteration, quality assurance purpose, flavor, contamination of food, and mining the food for natural products. GC is used as primary tool for the compositional analysis of sterols, alcohols, aroma profiles, and other food composition application and for analysis of volatile components in food. GC-TOF-MS method is used for the approach to know the separation efficiency (Steven and Jana 2002). Mass spectra is one of the promising techniques that includes automated searching for characteristic pattern such as sulphur-containing contaminants such as pesticides in foods and chlorinated- and brominated environmental samples based on the characteristic isotope (Kaushik et al. 2012). The main diterpenoid alkaloids derived from aconite root such as aconitine, mesaconitine, and hypaconitine have the property of analgesic, antipyretic, and local anesthetic functions and also have the beneficial effect against rheumatosis and rheumatoid arthritis. On the other hand, these are highly toxic and have a primary target on the heart and central nervous system. Both ^{1}H NMR- and gas chromatography/time-of-flight mass spectrometry (GC-TOF–MS)-based metabonomic approaches were used to investigate and compare the toxicities of the aconitum alkaloids (Bo et al. 2009).

The genus *Delphinium* plants are well known to be poisonous and contain components with chemical structure which are similar to aconitine, as major toxin. There are over 50 compounds reported and 42 are known compounds among them bicoloridine, condelphine, and delpheline, wherein GC-TOF–MS helps in identify these substances by accurately measuring their molecular weight (In et al. 2018). GC-TOF-MS has a wide scope of screening of samples in full-scan acquisition mode due to its high scan speed for both target and non-target analytes in food samples. GC-TOF-MS is usually operated in two approaches: high speed (100–500 spectra) with unit mass resolution, and high resolution with accurate-mass measurement. For quantitative analysis, high-speed GC-TOF-MS is often used for qualitative analysis of organic compounds. Organic pollutants such as PCBs, PAHs, PBDEs, pesticides, and related secondary metabolites are detected by using GC-TOF-MS and also it gives accurate-mass full-spectrum for qualitative analysis of around 200 target analytes. Pesticide residues in fruits, vegetables, and in other food samples were detected by GC-TOF-MS, which provides confirmation for the identification of several new toxic compounds in these samples (Xian et al. 2013) (Table 11.2).

11.4 LOW-PRESSURE (LP) GC-MS

Low-pressure gas chromatography (LPGC) is an efficient technique that has been employed for rapid examination of GC-amenable compounds for more than five decades.

Quantitative Determination by GC-MS

183

TABLE 11.2 Currently Used GC-MS Techniques

GC-MS technique	Abbreviation	Mass resolution	Application field
GC-single quadrupole	GC-MS	Nominal/low mass resolution	Screening and target analysis
GC-ion trap detector	GC-ITD	Low	Screening and target analysis, structure elucidation
GC-triple stage quadrupole	GC-QqQ-MS	Low	Highly sensitive target analysis, structure elucidation
GC-double focusing sector instruments/ tandem MS	–	High, R ≈ 10,000	Gas-phase studies, target analysis, structure elucidation
GC-isotope ratio MS	GC-IR-MS	R < 1000	Isotope ratio measurement
GC-time-of-flight MS	GC-ToF MS	a. Low b. High R ≈ 100,000	a. Fast GC, GC×GC b. Target analysis, structure elucidation
GC-hybride MS	GC-sector fields-Q	Depends on the analyzer	High sensitive target analysis, structure elucidation
GC-Orbitrap™	–	High R ≈ 100,000	High sensitive target analysis, research
GC-inductive coupled plasma MS (MS)	GC-ICP-MS (MS)	Depends on analyzer used	Element speciation, organometal compound analysis

GC-MS: gas chromatography-mass spectrometry, IR-MS: isotope ratio-mass spectrometry, ICP: inductive coupled plasma, ITD: ion trap detector, QqQ: triple-stage quadrupole, ToF: time-of-flight.
Reproduced with permission from Moeder (2014) with slight modification, Copyright © Springer-Verlag (2014).

Fastening analysis has long been a key objective for LPGC separations of various mixes. However, the extended detection and applications with LPGC are inextricably linked to mass spectroscopy (MS) (Al-Rubaye et al. 2017).

11.4.1 Working Principle and Instrumentation

In LPGC, the outlet of the analytical column (short and wide) is placed under vacuum conditions, which assists to reduce the viscosity of the carrier gas and thus to assure sophisticated flow rate and rapid separation. The diagnostic column is often linked to a thin, short, unmounted restriction capillary in order to maintain typical operating pressures at the inlet, where the capillary itself serves as a guard column. The quicker examination in LPGC often leads to lower separation efficacy as compared to traditional GC. Although the application of advanced selective detection practice mass spectrometry (MS) along with LPGC disables this disadvantage. Mass spectrometry (MS) is a very selective and sensitive technique, and generally all GC-MS equipment provides vacuum outlet settings for LPGC-MS(MS) deployment without the requirement of any modifications (Sapozhnikova and Lehotay 2015). LPGC/MS is also known as Rapid-MS. It is an intriguing way to quick analysis; where comparatively short (10 m) mega bore analytical column (0.53 mm) is employed. The MS's vacuum extends into the analytical column, resulting in a greater flow rate and discerning separation features. At the inlet end, a restriction capillary of suitable length (0.1–0.25 mm i.d.) is to be installed to assure positive input pressure which also enable typical GC injection. Due to bigger column capacity in LPGC/MS, larger injection volumes may be preferred for better outcomes (Maštovská et al. 2001). However, separation efficiency is lowered while using repeated injections, a limitation of this technique as compared to classic approaches.

11.4.2 Significances of LPGC-MS

A vast variety of compounds can be analyzed very accurately through this comprehensive technique. Some of the benefits of LPGC-MS are quick and easy separations, no need of instrumental amendments, advanced mega-bore columns for elevated injection volumes, peak widths as like conventional detection methods, increased peak heights and lower LOQ, improved peak shapes, and less possibility of thermal degradation of thermal-labile analytes. Other advantages of LPGC include less peak tailing, greater loadability, lower exposure confines, and added ruggedness that assists rapid data gathering rates (Lehotay and Hajšlová 2002). It is also possible to reduce GC separation times in a variety of methods as column shortening, cumulative carrier-gas movement, decreasing thickness of column-film, reducing viscosity of carrier-gas, elevating column diameter, and heating the column more rapidly. For instance, the use of a mega bore column showed enhanced analytical performances as contrast of using short microbore capillary column in LPGC-MS (Al-Rubaye et al. 2017) (Table 11.3).

11.4.3 Applications of LPGC-MS

LPGC-MS is effectively employed for the analysis of several sensitive bio-active compounds from diverse analytical samples. Currently, this technique is well optimized and extensively used for the quick detection of pesticides and a variety of other compounds as hydrocarbons especially polycyclic aromatic hydrocarbons (PAHs), volatile organic moieties as benzene, toluene, etc., volatile fragrances, and typical contaminants. However, alternations in column set-ups/MS systems may require for optimization of separation of each specific compound (Ravindra et al. 2008). For example, a rapid approach with LPGC combined mass spectrometry was developed and optimized to separate 27 typical pesticide components from grapes, musts, and wines. Where, a number of criteria as temperature of column, injection parameters, flow frequency, MS circumstances, and matrix properties, were investigated in order to accomplish the swiftest characterization in less than 20 minutes along maintaining the maximum sensitivity in MS detection (Cunha et al. 2009).

11.5 GC/SUPERSONIC MOLECULAR BEAM (SMB)-MS

The development of mass spectrometry–based advanced gas chromatography (GC) techniques has contributed immensely to the field of food toxicology, where the determination of toxins of plant origin is the most concerned issue in the food industry (Moazzen et al. 2018). Increase in the need of GC analysis led the manufacture industry to

TABLE 11.3 Analysis of Food Toxins by GC-MS

Analytes	Food matrix	Analysis method	Sensitivity	References
496 Pesticides	Dumplings	GC-MS	LODs: 0.01 mg g^{-1}	Matsuoka et al. (2011)
176 Pesticides	Tea, Chinese cabbage, human breast milk	GPC-GC-EI-MS	LODs: 1–20 ng ml^{-1}	Yu and Xu (2012)
Galaxolide, tonalide, triclosan	Barley, meadow fescue, carrot	GC-EI-MS-SIM	LOQs: 0.001–0.06 mg g^{-1}	Macherius et al. (2012)
8 Pesticides	Ginseng	GC-EI-Q-MS-SIM	LODs: 0.05–0.5 ng g^{-1}	Zhou et al. (2012)
Volatile hop constituents	Hops	HS-Trap GC-EI-Q-MS-SIM	LODs: 0.6 mg g^{-1}	Aberl and Coelhan (2012)
Biogenic amines	Beers	GC-EI-Q-MS-SIM	LODs: 2.9 ng ml^{-1}	Almeida et al. (2012)
130 Pesticides	Fruits and vegetables	GC-EI-QqQ-MS/MS	LODs: 0.1–50 ng g^{-1}	Cervera et al. (2010)
124 Pesticides	Rice	GC-EI-QqQ-MS/MS	LODs: 0.1–7.5 ng g^{-1}; LOQs: 0.4–26.3 ng g^{-1}	Hou et al. (2013)
PCBs	Meat and sea food	GC-EI-QqQ-MS/MS	LODs: 0.006–0.009 ng g^{-1}; LOQs: 0.02–0.03 ng g^{-1}	Huo et al. (2012)
Pyrethroids	Fruits and vegetables	GC-APCI-QqQ-MS/MS	LODs: 3 ng g^{-1}	Portolés et al. (2010)
PCBs	Chicken whole blood	GC-LVI-EI-IT-MS-MS/MS	LODs: 0.05–0.5 ng g^{-1}	Haskins et al. (2010)
PBBs	Fish	GC-EI-IT-MS/MS	LODs: 0.03–0.16 ng g^{-1}	Malavia et al. (2011)
PCBs and PBDEs	Milk	GC-EI-IT-MS/MS	LODs: 0.02–3.22 pg g^{-1}; LOQs: 0.07–10.72 pg g0^{-1}	Roszko et al. (2012)
Odorous compounds	Food packaging materials	GC-O-EI-IT-MS	LODs: 0.03–5.02 mg g^{-1}	Vera et al. (2012)
Organic pollutants	Drinking water, grape, fish, cucumber	GC-EI-TOF-MS	LODs: 5 ng g^{-1}	Ibáñez et al. (2012)
Organic contaminants	Marine salts	GC-EI-TOF-MS	–	Serrano et al. (2011)
PCBs, PBDEs, PAHs	Fish and shrimps	GC-EI-TOF-MS	LOQs: PCBs: 0.1–0.5 ng g^{-1}; PBDEs: 0.5 ng g^{-1}; PAHs: 0.05–0.25 ng g^{-1}	Kalachova et al. (2011)

(Continued)

TABLE 11.3 (Continued)

Analytes	Food matrix	Analysis method	Sensitivity	References
Pesticides	Fruits and vegetables	GC-EI-TOF-MS	LOQs: 0.01 mg g^{-1}	Cervera et al. (2012)
Pesticides	Orange, nectarine, and spinach	GC-APCI-Q-TOF-MS GC-EI-TOF-MS	–	Portolés et al. (2010)
Acrylic adhesives	Food packaging materials	GC-APCI-Q-TOF-MS GC-EI-Q-MS	LODs: 1.2–4.3 ng g^{-1}	Canellas et al. (2012)
68 Pesticides	Oil seeds	GC×GC-EI-TOF-MS	–	Wang et al. (2012)
Volatile compounds	Wines	GC×GC-EI-TOF-MS	–	Vestner et al. (2011)
Furans, lactones, volatile phenols, acetals	Wines	GC×GC-EI-TOF-MS	–	Perestrelo et al. (2011)
Cellular fatty acid	Foodborne bacteria	GC×GC-EI-TOF-MS	–	Gardner et al. (2011)
Pesticides, essential oil	Mandarin essential oil	GC×GC-EI-QqQ-MS	LODs: 3–21 ppb LOQs: 9–71 ppb	Tranchida et al. (2013)
Sandalwood essential oil	Sandalwood oil	GC×GC-EI-Q -MS	0.002%	Sciarrone et al. (2011)

Reproduced with permission from Wang et al. (2013) with slight modification, Copyright © Elsevier (2013).

construct a new mechanism with highly reliable, rapid, and sensitive GC analysis systems coupled with mass spectrometry (MS) instrument. There are many factors that help to achieve desired accuracy with pressure variation in the analysis of trace components in food materials. The microbore column in the GC-MS system is one of the factors that predominantly influence the pressure management, thereby limiting the practical applicability of the GC system (Maštovská 2008). However, recent innovations in the industry replaced the microbore column with a short megabore column system that furnishes a higher sample capacity (Q_s), better ruggedness, and reduced instrumentation issues (Maštovská et al. 2004). The newly introduced megabore columns can be operated at lower pressures along the entire length of the column. This system allows the solute to diffuse greatly maintaining lower column pressure in the gas phase, resulting in higher values of optimum mobile phase velocity and helps in faster GC separation (Leclercq and Cramers 1987).

To attain the desired high flow rates over the low-flow rates obtained in the conventional GC-MS instrument, a new improved technique was introduced: the supersonic molecular beam (SMB) GC-MS. The instrument was developed and the entire process was described by Dagan and Amirav (1994). The SMB system was consisted of a 50 cm long megabore column connecting the injector to the supersonic nozzle source with split-less sample injection system. It was demonstrated that the GC separation achieved was ultrafast and completed the separation of mixtures containing benzene, chlorobenzene, bromobenzene, and iodobenzene each in methanol (1%) solution (upper trace) in less than 1 second. This new technology exceptionally contributed to the sample analysis process in the field of food toxicology with the following advantages: (1) the newly constructed instrument was totally dependent on supersonic molecular beam (SMB) which is situated in between the vacuum and the atmospheric pressure inlet, (2) the sample preparation was easy, (3) non-volatile and thermolabile samples were also detected, (4) the high-flow rates of the machine helped to speed up the analysis process and repeatability of the sample. However, the study team observed increased complexity of gas consumption of MS instrument and the high cost, as the drawbacks (Dagan and Amirav 1994).

The development of supersonic molecular beam (SMB) GC-MS for biological application was demonstrated in several studies and is still in progress, while the lack of commercial availability is severely limiting the access and applicability in real-time practice (Maštovská 2008; Poliak et al. 2008). There are few studies that utilized the supersonic molecular beam (SMB) GC-MS system to analyze food toxins in the sample and are explained as follows. The family Cortinariaceae contains over 2,100 species of gilled mushrooms found all over the world. The compound orellanine is a deadly toxic compound found in more 34 species, particularly in *Cortinarius orellanus* Fr. species and usually inhabit Northern Europe to Southern Scandinavia. Cortinariaceae family mushrooms are quite common in East, North, and Central Europe and Northern America. The cases of intoxication owing to poisonous mushrooms are being reported in some parts of Asia and the compound orellanine was suspected as the prime foodborne toxin. A study was conducted employing supersonic molecular beam (SMB) GC-MS technique to analyze orellanine in the stomach of rats after ingestion (Brondz et al. 2012).

The study involved supersonic molecular beam (SMB) GC-MS system and was set up, maintaining the following conditions. The separation of compounds was performed with a VF-5HT column using helium as mobile phase with the column flow rate of 8 mL/min. About 1 µL of sample was injected with the programmed temperature of 120°C to 300°C at 30°C/min. This study was found that the use of SMB GC-MS can improve mass spectrum of the sample and allows molecules to stay intact producing visible mass ion signal (Brondz et al. 2012). Another study utilizing the same SMB GC-MS technique elucidated the structure of rubelline, another mushroom origin toxin extracted from *Cortinarius rubellus*. This study concluded that the supersonic molecular beam (SMB) GC-MS system assists in the faster separation and structural elucidation of any toxic compound of plant origin (Brondz 2013). The new SMB GC-MS technique also enables to record mass ion and a detailed fragmentation of a compound in the field of food toxicology.

11.6 GC-MS OF FOOD TOXINS OF PLANT ORIGIN

In the review article of Kopp et al. (2020), extracting and analyzing techniques of pyrrolizidine alkaloids among which GC MS are detailed.

A combined derivatization method for gas chromatographic/mass spectrometric (GC/MS) analysis of steroidal glycoalkaloid (SGA) aglycons was developed by Laurila J. et al. (1999), using both trimethylsilylation and pentafluoropropionylation. In comparison with underivatized or only silylated aglycons, this technique produces more specific and abundant fragmentation for compounds with a tomatidine-type structure. For example, the difference between solasodine and tomatidine, the former containing a double bond at position 5,6 in the steroidal skeleton, can be observed by their base peak fragments at m/z 417 ($C24H41O2Si2$) and m/z 419 ($C24H43O2Si2$). The method is well suited for the simultaneous determination of both solanidane- and spirosolane-type SGA aglycons from *Solanum* species and hybrids. The reproducibility of the method, including SGA extraction, hydrolysis, derivatization, and quantitative GC/MS analysis was <6% (CV) for the principal aglycons determined from a hybrid between a wild potato species, *Solanum brevidens* Phil., and a cultivated potato, *S. tuberosum* L. A single ion monitoring technique using specific fragments m/z 419 and 417 could be applied for the determination of minor stereoisomers, which are often overlapped by large amounts of tomatidine.

Ethanol was used to extract linear and angular furanocoumarins (Alehaideb Z. et al. 2017). Plant products were minced and reduced to fine powders using a Salton food processor (Dollard-des-Ormeaux, QC, Canada) model CG-1174. The plant powders were mixed with ethanol at 10 mg/mL and sonicated, using a Branson sonicator (Shelton, CT, USA), for 3 hours under atmospheric pressure at 40°C–60°C. The extracts were cooled to room temperature for 5–10 minutes. Ethanol was added to replace the volume lost during extraction. The extract was filtered using a Millipore Millex-LG filter unit (0.2 μm) and analyzed using GC-MS. The procedure used an Agilent HP-5 MS column (30 m × 0.25 mm, 0.25 μm). The following temperature program was used: 75°C (0–2 minutes), 75–250°C (2–20 minutes), 250–280°C (20–25 minutes), and 280°C (25–30 minutes). Injection volume was 1.0 μL. High-purity helium was used at a constant flow rate of 0.5 mL/min. The MS was set at full scan with a 50 to 600 m/z range.

11.7 CONCLUSION

GC-MS has been a widely used analytical method in food toxic analysis. Different analytical setups for GC-MS has been described for the analysis of various naturally occurring food toxins of plant origin. Optimization of the derivatization procedure and optimum sample preparation conditions using solid-phase extraction were found to be useful in analysis. GC-MS is the well-developed and validated method for quantitative analysis of the wide range of phytoestrogens and phytosterol phytotoxins. In the near future, GC-MS mediated detection system is expected to replace various current techniques of toxic food contaminant detection using selective GC detectors. GC-MS will be more beneficial especially when combined with fast-GC separations strategies. The enhanced sensitivity and selectivity of MS detectors help to achieve faster separations of lower chromatographic resolutions. The advanced GC-MS methods that may be useful in detection and quantification are included GC-TOF-MS, LP-GC-MS, and GC/SMB-MS. Hence, GC-MS-based detection and analysis of food toxins of plant origin will be a more interesting and reliable approach in food safety and standards studies.

REFERENCES

Aberl, A. and Coelhan, M. (2012). Determination of volatile compounds in different hop varieties by headspace-trap GC/MS – In comparison with conventional hop essential oil analysis. *Journal of Agricultural and Food Chemistry*, 60, 2785–2792.

Alajlouni, A. M., Al-Malahmeh, A. J., Isnaeni, F. N., Wesseling, S., Vervoort, J. and Rietjens, I. M. C. M. (2016). Level of alkenyl benzenes in parsley and dill based teas and associated risk assessment using the margin of exposure approach. *Journal of Agricultural and Food Chemistry*, 64, 8640–8646.

Alehaideb, Z., Sheriffdeen, M. and Law, F. C. (2017). Furanocoumarin bioactives in the Apiceae and Rutaceae families of plants. *Can. J. Appl. Sci.*, 11(2), 4157–4167.

Alexander, B. F., Mati, M. and Aviv, A. (2011). A low thermal mass fast gas chromatograph and its implementation in fast gas chromatography mass spectrometry with supersonic molecular beams. *J. Chromatogr. A*, 1218(52), 9375–9383. doi: 10.1016/j.chroma.2011.10.053

Almeida, C., Fernandes, J. O. and Cunha, S. C. (2012). A novel dispersive liquid–liquid microextraction (DLLME) gas chromatography-mass spectrometry (GC-MS) method for the determination of eighteen biogenic amines in beer. *Food Control*, 25, 380–388.

Al-Rubaye, A. F., Hameed, I. H. and Kadhim, M. J. (2017). A review: Uses of gas chromatography-mass spectrometry (GC-MS) technique for analysis of bioactive natural compounds of some plants. *International Journal of Toxicological and Pharmacological Research*, 9(1), 81–85.

Altameme, H. J., Hadi, M. Y. and Hameed, I. H. (2015). Phytochemical analysis of *Urtica dioica* leaves by fourier-transform infrared spectroscopy and gas chromatography-mass spectrometry. *Journal of Pharmacognosy and Phytotherapy*, 7(10), 238–252.

Ashish, C., Manish, K. G. and Priyanka, C. (2014). GC-MS technique and its analytical applications in science and technology. *Journal of Analytical and Bioanalytical Techniques*, 5(6), 222. doi: 10.4172/2155-9872.1000222

Baier, H. U. (2014). Shortening analysis time (Fast GC). In Dettmer-Wilde, K. and Engewald, W. (eds.). *Practical Gas Chromatography*. Berlin, Heidelberg: Springer, pp. 413–459. doi: 10.1007/978-3-642-54640-2_12

Bo, S., Ling, L., Shengming, W., et al. (2009). Metabolomic analysis of biofluids from rats treated with aconitum alkaloids using nuclear magnetic resonance and gas chromatography/time-of-flight mass spectrometry. *Analytical Biochemistry*, 395, 125–133.

Brondz, I. (2013). Structure elucidation of a new toxin from the mushroom *Cortinarius rubellus* using gas chromatography-mass spectrometry (GC-MS). *International Journal of Analytical Mass Spectrometry and Chromatography*, 1(2), 109–118. doi: 10.4236/ijamsc.2013.12014

Brondz, I., Nevo, E., Wasser, S. P. and Brondz, A. (2012). A direct gas chromatography-mass spectrometry (GC-MS) method for the detection of orellanine present in stomach content (Part I). *Journal of Biophysical Chemistry*, 3(1), 29–34. doi: 10.4236/jbpc.2012.31003

Canellas, E., Vera, P., Domeno, C., Alfaro, A. P. and Nerin, C. (2012). Atmospheric pressure gas chromatography coupled to quadrupole-time of flight mass spectrometry as a powerful tool for identification of non intentionally added substances in acrylic adhesives used in food packaging materials. *Journal of Chromatography A*, 1235, 141–148.

Cervera, M., Medina, C., Portolés, T., Pitarch, E., Beltrán, J. and Serrahima, E. (2010). Multi-residue determination of 130 multiclass pesticides in fruits and vegetables by gas chromatography coupled to triple quadrupole tandem mass spectrometry. *Analytical and Bioanalytical Chemistry*, 397, 2873–2891.

Cervera, M. I., Portoles, T., Pitarch, E., Beltran, J. and Hernandez, F. (2012). Application of gas chromatography time-of-flight mass spectrometry for target and non-target analysis of pesticide residues in fruits and vegetables. *Journal of Chromatography A*, 1244, 168–177.

Chang, S. S., Wu, M. L., Deng, J. F., Chin, T. F. and Liao, S. J. (1999). Poisoning by *Datura* leaves used as edible wild vegetables. *Vet. Hum. Toxicol.*, 41, 242–243.

Cunha, S. C., Fernandes, J. O., Alves, A. and Oliveira, M. B. P. P. (2009). Fast low-pressure gas chromatography – Mass spectrometry method for the determination of multiple pesticides in grapes, musts and wines. *Journal of Chromatography A*, 1216(1), 119–126.

Dagan, S. and Amirav, A. (1994). Fast, high temperature and thermolabile GC—MS in supersonic molecular beams. *International Journal of Mass Spectrometry and Ion Processes*, 133, 2187–2210. doi: 10.1016/0168-1176(94)03960-7

Edgar, J. A., Roeder, E. and Molyneux, R. J. (2002). Honey from plants containing pyrrolizidine alkaloids: A potential threat to health. *J. Agric. Food Chem.*, 50, 2719–2730.

Fu, P. P., Xia, Q., Lin, G. and Chou, M. W. (2004). Pyrrolizidine alkaloids–genotoxicity, metabolism, enzymes, metabolic activation, and mechanisms. *Drug Metab. Rev.*, 36, 1–55.

Gardner, J. Y., Brillhart, D. E., Benjamin, M. M., Dixon, L. G., Mitchell, L. M. and Dimandja, J. -M. D. (2011). The use of GCxGC/TOF MS with multivariate analysis for the characterization of foodborne pathogen bacteria profiles. *Journal of Separation Science*, 34, 176–185.

Guharoy, S. R. and Barajas, M. (1991). Atropine intoxication from the ingestion and smoking of jimson weed *Datura stramonium*. *Vet. Hum. Toxicol.*, 33, 588–589.

Hameed, I. H., Ibraheam, I. A. and Kadhim, H. J. (2015). Gas chromatography mass spectrum and fourier-transform infrared spectroscopy analysis of methanolic extract of *Rosmarinus oficinalis* leaves. *Journal of Pharmacognosy and Phytotherapy*, 7(6), 90–106.

Hartmann, T. and Witte, L. (1995). In Pelletier, S. W. (ed.). *Alkaloids: Chemical and Biological Perspectives*. Oxford: Pergamon Press, pp. 155–233.

Haskins, S. D., Kelly, S. D. and Weir, R. D. (2010). Novel pressurized solvent extraction vessels for the analysis of polychlorinated biphenyl congeners in avian whole blood. *Analytica Chimica Acta*, 677, 19–23.

Hayman, J. (1985). *Datura arborea* poisoning the angels trumpet. *Pathology*, 17, 465–466.

Hou, X., Han, M., Dai, X., Yang, X. and Yi, S. (2013). A multi-residue method for the determination of 124 pesticides in rice by modified QuEChERS extraction and gas chromatography-tandem mass spectrometry. *Food Chemistry*, 138, 1198–1205.

Huo, L. G., Li, H. D., Zhao, C. L., et al. (2012). The determination of PCBs in meat and sea food by GC-QqQ-MS/MS. *Food Analytical Methods*, 5, 1481–1491.

Ibáñez, M., Portoles, T., Rubies, A., et al. (2012). The power of hyphenated chromatography/time-of-flight mass spectrometry in public health laboratories. *Journal of Agricultural and Food Chemistry*, 60, 5311–5323.

In, S. H., Choi, S., Jeon, Y. J. and Seol, I. (2018). A case report on the analysis of poisonous alkaloids in Delphinium plant by liquid chromatography-quadrupole time-of-flight mass spectrometry. *J Forensic Toxicol Pharmacol.*, 7, 1. doi: 10.4172/2325-9841.1000159

Jasim, H., Hussein, A. O., Hameed, I. H. and Kareem, M. A. (2015). Characterization of alkaloid constitution and evaluation of antimicrobial activity of *Solanum nigrum* using gas chromatography mass spectrometry (GC-MS). *Journal of Pharmacognosy and Phytotherapy*, 7(4), 56–72.

José, G. V. and Sonia, O. (2016). Simultaneous determination of plant hormones by GC-TOF-MS. In Jose, R. B. and Miguel, A. B. (eds.). *Plant Signal Transduction: Methods and Protocols*. New York: Springer, pp. 1064–3745.

Kadhim, M. J., Sosa, A. A. and Hameed, I. H. (2016). Evaluation of anti-bacterial activity and bioactive chemical analysis of *Ocimum basilicum* using Fourier transform infrared (FT-IR) and gas chromatography-mass spectrometry (GC-MS) techniques. *International Journal of Pharmacognosy and Phytochemical Research*, 8(6), 127–146.

Kalachova, K., Pulkrabova, J., Drabova, L., et al. (2011). Simplified and rapid determination of polychlorinated biphenyls, polybrominated diphenyl ethers, and polycyclic aromatic hydrocarbons in fish and shrimps integrated into a single method. *Analytica Chimica Acta*, 707, 84–91.

Kateřina, M. and Steven, J. L. (2003). Practical approaches to fast gas chromatography–mass spectrometry. *Journal of Chromatography A*, 1000(1–2), 153–180. doi: 10.1016/s0021-9673(03)00448-5

Kaushik, B., Soma, D. and Sagar, C. U. (2012). Application of GC–TOF-MS for pesticide residue analysis in grapes. In Amadeo, R. and Fernandez-Alba (eds.). *TOF-MS within Food and Environmental Analysis*. Amsterdam, Netherlands: Elsevier, pp. 367–413.

Kawana, S., Nakagawa, K., Hasegawa, Y., Kobayashi, H. and Yamaguchi, S. (2008). Improvement of sample throughput using fast gas chromatography mass-spectrometry for biochemical diagnosis of organic acid disorders. *Clin. Chim. Acta.*, 392(1–2), 34–40. doi: 10.1016/j.cca.2008.02.025

Kempf, M., Beuerle, T., Bhringer, M., Denner, M., Trost, D., Von der Ohe, K., Bhavanam, V. B. R. and Schreier, P. (2008). Pyrrolizidine alkaloids in honey: Risk analysis by gas chromatography-mass spectrometry. *Mol. Nutr. Food Res.*, 52(10), 1193–1200.

Kopp, T., Abdel-Tawab, M., and Mizaikoff, B. (2020). Extracting and analyzing pyrrolizidine alkaloids in medicinal plants: A review. *Toxins*, 12(5), 320. doi: 10.3390/toxins12050320

Kowalczyk, E., Sieradzki, Z. and Kwiatek, K. (2018). Determination of pyrrolizidine alkaloids in honey with sensitive gas chromatography-mass spectrometry method. *Food Anal. Methods*, 11, 1345–1355.

Laurila, J., Laakso, I., Väänänen, T., Kuronen, P., Huopalahti, R., and Pehu, E. (1999). Determination of solanidine-and tomatidine-type glycoalkaloid aglycons by gas chromatography/mass spectrometry. *Journal of Agricultural and Food Chemistry*, 47(7), 2738–2742. doi: 10.1021/jf981009b

Leclercq, P. A. and Cramers, C. A. (1987). Minimum analysis time in capillary gas chromatography vacuum- versus atmospheric-outlet column operation. *Journal of High Resolution Chromatography*, 10(5), 269–272. doi: 10.1002/jhrc.1240100511

Lehotay, S. J. and Hajšlová, J. (2002). Application of gas chromatography in food analysis. *TrAC Trends in Analytical Chemistry*, 21(9-10), 686–697.

Lopez, P., Van Sisseren, M., De Marco, S., Jekel, A., De Nijs, M. and Mol, H. G. J. (2015). A straightforward method to determine flavouring substances in food by GC-MS. *Food Chemistry*, 174, 407–416.

Luigi, M. and Hans-Ulrich, B. (2022). Fast GC and GC/MS using narrow-bore columns: Principles and applications. *Application Handbook Fast GC/GCMS Chromatography Application, Vol. 2*, Deutschland, Germany: Shimadzu. https://www.shimadzu.eu/sites/shimadzu.seg/files/Application_Book_FastGCvol2.pdf.

Macherius, A., Eggen, T., Lorenz, W. G., Reemtsma, T., Winkler, U. and Moeder, M. (2012). Uptake of galaxolide, tonalide and triclosan by carrot, barley and meadow fescue plants. *Journal of Agriculture and Food Science*, 60, 7785–7791.

Malavia, J., Santos, F. J. and Galceran, M. T. (2011). Simultaneous pressurized liquid extraction and clean-up for the analysis of polybrominated biphenyls by gas chromatography-tandem mass spectrometry. *Talanta*, 84, 1155–1162.

Maštovská, K. (2008). Recent developments in chromatographic techniques. In Picó, Y. (ed.). *Comprehensive Analytical Chemistry*. Amsterdam, Netherlands: Elsevier, pp. 175–200. doi: 10.1016/S0166-526X(08)00006-8

Maštovská, K., Hajšlová, J. and Lehotay, S. J. (2004). Ruggedness and other performance characteristics of low-pressure gas chromatography – Mass spectrometry for the fast analysis of multiple pesticide residues in food crops. *Journal of Chromatography A*, 1054(1), 335–349. doi: 10.1016/j.chroma.2004.08.061

Maštovská, K., Lehotay, S. J. and Hajšlová, J. (2001). Optimization and evaluation of low-pressure gas chromatography–mass spectrometry for the fast analysis of multiple pesticide residues in a food commodity. *Journal of Chromatography A*, 926(2), 291–308.

Mathieu, P. E., Leonine, E. E. and Mark, G. B. (2013). Keeping pace with NPS releases: Fast GC-MS screening of legal high products. *Drug Test Anal.*, 5(5), 281–290. doi: 10.1002/dta.1434

Matisová, E. and Hrouzková, S. (2012). Analysis of endocrine disrupting pesticides by capillary GC with mass spectrometric detection. *International Journal of Environmental Research and Public Health*, 9(9), 3166–3196. 10.3390/ijerph9093166

Matsuoka, T., Akiyama, Y. and Mitsuhashi, T. (2011). Application of multi-residue analytical method for determination of 496 pesticides in frozen gyoza dumplings by GC-MS and LC-MS. *Journal of Pesticide Science*, 36, 486–491.

Michal, K., Eva, M., Svetlana, H. and de Jaap, Z. (2005a). Possibilities and limitations of quadrupole mass spectrometric detector in fast gas chromatography. *Journal of Chromatography A*, 1090(1–2), 126–132. doi: 10.1016/j.chroma.2005.06.090

Michal, K., Eva, M., Robert, O., Andrea, H. and de Jaap, Z. (2005b). Search on ruggedness of fast gas chromatography – Mass spectrometry in pesticide residues analysis. *Journal of Chromatography A*, 1084(1–2), 63–70. doi: 10.1016/j.chroma. 2004.10.043

Mikolich, J. R., Paulson, G. W. and Cross, C. J. (1975). Acute anticholinergic syndrome due to jimson seed ingestion. clinical and laboratory observation in 6 cases. *Ann. Intern. Med.*, 83, 321–325.

Moazzen, M., Amir, H. M., Nabi, S., Gholamreza, J. K., Shahrokh, N., Mahmood, A. et al. (2018). Determination of phthalate acid esters (PAEs) in carbonated soft drinks with MSPE/GC–MS method. *Toxin Reviews*, 37(4), 319–326. doi: 10.1080/155 69543.2017.1378234

Moeder, M. (2014). Gaschromatography-massspectrometry, Practical Gas Chromatography. Springer: Berlin, Heidelberg, pp. 303–350.

Nogue, S., Pujol, L., Sanz, P. and Torre, R. (1995). *Datura stramonium* poisoning: Identification of tropane alkaloids in urine by gas chromatography–mass spectrometry. *J. Int. Med. Res.*, 23, 132–137.

Perestrelo, R., Barros, A. S., Camara, J. S. and Rocha, S. M. (2011). In-depth search focused on furans, lactones, volatile phenols, and acetals as potential age markers of Madeira wines by comprehensive two-dimensional gas chromatography with time-of-flight mass spectrometry combined with solid phase microextraction. *Journal of Agricultural and Food Chemistry*, 59, 3186–3204.

Peter, Q. T., Mariosimone, Z. and Luigi, M. (2020). High-speed GC-MS: Basic theory, practical aspects, and applications. In Peter, Q. T. and Luigi, M. (eds.). *Hyphenations of Capillary Chromatography with Mass Spectrometry*. Amsterdam, Netherlands: Elsevier, pp. 109–132. doi: 10.1016/B978-0-12-809638-3.00003-X

Pierangeli, G., Vital, G. and Rivera, W. (2009). Antimicrobial activity and cytotoxicity of *Chromolaena odorata* (L.) King and Robinson and *Uncaria perrottetii* (A. Rich) Merr. extracts. *J. Medicinal Plants Res*, 3(7), 511–518.

Poliak, M., Fialkov, A. B. and Amirav, A. (2008). Pulsed flow modulation two-dimensional comprehensive gas chromatography – Tandem mass spectrometry with supersonic molecular beams. *Journal of Chromatography A*, 1210(1), 108–114. doi: 10.1016/j.chroma.2008.09.039

Portolés, T., Sancho, J. V., Hernandez, F., Newton, A. and Hancock, P. (2010). Potential of atmospheric pressure chemical ionization source in GC-QTOF MS for pesticide residue analysis. *Journal of Mass Spectrometry*, 45, 926–936.

Ravindra, K., Dirtu, A. C. and Covaci, A. (2008). Low-pressure gas chromatography: Recent trends and developments. *TrAC Trends in Analytical Chemistry*, 27(4), 291–303.

Roszko, M., Rzepkowska, M., Szterk, A., Szymczyk, K., Jedrzejczak, R. and Bryla, M. Y. (2012). Application of semi-permeable membrane dialysis/ion trap mass spectrometry technique to determine polybrominated diphenyl ethers and polychlorinated biphenyls in milk fat. *Analytica Chimica Acta*, 748, 9–19.

Sapozhnikova, Y. and Lehotay, S. J. (2015). Review of recent developments and applications in low-pressure (vacuum outlet) gas chromatography. *Analytica Chimica Acta*, 899, 13–22.

Sciarrone, D., Costa, R., Ragonese, C., Tranchida, P. Q., Tedone, L., Santi, L., Dugo, P., Dugo, G., & Mondello, L. (2011). Application of a multidimensional gas chromatography system with simultaneous mass spectrometric and flame ionization detection to the analysis of sandalwood oil. *Journal of Chromatography A*, 1218(1), 137–142.

Serrano, R., Nacher-Mestre, J., Portoles, T., Amat, F. and Hernandez, F. (2011). Non-target screening of organic contaminants in marine salts by gas chromatography coupled to high-resolution time-of-flight mass spectrometry. *Talanta*, 85, 877–884.

Singleton, V. L. (1981). Naturally occurring food toxicants: Phenolic substances of plant origin common in foods. *Advances in Food Research*, 27, 149–242.

Stegelmeier, B. L., Edgar, J. A., Colegate, S. M., Gardner, D. R., et al. (1999). Pyrrolizidine alkaloid plants, metabolism and toxicity. *J. Nat. Toxins.*, 8, 95–116.

Steven, J. L. and Jana, H. (2002). Application of gas chromatography in food analysis. *Trends in Analytical Chemistry*, 21, 686–697. doi: 10.1016/B978-0-444-53810-9.00005-5

Stuart, P. C., Terence, H. R., Larry, R. F. and Wei-Lu, Y. (1980). Gas chromatography. *Analytical Chemistry*, 52(5), 324–360. doi: 10.1021/ac50055a030

Thabet, H., Brahmi, N., Amamou, M., Salah, N. B. and Yacoub, M. (1999). *Datura stramonium* poisonings in humans. *Vet. Hum. Toxicol.*, 41, 320–321.

Thomas, C. and Hajšlová, J. (2007). Gas chromatography–time-of-flight mass spectrometry in food analysis. *LCGC Europe*, 20(1), 25.

Tiago, F. J., Ana, T. M. and Carla. A. (2016). Mass spectrometry as a quantitative tool in plant metabolomics. *Philosophical Transaction of the Royal Society A*, 374(2079), 20150370. doi: 10.1098/rsta.2015.0370

Tranchida, P. Q., Franchina, F. A., Zoccali, M., Pantò, S., Sciarrone, D., Dugo, G. and L Mondello, L. (2013). Untargeted and targeted comprehensive two-dimensional GC analysis using a novel unified high-speed triple quadrupole mass spectrometer. *Journal of Chromatography A*, 1278, 153–159.

Urich, R. W., Bowerman, D. L., Levisky, J. A. and Pflug, J. L. (1982). *Datura stramonium*: A fatal poisoning. *J. Forensic Sci.*, 27, 948–954.

Vera, P., Uliaque, B., Canellas, E., Escudero, A. and Nerín, C. (2012). Identification and quantification of odorous compounds from adhesives used in food packaging materials by headspace solid phase extraction and headspace solid phase microextraction coupled to gas chromatography olfactometry mass spectrometry. *Analytica Chimica Acta*, 745, 53–63.

Vestner, J., Malherbe, S., Toit, M. D., et al. (2011). Investigation of the volatile composition of pinotage wines fermented with different malolactic starter cultures using comprehensive two-dimensional gas chromatography coupled to time-off-light mass spectrometry (GCxGC-TOF-MS). *Journal of Agricultural and Food Chemistry*, 59, 12732–12744.

Wang, X., Li, P., Zhang, W., Zhang, Q., Ma, F., Yu, L. and Wang, L. (2012). Screening for pesticide residues in oil seeds using solid-phase dispersion extraction and comprehensive two-dimensional gas chromatography time-of-flight mass spectrometry. *Journal of Separation Science*, 35, 1634–1643.

Wang, X., Wang, S. and Cai, Z. (2013). The latest developments and applications of mass spectrometry in food-safety and quality analysis. *TrAC Trends in Analytical Chemistry*, 52, 170–185. doi: 10.1016/j.trac.2013.08.005

Wedad, Q. A., Anwar, N., Aseela, S. Q., et al. (2017). Gas chromatography: Principles, advantages and applications in food analysis. *International Journal of Agriculture Innovations and Research*, 6(1), 1473–2319.

Xian, W., Shujuan, W. and Zongwei, C. (2013). The latest developments and applications of mass spectrometry in food-safety and quality analysis. *Trends in Analytical Chemistry*, 52, 170–185. doi: 10.1016/j.trac.2013.08.005

Yu, S. and Xu, X. M. (2012). Study of matrix-induced effects in multi-residue determination of pesticides by online gel permeation chromatography-gas chromatography/mass spectrometry. *Rapid Communication in Mass Spectrometry*, 26, 963–977.

Zhou, T., Xiao, X. and Li, G. (2012). Microwave accelerated selective Soxhlet extraction for the determination of organophosphorus and carbamate pesticides in ginseng with gas chromatography/mass spectrometry. *Analytica Chemistry*, 84, 5816–5822.

CHAPTER 12

High-Performance Thin-Layer Chromatography

Semih Ötles[1] and Vasfiye Hazal Özyurt[2]

[1]Food Engineering Department, Ege University, Izmir, Turkey
[2]Gastronomy and Culinary Arts, Mugla Sitki Kocman University, Mugla, Turkey

CONTENTS

12.1	Introduction	195
12.2	Principles	195
12.3	Natural Toxins in Food	196
References		197

12.1 INTRODUCTION

Natural toxins are natural defense chemicals which are found in plant to protect itself against attacks and stressors (Hickman et al. 2021). There are more than 20,000 natural toxic compounds to humans and their toxicity are different. Natural toxins may also be produced in high quantities which are some exceeding 10 g/m^2/year (Hansen et al. 2021). It is important to determine the current toxin levels in plant in order to determine any potential contamination. Moreover, since their concentrations are in the trace and ultra-trace range, analytical experience is needed for appropriate sampling, sample preparation, subsequent measurement by a suitable method and finally the evaluation and interpretation of the data obtained (Charegaonkar 2011; Patel et al. 2011). For natural toxin analysis, several methods exist. The largest effort focuses on high-performance liquid chromatography (HPLC) with fluorescence or ultraviolet detection and liquid chromatography coupled to mass spectrometry for detection (LC-MS) (Otero et al. 2013). This chapter describes HPTLC applications for the analysis of natural toxins with different chemical characteristics.

12.2 PRINCIPLES

Thin-layer chromatography (TLC) contains a layer of silica poured onto a glass plate as a stationary phase. There are two phases in TLC: the stationary phase and the mobile phase. The mobile phase is placed in the bottom of a glass container and the glass plate, to which the test samples have been applied, is placed into this. As the mobile phase passes up through the glass plate, the compounds will adhere to the stationary phase at different

DOI: 10.1201/9781003222194-15

rates leading to separation of components (Patel et al. 2011). HPTLC is a modern technique which is adapted from thin-layer chromatography. Thin-layer chromatography is used to isolate and to analyse individual compounds (Şeremet et al. 2013). HPTLC also include separation and detection (Meisen et al. 2005) and is used for fingerprinting of extracts and determination and identification of new compounds (Kokotkiewicz et al. 2016; Scrob et al. 2019). HPTLC is a method which is suitable for both qualitative and quantitative analytical tasks (Loescher et al. 2014). It is an offline process (Rashmin et al. 2012). HPTLC have much greater resolution and separation of components capacity than TLC. It is a simple, low cost, fast method, as well as it has a chance to detect multiple components. It is applied for analysis in food, drug, and environmental samples (Ramachandra 2017). HPTLC is composed of a sample applicator, development chamber, derivatizer, scanner, mass spectrometers, and software. A sample applicator is used to apply sample to stationary phase. A development chamber is used to separate components. A derivatizer help to detect compounds unvisible under white/UV light. A scanner is used to visualization. Mass spectrometer is optional. Software is obligatory for the analysis (Charegaonkar 2011).

12.3 NATURAL TOXINS IN FOOD

Cassava, sorghum, stone fruits, bamboo roots, and almonds are especially important foods containing cyanogenic glycosides. The potential toxicity of a cyanogenic plant depends primarily on the potential that its consumption will produce a concentration of cyanide that is toxic to exposed humans. Death due to cyanide poisoning can occur when the cyanide level exceeds the limit an individual is able to detoxify. Linamarin is correlated with bitter taste. Lebot and Kaoh (2017) developed a high-performance thin-layer chromatography method for the quantitation of linamarin in the sweet-type cultivars of cassava. (Lebot and Kaoh 2017). Bodart et al. (2016) presented a method for the determination of linamarin using HPTLC in cassava (Bodart et al. 2016).

Furocoumarins are polyphenolic compounds, synthesized from L-phenylalanine, which may occur in a linear form with the furan ring attached to the 6, 7 position of the benzo-2-pyrene nucleus (Rawat et al. 2013). These toxins are present in many plants such as parsnips (closely related to carrots and parsley), celery roots, citrus plants (lemon, lime, grapefruit, bergamot) and some medicinal plants. Some of these toxins can cause gastrointestinal problems in susceptible people. Rawat et al. (2013) developed a simple high performance thin layer chromatography (HPTLC) method for the simultaneous determination of psoralen and heraclenol in the fruits of *Heracleum candicans* (Rawat et al. 2013).

Mycotoxins are naturally occurring toxic compounds produced by certain types of molds. Molds that can produce mycotoxins grow on numerous foodstuffs such as cereals, dried fruits, nuts, and spices. Most mycotoxins are chemically stable and survive food processing. Long-term effects on health of chronic mycotoxin exposure include the induction of cancers and immune deficiency. Khatoon et al. (2012) assayed for 14 toxicologically significant mycotoxins such as aflatoxin B1 (AfB1), aflatoxin B2 (AfB2), aflatoxin G1 (AfG1), aflatoxin G2 (AfG2), zearalenone (ZON), deoxynivalenol (DON), 3acetyl-deoxynivalenol (3A DON), 15acetyl-deoxynivalenol (15A-DON), nivalenol (NIV), T-2 toxin (T-2), HT-2 toxin (HT-2), diacetoxyscirpenol (DAS), neosolaniol (NEOS), and fusarenone-x (Fus-x) simultaneously in maize from main maize growing areas of Pakistan. High-performance thin-layer chromatography (HPTLC) was used for quantification (Khatoon et al. 2012). Vega-Herrera et al. (2017) used HPTLC/MS to confirm the presence of deoxynivalenol in wheat crops (Vega-Herrera et al. 2017). Priyanka et al. (2014) investigated the different mycotoxigenic *Aspergillus* species from the major food grains of southern states of India viz. maize, paddy, groundnut, and sorghum using HPTLC. Among the different grain samples tested 83% of ground nut, 69% of maize, 57% of sorghum, and 29% of paddy had aflatoxin B1 beyond the permitted limits, where 82% of maize, 70% of sorghum, 42% of paddy, and 17% of groundnut had higher concentrations of ocharatoxin A of the permitted threshold (5 µg kg-1) (Priyanka et al. 2014). Rao et al. (2016) examined the degradation of aflatoxin B1 using HPTLC (Rao et al. 2016). Pradhan and Ananthanarayan (2021) validated a method for the analysis of Aflatoxin B1 using a HPTLC with respect to linearity, sensitivity, precision, and accuracy. The optimized HPTLC method was then used to assess the AFB1 content in marketed samples of groundnut, corn, rice, wheat and dried chillies from the Mumbai region and it was found that eight groundnut samples, five corn samples, five wheat samples, and seven chilli samples out of 20 each had AFB1 concentrations above the Indian permissible regulatory limit of 30 ppb (Pradhan and Ananthanarayan 2021). Soylemez and Yamac (2021) screened 94 macro fungi isolates for aflatoxin B1 and ocratoxin A degradation potantial using a semi-quantitative HPTLC method (Söylemez and Yamac 2021). Arunachalam et al. (2012) attempted to evaluate the inhibition of *Aspergillus flavus* growth and aflatoxin production using *Cassia alata*. The aflatoxin production was observed in HPTLC (Arunachalam et al. 2012).

All solanacea plants, which include tomatoes, potatoes, and eggplants, contain natural toxins called solanines and chaconine (which are glycoalkaloids). While levels are generally low, higher concentrations are found in potato sprouts and bitter-tasting peel and green parts, as well as in green tomatoes. The plants produce the toxins in response to stresses like bruising, UV light, microorganisms, and attacks from insect pests and herbivores. To reduce the production of solanines and chaconine it is important to store potatoes in a dark, cool, and dry place, and not to eat green or sprouting parts. Skarkova et al. 2008 used a method for quantification of glycoalkaloids (α-solanine and α-chaconine) in peeled potato tubers using HPTLC (Skarkova et al. 2008). Mader et al. (2009) and Mader et al. (2009) used an HPTLC method for analysis of α-solanine and α-chaconine in potatoes during industrial processing (Mader et al. 2009; Mäder et al. 2009).

Pyrrolizidine alkaloids (PA) are stable during processing, and have been detected in herbal teas, honey, herbs and spices and other food products, such as cereals and cereal products. Human exposure is estimated to be low, however. Due to the complexity of the subject and the large number of related compounds, the overall health risk has not been fully evaluated yet. Şeremet et al. (2013) developed a HPTLC method using a new mobile phase (chloroform: methanol: ammonia 25%: hexane – 82: 14: 2.6: 20 v/v) that permited the separation of senecionine (Şeremet et al. 2013). Smyrska-Wieleba et al. (2016) utilized HPTLC for the determination of pyrrolizidine alkaloids (PAs) and their N-oxides (PANOs) in *Tussilago farfara* L and *Arnebia euchroma* (Royle) I.M. Johnst. HPTLC confirmed the postulated presence of PAs (saturated and unsaturated) or PANOs in the tested extracts (Smyrska-wieleba et al. 2016).

REFERENCES

Arunachalam, C., Arunkumar, S., Murugan, A. M., Wainwright, M. and Zayed, M. E. (2012). Efficacy of cassia alata leaves powder on inhibition of aspergillus flavus growth and aflatoxin production. *Biosci. Biotechnol. Res. Asia*, 9, 223–227.

Bodart, P., Penelle, J., Angenot, L. and Noirfalise, A. (2016). Direct quantitative analysis of linamarin in cassava by high-performance thin layer chromatography. *J. Planar Chromatogr.*, 11, 38–42.

Charegaonkar, D. (2011). High-performance thin-layer chromatography: Excellent automation. In *High-performance thin-layer chromatography (HPTLC)*. Springer, Berlin, Heidelberg. pp. 55–65.

Hansen, H. C. B., Hilscherova, K. and Bucheli, T. D. (2021). Natural toxins: Environmental contaminants calling for attention. *Environmental Sciences Europe*, 33(1), 1–8. 10.1186/s12302-021-00543-6

Hickman, D. T., Rasmussen, A., Ritz, K., Birkett, M. A. and Neve, P. (2021). Review: Allelochemicals as multi-kingdom plant defence compounds: Towards an integrated approach. *Pest Manag. Sci.*, 77, 1121–1131. 10.1002/ps.6076

Khatoon, S., Hanif, N., Tahira, I., Sultana, N., Sultana, K. and Ayub, N. (2012). Natural occurrence of aflatoxins, zearalenone and trichothecenes in maize grown in Pakistan. *Pak. J. Bot.*, 44, 231–236.

Kokotkiewicz, A., Migas, P., Stefanowicz, J. and Luczkiewicz, M. (2016). Densitometric TLC analysis for the control of tropane and steroidal alkaloids in Lycium barbarum. *Food Chemistry*, 221, 535–540. 10.1016/j.foodchem.2016.11.142

Lebot, V. and Kaoh, J. (2017). Estimating sugars in 212 landraces and hybrids of sweet type cassava (Manihot esculenta Crantz, Euphorbiaceae). *Genet Resour Crop Evol*, 64, 2093–2103.

Loescher, C. M., Morton, D. W., Razic, S. and Agatonovic-Kustrin, S. (2014). High performance thin layer chromatography (HPTLC) and high performance liquid chromatography (HPLC) for the qualitative and quantitative analysis of calendula officinalis-advantages and limitations. *J. Pharm. Biomed. Anal.*, 98, 52–59. 10.1016/j.jpba.2014.04.023

Mäder, J., Fischer, W., Schnick, T. and Kroh, L. W. (2009). Changes in glycoalkaloid composition during potato processing: Simple and reliable quality control by HPTLC. *JPC–Journal of Planar Chromatography–Modern TLC*, 22(1), 43–47. 10.1556/JPC.22.2009.1.8

Mader, J., Rawel, H. and Kroh, L. (2009). Composition of phenolic compounds and glycoalkaloids a-solanine and a-Chaconine during commercial potto processing. *J. Agric. Food Chem.*, 57, 6292–6297.

Meisen, I., Friedrich, A. W., Karch, H., Witting, U., Peter-katalinic, J. and Mu, J. (2005). Application of combined high-performance thin-layer chromatography immunostaining and nanoelectrospray ionization quadrupole time-of-flight tandem mass spectrometry to the structural characterization of high- and low-affinity binding ligands of Shiga toxi. *Rapid Commun. Mass Spectrom.*, 1, 3659–3665. 10.1002/rcm.2241

Otero, P., Rodríguez, P., Botana, A. M., Alfonso, A. and Botana, L. M. (2013). Analysis of natural toxins. In *Liquid chromatography: Applications*, pp. 411–430. Elsevier. 10.1016/B978-0-12-415806-1.00015-2

Patel R. B., Patel M. R. and Batel B. G. (2011). Experimental aspects and implementation of HPTLC. In Srivastava, M. (ed.). *High-performance thin-layer chromatography (HPTLC)*. pp. 41–55, Springer, Berlin, Heidelberg.

Pradhan, S. and Ananthanarayan, L. (2021). Standardization and validation of a high - performance thin - layer chromatography method for the quantification of aflatoxin B1 and its application in surveillance of contamination level in marketed food commodities from the Mumbai region. *JPC – J. Planar Chromatogr. – Mod. TLC*, 33, 617–630. 10.1007/s00764-020-00073-6

Priyanka, S. R., Venkataramana, M., Kumar, P., Rao, K., Sripathi, C. and Vardhan, H. (2014). Occurrence and molecular detection of toxigenic aspergillus species in food grain samples from India. *J Sci Food Agri*, 94, 537–543. 10.1002/jsfa.6289

Ramachandra, B. (2017). Development of impurity profiling methods using modern analytical techniques. *Critical Reviews in Analytical Chemistry*, 47(1), 24–36.

Rao, K. R., Vipin, A. V., Hariprasad, P., Appaiah, K. A. A. and Venkateswaran, G. (2016). Biological detoxification of aflatoxin B1 by Bacillus licheniformis CFR1 K. *Food Control*, 71, 234–241. 10.1016/j.foodcont.2016.06.040

Rashmin, P., Mrunali, P., Nitin, D., Nidhi, D. and Bharat, P. (2012). HPTLC method development and validation: Strategy to minimize methodological failures. *J. Food Drug Anal.*, 20, 794–804. 10.6227/jfda.2012200408

Rawat, A. K. S., Singh, A. P., Singh, D. P., Pandey, M. M., Govindarajan, R. and Srivastava, S. (2013). Separation and identification of furocoumarin in fruits of heracleum candicans DC. by HPTLC. *Journal of Chemistry*.

Scrob, T., Hosu, A. and Cimpoiu, C. (2019). Trends in analysis of vegetables by high performance TLC. *J. Liq. Chromatogr. Relat. Technol.*, 1–9. 10.1080/10826076.2019.1585611

Şeremet, O. C., Olaru, O. T., İlie, M., Negreş, S. and Balalau, D. (2013). HPTLC evaluation of the pyrollizidine alkaloid senecionine in certain. *Farmacia*, 61, 756–763.

Skarkova, J., Ostry, V. and Ruprich, J. (2008). Instrumental HPTLC determination of α -solanine and α -chaconine in peeled potato tubers. *J. Planar Chromatogr.*, 21, 113–117. 10.1556/JPC.21.2008.2.7

Smyrska-wieleba, N., Wojtanowski, K. K. and Mroczek, T. (2016). Comparative HILIC/ ESI-QTOF-MS and HPTLC studies of pyrrolizidine alkaloids in flowers of Tussilago farfara and roots of Arnebia euchroma. *Phytochemistry Letters*, 20, 339–349. 10.1016/j.phytol.2016.11.009

Söylemez, T. and Yamac, M. (2021). Screening of macrofungi isolates for Aflatoxin B1 and ochratoxin a degradation. *Biol. Bullettin*, 48, 122–129.

Vega-Herrera, M., Madariaga, R., Aranda, M. and Morlock, G. E. (2017). Confirmation of deoxynivalenol presence in chilean wheat by high-performance thin-layer chromatography-mass spectromtry. *J. Chil. Chem*, 62, 3435–3437.

CHAPTER 13

Capillary Electrophoresis

Suraj Singh S. Rathod[1], Gagandeep Kaur[1], Neha Sharma[1], Navneet Khurana[1], and Awanish Mishra[2]

[1]Department of Pharmacology, School of Pharmaceutical Sciences, Lovely Professional University, Phagwara, India
[2]Department of Pharmacology and Toxicology, National Institute of Pharmaceutical Education and Research (NIPER) – Guwahati, India

CONTENTS

13.1	Introduction	201
13.2	Types of Capillary Electrophoresis	202
	13.2.1 Capillary Zone Electrophoresis (CZE)	202
	13.2.2 Capillary Electrochromatography (CEC)	203
	13.2.3 Micellar Electrokinetic Capillary Chromatography (MEKC)	203
	13.2.4 Capillary Gel Electrophoresis (CGE)	203
	13.2.5 Capillary Isoelectric Focusing (CIEF)	203
	13.2.6 Capillary Isotachorphoresis (CITP)	204
13.3	Analysis of Cyanogenic Glycosides, Pyrrolizidine Alkaloids, and Glycoalkaloid	204
	13.3.1 Detection Methods of Food Toxins	207
13.4	Conclusion and Future Perspective	212
References		212

13.1 INTRODUCTION

Capillary electrophoresis (CE) is a technique used for separating charged particles. Due to its flexibility, CE has recently gained popularity as a separation technique. It was originally employed in the 1980s. CE has a high extraction speed and good resolution. The charge, isoelectric point, hydrodynamic radius, and molecular weight of analytes are all factors that influence the separation method. When utilized in combination with mass spectrometry (MS), this advanced analytical approach offers mass information and allows for molecular characterization based on fragmentation (Stolz et al. 2019). Food assessment may be thought of as a series of steps that begins with isolation and ends with identification. The success of food analysis depends on the precision and reliability of these techniques. Nutritional content, quality aspects, and the detection of dangerous chemicals such as toxins are all evaluated in food components and supplies (Juan-Garca et al. 2005) (Figure 13.1).

DOI: 10.1201/9781003222194-16

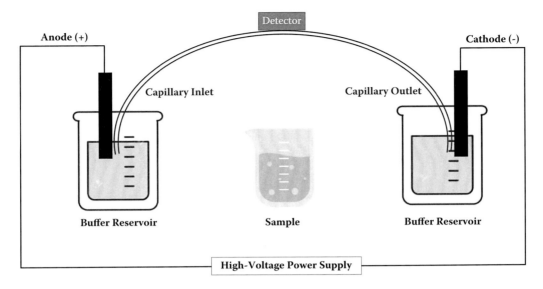

FIGURE 13.1 Instrumentation of capillary electrophoresis.

13.2 TYPES OF CAPILLARY ELECTROPHORESIS

Different biomolecules are analyzed by different CE methods like capillary gel electrophoresis (CGE), capillary electrochromatography (CEC), electrokinetic capillary chromatography (MEKC), capillary zone electrophoresis (CZE), micellar capillary isoelectric focusing (CIEF), and capillary isotachophoresis (CITP) for food quality and toxicant analysis (Schmitt-Kopplin, P. 2005) (Figure 13.2).

13.2.1 Capillary Zone Electrophoresis (CZE)

The most used technique among the CEs is capillary zone electrophoresis (CZE), also known as free solution CE. A combination in a solution can be instantly and easily

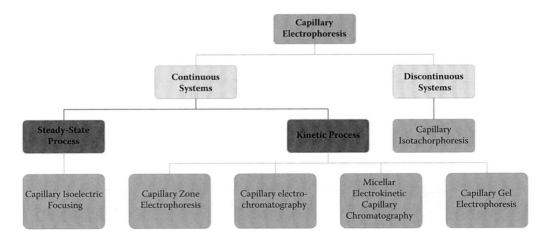

FIGURE 13.2 Types of capillary electrophoresis.

separated into its constituents. The separation is based on variations in electrophoretic mobility, which are proportionate to molecule charge and inversely proportionate to solvent viscosity and atom radius. Increased selectivity in the separation of charged or neutral analytes can be achieved by modifying the quality of the separation buffer and optimizing interactions between the sample and specific buffer constituents. CZE permits anions and cations to be separated based on their effective charge-to-size ratio (Schmitt, P. 1997). The rate at which the ion moves is precisely proportionate to the electrophoretic mobility and the magnitude of the electric field (Castagnola 1993).

13.2.2 Capillary Electrochromatography (CEC)

Capillary electrochromatography is a separating technique that is like packed column chromatography. The mobile liquid travels through the particles and over the silica wall. Because of the charges on the stationary surface, electroosmosis flow ensues. In comparison to the pumped parabolic flow that enhances band broadening, CEC and CZE both feature a plug-type flow.

13.2.3 Micellar Electrokinetic Capillary Chromatography (MEKC)

It is a separation technique based on the splitting of solutes among micelles and the solvent. Micelles are surface-active agent molecule aggregates that develop when a surface-active agent is applied to a solution at a concentration higher than the critical micelle concentration. The granules are naturally attracted to the positively charged anode since they have polar negatively charged surfaces. The micelles are represented at the cathode as well, although at a slower rate due to the electroosmotic flow toward the cathode. Hydrophobic molecules will spend most of their time in the micelle, whereas hydrophilic molecules will move beyond the solvent more instantly. When micelles are absent, neutral molecules move with the electroosmotic flow, resulting in no separation. There are several factors which influences the electroosmotic flow in MEKC are pH, surfactant content, additives, and capillary wall polymer (Castagnola 1993).

13.2.4 Capillary Gel Electrophoresis (CGE)

Capillary gel electrophoresis is used as a separation technique based on a difference in solute size as the particles migrate through the gel. Gels are beneficial because they reduce solute diffusion, which promotes zone expansion, and prevents the solute from being absorbed by the capillary walls. They also slow down the molecules, limiting heat transmission. Capillary SDS-PAGE is a regularly used gel device for separating proteins. It's a very sensitive system that only needs a few samples to work.

13.2.5 Capillary Isoelectric Focusing (CIEF)

This is a method for separating peptides and proteins that are widely utilized. Because they contain both positive and negative charges, these molecules are referred to as zwitterionic compounds. The charge is determined by the functional groups linked to the main chain as well as the pH of the surrounding environment. Furthermore, each molecule has its isoelectric point (pI). The molecule has no net charge when the surrounding pH is equal to its pI. To be clear, this isn't the pH at which all bases are deprotonated, and all acids are protonated in a protein; rather, it's the pH at which positive and negative charges cancel out to zero (Weston and Brown 1997). When the pH is below the pI, the molecule is positive, and when the pH is above the pI, the molecule is negative. A pH gradient can be used to separate molecules in a mixture because the charge varies with pH. During a CIEF separation, the capillary is filled with the sample in solution, and no EOF is normally employed (EOF is removed by using a coated capillary). When the voltage is applied, the ions will travel to an area where they will become neutral (pH = pI). The capillary's anodic end is immersed in an acidic solution (low pH), while the cathodic end is immersed in a basic solution (high pH). Compounds with equal isoelectric points are "focused" into sharp segments and remain in their zone, allowing them to be distinguished.

13.2.6 Capillary Isotachorphoresis (CITP)

In a discontinuous system, only capillary isotachophoresis (CITP) can be employed. The analyte migrates in zones, and the length of each zone may be quantified to determine the amount of sample present.

Generally, CE has represented its utility in the analysis of low molecular weight organic acids, amino acids, sea products, wines, proteins, polyphenols, and vitamins available in food as well as for detection of microbial contaminations and food quality.

13.3 ANALYSIS OF CYANOGENIC GLYCOSIDES, PYRROLIZIDINE ALKALOIDS, AND GLYCOALKALOID

Cyanogenic glycosides (CGs) are alpha-hydroxy nitriles derivatives. Linamarin, prunasin, amygdalin all are examples of CGs. These compounds release cyanide in high concentrations when ingested, which is toxic to human beings. Numerous studies have been performed to find reliable methods for cyanide detection. Initially, these are usually achieved by converting these compounds to cyanide via enzymatic or chemical hydrolysis, followed by detection using enzyme membrane reactors and titration with a standard AgNO3 solution (Campa et al. 2000). Among all the different methods, CE can also help to analyze CGs. Out of all CE methods, micellar capillary electrophoresis was found to have more selectivity towards CGs detection (Campa et al. 2000). Craig et al. detected amygdalin, a source of cyanide in a sample using capillary electrophoresis. The concentration of amygdalin was found 1 µM in the sample injected into the apparatus (Craig and Guimond 2022). Using CE methods, Marchart et al. detected flavone-O- and C-glycosides in *Achillea setacea*. Seigler et al. reported the presence of prunasin, sambunigrin in *Passiflora edulis*, and *Carica papaya* plants when performed detection for both in CE (Seigler et al. 2002).

MEKC was used to determine the concentrations of prunasin and amygdalin in peach seeds, which were found to be 50 µg/g and 90 µg/g (dry weight), respectively (Campa et al. 2000). Apple seeds contain amygdalin, which can be identified at a concentration of 1 µM (Craig and Guimond 2022). CZE tested 0.5–1.5 mg of prunasin and sambunigrin from *Passiflora edulis* and *Carica papaya* mentioned in Table 13.1 (Seigler et al. 2002).

Sr. no.	Method	Category	Sample	Reagents	Compound detected	References
1.	Micellar capillary electrophoresis	Cyanogenic Glycosides	Peach seeds	sodium dodecyl sulfate, acetic acid, acetate, sodium hydroxide	Prunasin and Amygdalin	Campa et al. 2000
2.	Capillary electrophoresis	Cyanogenic Glycosides	Apple seeds	fluorogenic reagent 3-(2-furoyl) quinoline-2-carboxaldehyde (FQ) and glutamic acid	Amygdalin	Craig and Guimond 2022
3	Capillary electrophoresis	Cyanogenic Glycosides	*Achillea setacea*	sodium borate, methanol	Schaftoside, iso-schaftoside and vicenin-2	Marchart and Kopp 2003
4	Capillary zone electrophoresis	Cyanogenic Glycosides	*Passiflora edulis*	23% TFA solution, S– (-)–1–phenylethylamine derivatives, Na2B4O7, NaBH3CN etc	Prunasin, sambunigrin,	Seigler et al. 2002
5	Capillary zone electrophoresis	Cyanogenic Glycosides	*Carica papaya*	23% TFA solution, S– (-)–1–phenylethylamine derivatives, Na2B4O7, NaBH3CN etc	Prunasin	Seigler et al. 2002
6	Capillary zone electrophoresis	Cyanogenic Glycosides	*Passiflora edulis*	23% TFA solution, S–(-)–1–phenylethylamine derivatives, Na2B4O7, NaBH3CN etc	Cyanogenic Glycosides (2R)–β-d-allopyransosyloxy-2-phenylacetonitrile	Seigler et al. 2002
7	Micellar electrokinetic capillary chromatography	Pyrrolizidine Alkaloids	*Tussilago farfara*	Sodium tetraborate, SDS, methanol, hydrochloric 7 acid, and sodium hydroxide	Senkirkine and Senecionine	Cao et al. 2016
8	Micellar electrokinetic capillary chromatography	Pyrrolizidine Alkaloids	*Gynura segetum*	Lauric acid, Methanol, acetonitrile, ethanol, isopropyl alcohol, acetone.	Seneciphylline and Senecionine	Qi et al. 2009
9	Capillary electrophoresis	Pyrrolizidine Alkaloids	*Tussilago farfara*	Water, methanol, ammonia	Senkirkine and Senecionine	Lebada, R. et al. 2000

(*Continued*)

TABLE 13.1 (Continued)

Sr. no.	Method	Category	Sample	Reagents	Compound detected	References
10	Capillary electrophoresis coupled with electrospray ionization-ion trap mass spectrometry	Glycoalkaloids	Wild potatoes	MeCN-MeOH containing 50 mM ammonium acetate and 1.2 M acetic acid	Solanidine, Alpha-chaconine, Alpha-solanine.	Bianco, G. et al. 2003
11	Capillary electrophoresis coupled with electrospray ionization-ion trap mass spectrometry	Glycoalkaloids	Tomato	MeCN-MeOH containing 50 mM ammonium acetate and 1.2 M acetic acid	Solanidine, Tomatidine, Alpha-chaconine, Alpha-solanine, and Alpha-tomatine	Bianco, G. et al. 2003
12	Nonaqueous capillary electrophoresis	Glycoalkaloids	Solanaceae Plants	600 mM acetic acid, 10 mM sodium hydroxide in methanol, 30 kV	Solasodine, Solasonine, Tomatidine, and Demissidine	Cataldi and Bianco 2008
13	Nonaqueous capillary electrophoresis	Glycoalkaloids	Transgenic potatoes	Ammonium acetate, acetic acid, sodium hydroxide, HPLC-grade methanol, and acetonitrile	α-chaconine, α-solanine	Bianco, G. et al. 2002

Pyrrolizidine alkaloids (PAs) are found in a wide range of plants across the world, including many species that are consumed by humans. In addition to the toxicity that these compounds can produce in people and cattle, they are renowned for their vast spectrum of pharmacological characteristics, which may be utilized in drug development programs. PAs are a class of alkaloids generated from ornithine that are present in plants of many taxa, as well as insects that consume them to protect themselves from predators. It is found as esters (mono-, di-, or macrocyclic diesters) generated by a necine base (amino alcohols) and one or more nucleic acids (mono-or dicarboxylic aliphatic acids), which are responsible for their structural variety (Moreira et al. 2018).

CE is used for the detection of PA in food as well as in herbal plants. Cao et al. (2016) detected the presence of senkirkine and senecionine in *Tussilago farfara* using the micellar electrokinetic chromatography method (MKEC). The limit for the detection of senkirkine and senecionine in *Tussilago farfara* is 2–5 microg/L (Cao et al. 2016). MKEC detected PA responsible for hepatotoxicity in *Gynura segetum*. Seneciphylline and senecionine are present in *Gymura segetum*, a herbal medicine used for hemostasis and detumescence. The contents of seneciphylline and senecionine in *Gynura segetum* were 526.1 ± 7.2 microg/1 and 218.2 ± 2.8 microg/1 (Qi et al. 2009). Using CE methods senkirkine and senecionine were detected in *Tussilago fafara*. The concentration for senkirkine is 4.5–664.1 µg/mL and for senecionine.4.9–593.0 µg/mL mentioned in Table 13.1.

Secondary natural toxic metabolites generated by plants in the Solanaceae family are known as glycoalkaloids. Peppers (*Capsicum annum*), potatoes (*Solanum tuberosum* L.), eggplant (*Solanum melongena*), tomatoes (*Lycopersicon esculentum*), nightshade, thorn apple, and capsicum are all examples of these plants. They play a crucial function in plants because of their toxicity and the potential for negative effects on humans if taken in large quantities. They provide taste, but in greater concentrations, they can cause bitterness as well as a burning sensation when swallowed. Glycoalkaloids are found in every organ of the plant. Unripe berries, young leaves, blossoms, branches, and sprouts have the greatest amounts of (metabolically active parts). Worms, fungi, insects, bacteria, and viruses are among the diseases and predators that allelochemicals protect against (Omayio et al. 2016).

Cataldi and Bianco determined glycoalkaloids in potatoes and tomato extract. α-solanine, α-chaconie, α-tomatine, solanidine, and tomatidine are the glycoalkaloids detected in the sample by non-aqueous capillary electrophoresis (NACE) (Cataldi and Bianco 2008). Solanidine, alpha-chaconine, alpha-solanine were also detected in wild potatoes and tomatoes, belonging to the family. In combination with online electrospray ionization–mass spectrometry, NACE identifies glycoalkaloid in virus-resistant potatoes. When compared to tubers from control, intermediate, and susceptible line plants, tubers from the resistant line had a slightly greater concentration of solanine in the peel and meat. It is well established that chaconine is more embryotoxic and induces more liver enzymes than solanine (Bianco et al. 2003). The concentration 0.32–0.34 mg/g for solanidine, 0.21–0.24 mg/g for tomatidine, alpha-chaconine, alpha-solanine, and alpha-tomatine is analyzed by capillary electrophoresis coupled with electrospray ionization-ion trap mass spectrometry.

13.3.1 Detection Methods of Food Toxins

Cyanogenic glycosides, pyrrolizidine alkaloids, and glycoalkaloids have all been linked to food toxins. Table 13.2 lists the food toxins that have been found and may be detected using analytical techniques. Prunasin, Amygdalin, Schaftoside, Iso-Schaftoside, Vicenin,

TABLE 13.2 Structural Characterization and Detection Methods of Food Toxins

S. no.	Chemical name	Structure	CE mode	Detection methods	Reference
1	Prunasin		MEKC and CZE	UV = 214 nm, Nuclear Magnetic Resonance (NMR)	Campa et al. 2000; Seigler et al. 2002
2	Amygdalin		MEKC and CE	UV = 214 nm, Laser-induced fluorescence	Campa et al. 2000; Craig and Guimond 2022
3	Schaftoside		CZE	UV = 275 nm	Marchart and Kopp 2003
4	Iso-Schaftoside		CZE	UV = 275 nm	Marchart and Kopp 2003

5	Vicenin		CZE	UV = 275 nm	Marchart and Kopp 2003
6	Sambunigrin		CZE	NMR	Seigler et al. 2002
7	Senkirkine		CE, MEKC	Electrospray ionization multistage mass spectrometry (ESI-MS), UV = 220 nm	Lebada et al. 2000; Cao et al. 2016
8	Senecionine		CE, MEKC	ESI-MS, UV = 220 nm	Lebada et al. 2000; Qi et al. 2009; Cao et al. 2016

(*Continued*)

TABLE 13.2 (Continued)

S. no.	Chemical name	Structure	CE mode	Detection methods	Reference
9	Seneciphylline		MEKC	ESI-MS, UV = 220 nm	Qi et al. 2009
10	Solanidine		Non-aqueous Capillary electrophoresis (NACE)	UV = 205 nm ESI-MS	Bianco et al. 2003
11	Alpha-chaconine		NACE	UV = 205 nm ESI-MS	Bianco et al. 2003; Bianco et al. 2002
12	Alpha-tomatine		NACE	UV = 205 nm ESI-MS	Bianco et al. 2003; Bianco et al. 2002

13	Tomatidine		NACE	UV = 205 nm ESI-MS	Bianco et al. 2003; Cataldi and Bianco 2008
14	Alpha-solanine		NACE	UV = 205 nm ESI-MS	Bianco et al. 2003; Bianco et al. 2002
15	Solasodine		NACE	UV = 195 nm ESI-MS	Cataldi and Bianco 2008
16	Solasonine		NACE	UV = 195 nm ESI-MS	Cataldi and Bianco 2008
17	Demissidine		NACE	UV = 195 nm ESI-MS	Cataldi and Bianco 2008

Sambunigrin, Senkirkine, Senecionine, Seneciphylline, Solanidine, Alpha-chaconine, Alpha-tomatine, Tomatidine, Alpha-solanine, Solasodine, Solasonine, and Demissidine were identified using different capillary electrophoresis methods like MEKC, CZE, and NACE with the help detection technique UV, ESI-MS, and NMR, respectively.

13.4 CONCLUSION AND FUTURE PERSPECTIVE

Capillary electrophoresis over chromatographic approaches in determining toxins in food, it has become more common. However, most of the publications discuss methodological advancements that are insignificant in terms of using the approach to tackle specific issues. Because the CE's effectiveness is strongly dependent on the physical-chemical characteristics of the solutes, even small changes in the composition can have a significant impact. Micellar electrokinetic capillary chromatography (MEKC) has been identified as the most important capillary electrophoresis for the detection of food toxins, and it has been examined using UV, ESI-MS, and NMR. For the identification and characterization of food toxins, there is a lot of literature available on cyanogenic glycosides, pyrrolizidine alkaloids, and glycoalkaloids. Furocoumarins and lectins should be considered in future studies.

REFERENCES

Bianco, G., Schmitt-Kopplin, P., Crescenzi, A., Comes, S., Kettrup, A. and Cataldi, T. R. (2003). Evaluation of glycoalkaloids in tubers of genetically modified virus Y-resistant potato plants (var. Désirée) by non-aqueous capillary electrophoresis coupled with electrospray ionization mass spectrometry (NACE-ESI-MS). *Analytical and Bioanalytical Chemistry*, 375(6), 799–804. 10.1007/s00216-003-1831-3

Bianco, G., Schmitt-Kopplin, P., De Benedetto, G., Kettrup, A. and Cataldi, T. R. (2002). Determination of glycoalkaloids and relative aglycones by nonaqueous capillary electrophoresis coupled with electrospray ionization-ion trap mass spectrometry. *Electrophoresis*, 23(17), 2904–2912. 10.1002/1522-2683(200209)23:17<2904::AID-ELPS2904>3.0.CO;2-1

Brown, P. R. and Weston, A. (1997). *HPLC and CE: Principles and practice*. Academic Press.

Campa, C., Schmitt-Kopplin, P., Cataldi, T. R., Bufo, S. A., Freitag, D. and Kettrup, A. (2000). Analysis of cyanogenic glycosides by micellar capillary electrophoresis. *Journal of Chromatography. B, Biomedical Sciences and Applications*, 739(1), 95–100. 10.1016/s0378-4347(99)00375-8

Cao, K., Xu, Y., Mu, X., Zhang, Q., Wang, R. and Lv, J. (2016). Sensitive determination of pyrrolizidine alkaloids in tussilago farfara L. by field-amplified, sample-stacking, sweeping micellar electrokinetic chromatography. *Journal of Separation Science*, 39(21), 4243–4250. 10.1002/jssc.201600762

Castagnola, M. (1993). Capillary electrophoresis: Principles, practice, and application S. F. Y. LI (ED). Elsevier science publishers, Amsterdam, The Netherlands, 1992. *Journal of Chromatography Library, vol. 52. Biomedical Chromatography*, 7(2), 1–581. 10.1002/BMC.1130070215

Cataldi, T. R. and Bianco, G. (2008). Capillary electrophoresis of tropane alkaloids and glycoalkaloids occurring in Solanaceae plants. *Methods in Molecular Biology (Clifton, N.J.)*, 384, 171–203. 10.1007/978-1-59745-376-9_9

Craig, D. B. and Guimond, M. S. (2022). Analysis of cyanide using fluorogenic derivatization and capillary electrophoresis. *Food Chemistry*, 370, 131377. 10.1016/j.foodchem.2021.131377

Juan-García, A., Font, G. and Picó, Y. (2005). Determination of organic contaminants in food by capillary electrophoresis. *Journal of Separation Science*, 28(9–10), 793–812. 10.1002/jssc.200500041

Lebada, R., Schreier, A., Scherz, S., Resch, C., Krenn, L. and Kopp, B. (2000). Quantitative analysis of the pyrrolizidine alkaloids senkirkine and senecionine in tussilago farfara L. by capillary electrophoresis. *Phytochemical Analysis*, 11(6), 366–369.

Marchart, E. and Kopp, B. (2003). Capillary electrophoretic separation and quantification of flavone-O- and C-glycosides in Achillea setacea W. et K. *Journal of Chromatography. B, Analytical Technologies in the Biomedical and Life Sciences*, 792(2), 363–368. 10.1016/s1570-0232(03)00262-9

Moreira, R., Pereira, D. M., Valentão, P. and Andrade, P. B. (2018). Pyrrolizidine alkaloids: Chemistry, pharmacology, toxicology and food safety. *International Journal of Molecular Sciences*, 19(6), 1668. 10.3390/ijms19061668

Omayio, D. G., Abong, G. O. and Okoth, M. W. (2016). A review of the occurrence of glycoalkaloids in potato and potato products. *Current Research in Nutrition and Food Science*, 4(3), 195–202. 10.12944/CRNFSJ.4.3.05

Qi, X., Wang, S., Wu, B. and Qu, H. (2009). Determination of hepatotoxic pyrrolizidine alkaloids in gynura segetum by MEKC. *Chromatographia*, 70(1–2), 281–285. 10.1365/S10337-009-1149-1

Schmitt, P., Poiger, T., Simon, R., Freitag, D., Kettrup, A. and Garrison, A. W. (1997). Simultaneous determination of ionization constants and isoelectric points of 12 hydroxy-s-triazines by capillary zone electrophoresis and capillary isoelectric focusing. *Analytical Chemistry*, 69(13), 2559–2566. 10.1021/ac9609

Schmitt-Kopplin, P. (2005). Capillary electrophoresis: From small ions to macromolecules. Humana Press Totowa: NJ.

Seigler, D. S., Pauli, G. F., Nahrstedt, A. and Leen, R. (2002). Cyanogenic allosides and glucosides from passiflora edulis and carica papaya. *Phytochemistry*, 60(8), 873–882. 10.1016/s0031-9422(02)00170-x

Stolz, A., Jooß, K., Höcker, O., Römer, J., Schlecht, J. and Neusüß, C. (2019). Recent advances in capillary electrophoresis-mass spectrometry: Instrumentation, methodology and applications. *Electrophoresis*, 40(1), 79–112. 10.1002/ELPS.201800331

CHAPTER **14**

Enzyme-Linked Immunosorbent Assay

Ritam Bandopadhyay[1], Suraj Singh S. Rathod[1],
and Awanish Mishra[1,2]

[1]Department of Pharmacology, School of Pharmaceutical Sciences, Lovely Professional
University, Phagwara, India
[2]Department of Pharmacology and Toxicology, National Institute of Pharmaceutical
Education and Research (NIPER) – Guwahati, India

CONTENTS

14.1	Introduction	215
14.2	Enzyme-Linked Immunosorbent Assay	216
	14.2.1 Working Principle of ELISA	217
	14.2.2 Types of ELISA	217
	14.2.2.1 Direct ELISA	218
	14.2.2.2 Indirect ELISA	218
	14.2.2.3 Sandwich ELISA (Antibody Screening)	219
	14.2.2.4 Competitive ELISA (Antigen/Antibody Screening)	219
14.3	ELISA in the Structural Characterization and Identification of Naturally Occurring Food Toxins	219
	14.3.1 Cyanogenic Glycosides	219
	14.3.2 Pyrazolidine Alkaloids	222
	14.3.3 Furocoumarins	222
	14.3.4 Glycoalkaloids	223
	14.3.5 Lectin	225
14.4	Structural Characterization and Detection Methods of Food Toxins	226
14.5	Conclusion	235
References		235

14.1 INTRODUCTION

The basic principle of Enzyme-Linked Immunosorbent Assay (ELISA) was established in 1941, simultaneous to the development of radioimmunoassay (RIA) [1]. Both techniques, ELISA and RIA, follow the same principle, but are different in some respects. The first practical application of RIA was evident in the detection of endogenous insulin and was implicated by two scientists namedYalow and Berson [1]. Swiss scientists Engvall and Perlmann were the innovators of ELISA development [2]. In 1971, they modified the RIA

DOI: 10.1201/9781003222194-17

technique by coupling enzymes with radiolabeled antigen and antibodies used in RIA instead of conjugating them with I^{125}. This modification was implicated first in the identification of serum I_gG in rabbits [2]. In that same year, another application of ELISA was efficaciously implemented for the quantification of HCG (human chorionic gonadotropin) existing in human urine. Horseradish peroxidase (EC 1.11.17) enzyme was used as a linked enzyme in this study [3]. Along with this, ELISA was also used in various other aspects. Detection of trichinosis, malaria diagnosis, diagnostic microbiology, and discovery of infections caused by influenza, parainfluenza, and mumps viruses are a few examples of it [4]. Then comes the incorporation of microtitration plates in the ELISA technique, which revolutionized the quantitative application of the ELISA process and gave ELISA an entry inside the hormones, peptides, and proteins concentration measurement field [5]. Till then, ELISA has evolved in many fronts, and it is currently indivisibly involved in various fields. Nowadays, it is a routinely practiced technique in research and diagnostic laboratories.

Food is a biomedical substrate and, as such, it can enable the proliferation of microorganisms that could create harmful chemicals. Mycotoxins, marine biotoxins, plant toxins, cyanogenic glycosides, and toxins found in toxic mushrooms are examples of environmental toxic elements. Environmental poisons not only endanger animal and personal welfare, but they also have an influence on agricultural and nutritional availability by limiting people's accessibility to nutritious foods. Food manufacturers, wholesalers, transporters, and merchants have a main duty for monitoring and detecting environmental poisons in foods all the way back to their original source. Inspections should be carried out by government organizations to guarantee that quantities of the most significant environmental toxins in food conform with both national and global prescribed limits or applicable toxicological parameters.

The grade of chosen immunoreagents, particularly the antibodies, has a significant impact on the ultimate outcomes of immunoassays. As a result, existing investigations continue to place a high priority on the evaluation and discovery of novel antibodies. Based on the discovery of two unique monoclonal antibodies, Maragos et al. suggested a new competitive enzyme-linked immunosorbent test (ELISA) to check for citreoviridin and its geometric isomer, iso-citreoviridin, in white rice. The devised tests were moderately resistant to methanol and acetonitrile, and they gave sufficient quantitative results in terms of sensitivity, correctness, and specificity, enabling the screening of citreoviridin and iso-citreoviridin at toxicologically relevant levels [6]. Research by Bever et al. focuses on the development of new monoclonal antibodies and their use in competitive ELISAs and lateral flow immunoassays (LFIAs) to detect amatoxins, which are fatal poisons present in a range of mushroom taxa. The proposed tests are aimed for analyzing wild mushrooms (ELISA) and urine samples (LFIA), offering supplementary methods for analyzing amatoxins incidence and geographical spread as well as effectively detecting human and animal amatoxin exposure [6]. This chapter discusses the use of the ELISA technique in the detection and quantification of food toxins such as cyanogenic glycosides, pyrrolizidine alkaloids, furocoumarins, lectins, and glycoalkaloids.

14.2 ENZYME-LINKED IMMUNOSORBENT ASSAY

ELISA is an identification and quantification technique that works upon antigen-antibody interaction. It is subdivided into various types, such as direct ELISA, indirect ELISA, competitive ELISA, and sandwich ELISA. The working principle and subtypes are elaborated further in this section.

14.2.1 Working Principle of ELISA

The basic working principle of ELISA is based on the antibody-antigen interaction. The antibodies developed against an analyte are first applied on a solid surface or scientifically speaking, a solid phase, which subsequently binds themselves to the molecules in the solid phase. These solid phases are generally tube or microplates comprising of polypropylene, polystyrene, or polyvinyl. The solid phase is chosen in such a way that it can absorb the antigens and antibodies in appropriate amounts, but it lacks absorption power towards other phases [1,7]. After successful mounting over the solid phase, the analyte is applied over it. The analytical antigens then becomes attached to the antibodies present on the solid surface, forming an antigen-antibody complex. This step is followed by a few rounds of washing to confiscate other contaminants from the medium [4].

The enzymes linked to the immunogens play a very crucial role in the workings of ELISA. They get activated subsequently to the antigen-antibody interaction and convert the substrates applied, which denotes the analyte presence by inducing a color change, peroxidase, glucose oxidase, beta galactosidase, and alkaline phosphatase are some prime examples of linked enzymes. Alkaline phosphate and P-nitro-phenyl phosphate are the substrates of alkaline phosphatase which, upon conversion, induce a yellow color in the medium. Alkaline phosphate and P-nitro-phenyl phosphate are easily available in tablet form. Similarly, five amino salicylic acid and orthophenylenediamine are the substrates of peroxidase, which induce a brown color upon enzymatic conversion. If beta galactosidase is used as a linked enzyme in the experimental setup, colorimetric detection is less feasible. In this case, fluorometric detection is desired for proper detection. However, the specificity and acceleration of the enzyme-substrate reaction exclusively depend on the catabolic properties of the enzyme [8]. For the completion of enzyme-substrate reaction, 30–60 minutes time is required. Alternatively it can also be terminated using sodium hydroxide, hydrochloric acid, or sulfuric acid [9]. The results are generally read by using a spectrophotometer, at a wavelength of 400–600 nm, depending on the characteristics of the conjugate [9].

14.2.2 Types of ELISA

Enzymatic immunoassay methods are divided into two generalized classes, i.e. homogeneous enzymatic immunoassay and heterogeneous enzymatic immunoassay methods [10]. The homogeneous method is characterized by inactivation of a linked enzyme upon binding to the antibody, thus, eliminating the washing stage (separation of antigen from the medium). The homogeneous method is implicated for the measurement of analytes in small quantities i.e., therapeutic drugs. This method is very easy to use, however, high cost and lower sensitivities are some major drawbacks of this method [10].

On the other hand, heterogeneous methods are more commonly used, and they are subdivided into many classes. Working and types of this method are described later inside this chapter. The main characteristic feature of this method is the bound nature of the antigen-antibody complex with the experimental wall and a separate washing stage to

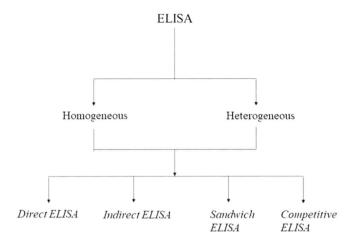

FIGURE 14.1 Types of ELISA.

prevent contamination with other components of the analyte, as well as to remove the unbound antigens from the medium. This washing stage is necessary to separate the bound antigen from the free antigen, post interaction of antigen and antibody. Due to its increased sensitivity, the heterogeneous method is more commonly used. ELISA is a heterogeneous method implied for the detection of explicit antibodies and decipherable antigens. However, as structures and characteristics of different antigens are different, a wide range to ELISA techniques were developed to increase the sensitivity and specificity of ELISA techniques [4,10]. Various types of ELISA techniques are described in the following section. Different types of ELISA techniques are presented in Figure 14.1.

14.2.2.1 Direct ELISA

Engvall and Perlmann [11] and Van Weemen and Schuurs [12] invented the direct ELISA technology concurrently in 1971, and it became the forerunner of subsequent ELISA kinds. This ELISA approach is ideal for quantification of high molecular weight antigens. This method starts with a direct coating of the antibody or antigen on the surface of the solid phase. The assessment is possible due to an enzyme-linked immunogens (antigen or antibody). After incubating, the unattached immunogens are removed from the media by rinsing or washing. Then, a suitable substrate is added to the medium, which is converted to a colored compound. This conversion imparts signatural color change into the mixture. The quantity of immunogen is determined by measuring the signal or color change [8,13].

14.2.2.2 Indirect ELISA

After being inspired by the direct ELISA method, Lindström and Wager, in 1978, invented the indirect ELISA method for the purpose of measuring porcine IgG [14]. The indirect technique gets its name from the fact that another additional antibody is used in the matrix that identifies and isolates the target to be tested, rather than the main antibody. The process is started with the addition of analyte serum to the antigen-coated solid phase plates or wells. After incubation, an antigen antibody complex is formed, which gets attached to the solid phase surface. An additional antibody is then added to the medium to make the antigen–antibody complex observable. After that, the enzyme's substrates are introduced to the mixture to generate color, and subsequently the analyte is quantified. This approach of detecting antigens is more widely employed in endocrinology [4].

14.2.2.3 Sandwich ELISA (Antibody Screening)

Kato and his colleagues invented the approach in 1977 [15]. In this procedure the surface of the solid phase is coated with antibodies and those antibodies are fixed on the surface. The specimen is then placed on the antibody-coated plate, which is then cultured for a long period before being cleaned. The antigens that are not attached to the coated antibodies are then washed away. These antigens cannot be eliminated after the antigen unique to the attached antibody is identified. After washing, antibodies that have been labeled with the antigen-specific enzyme are introduced and cultured. If there are antigens in the mixture after culture and rinsing, they cannot be eliminated because the enzyme-tagged antibodies are linked to them. Enzyme reagent is introduced to the medium and coloring is provided to demonstrate enzymatic functioning. A successful outcome is shown by coloring, whereas an absence of coloring implies a negative result. Sandwich ELISA gets its name from the fact that the required peptide is sandwiched within two antibody units. Sandwich ELISAs are two to five times more sensitive than any other ELISAs, according to studies [15].

14.2.2.4 Competitive ELISA (Antigen/Antibody Screening)

Yorde and his colleagues [16] invented this approach in 1976. The antigen-specific antibody or antibody-specific antigen is adsorbed on the exterior of the solid phase in this approach. The enzyme-tagged antigen or antibody, and the specimen to be analyzed, are both inserted in the well at the same time. To attach to the bound antibody or antigen, the labeled and unlabeled antigen (patient antigen) or antibody units contend within themselves. The coloring that results after the wells have been rinsed and enzyme substrate has been introduced allows the amount of the analyte to be quantified. The sample quantity and the strength of the resultant coloring are inversely proportional [16]. Competitive ELISA is the most preferred technique in the detection and quantification of food toxins (Figure 14.2).

14.3 ELISA IN THE STRUCTURAL CHARACTERIZATION AND IDENTIFICATION OF NATURALLY OCCURRING FOOD TOXINS

Immunoassays are often used in food analysis to identify food pollutants, as well as in the agricultural sector to identify hazardous remnants in plants, water, and soil [17–19]. For the identification and characterization of dangerous chemicals in food, many ELISA approaches have been developed. For the identification of fumonisin B1 (FB1) in food and animal feeds, Quan et al. [20] developed a sensitive enhanced chemiluminescence enzyme linked immunosorbent assay (ECL-ELISA). Pastor-Navarro et al. [21] used an enzyme-linked immunosorbent test (ELISA) to identify various tetracycline compounds in honey. Herbicides in water [22], pesticides, and estrogenic chemicals in agricultural goods [22,23] have all been monitored using ELISA techniques.

14.3.1 Cyanogenic Glycosides

Cyanogenic glycosides are a naturally occurring food toxin/chemical weapon/defense mechanism of plants against herbivorous animals. It is found in approximately 2,500 plant

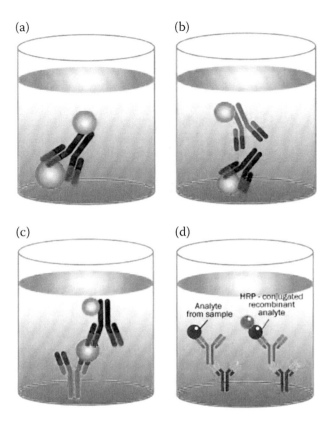

FIGURE 14.2 Different types of ELISA: (a) direct ELISA, (b) indirect ELISA, (c) sandwich ELISA, and (d) competitive ELISA.

species. The main mechanistic feature of its chemical weaponry is generation of hydrogen cyanide, which is very toxic to herbivorous animals and humans [24,25]. Cyanogenic glycosides are stored inside the plant vacuoles and are inaccessible under normal conditions. It is released form the vacuoles after bruising, chewing, or any kind of mechanical damage. After release, cyanogenic glycosides encounter enzymes like β-glucosidases and α-hydroxynitrile lyases and get converted into benzaldehyde and hydrogen cyanide [26]. This produced hydrogen cyanide causes cyanide toxicity in humans, involving symptoms such as "vomiting, diarrhoea, abdominal cramps, headache, drowsiness, confusion, nausea, hypotension, paralysis and coma" [27–31]. After ingestion, naturally occurring gut enzymes (produced by gut microflora), can also play a critical role in the hydrolysis of intact cyanogenic glycosides into benzaldehyde and hydrogen cyanide inside the human GIT [26]. The 0.5–3.5 mg/kg body weight of cyanide (generated from food) is enough to cause cyanide toxicity [27].

Generation of cyanide toxicity symptoms are reported after the congestion of apricot kernels, tapioca cake, cassava-based meal, and almonds [27–31]. Additionally, other plant crops such as "cassava, sorghum, millet, barley, almond, cherry, macadamia nut, peach, lima beans, and kidney beans", were also investigated to be biosynthesizing and accumulating cyanogenic glycosides inside their cells [17,32–34]. Amygdalin (D-mandelonitrile-β-D-gentiobioside) is the most abundant cyanogenic glycoside and exists in the kernels and seeds of various fruits like "apricot, almond, apple, cherry, plum, peach and nectarine" [35]. Even though all the above-mentioned fruits are not consumed directly, the edible portion of these fruits are the main store house for cyanogenic glycosides [35]. The HCN equivalents in these plants were reported to be ranging between 3–4 mg/kg and some canned stone fruits were also reported to contain 4 mg/kg of HCN equivalents [36–38].

Conventional methods for cyanogenic glycoside detection involves the use of HPLC and GC-MS. Even though these processes are sensitive, high cost and complex sample pretreatment procedures make them difficult to pursue [39–42]. Thus, there is a need for a cost-effective, simple, rapid, and sensitive detection technique. ELISA can fill this need even after providing a high sample throughput. Antibodies against the neurotransmitter amygdalin have been discovered earlier [43,44]; however, they have yet to be properly described or statistically used. The creation, characterization, and (for the first time) implementation of an ELISA for the measurement of amygdalin in commercially accessible food items was presented for the first time by Bolarinwa et al. [35].

Production of antibodies against the low molecular weight compounds such as cyanogenic glycosides is the main crucial step in the detection and characterization of them through ELISA, as they are unable to directly excite the immune system to produce antibodies. Thus, hapten molecules are used for this purpose. Haptens are small molecules that upon binding to a non-immunogenic protein (carrier protein), can elicit immunogenic response against themselves. Thus, it turns a non-immunogenic protein to an immunogenic one. This hapten-protein complex is known as a conjugate. As the sensitivity and specificity of ELISA solely depends on the hapten-assisted production of antibodies, design of this conjugate is the most crucial step in the development of immunological identification techniques, such as ELISA [45].

There are some methods mentioned in the literature that are used to produce conjugates, like the "succinic anhydride conjugation method" [46,47] and "cyanuric chloride conjugation method" [48]. The periodate cleavage assisted protein-hapten coupling is another very useful procedure for conjugate production [45]. In these methods, bovine serum albumin (BSA) is the most preferred hapten. Keyhole limpet hemocyanin (KLH) is also used in these methods as a hapten charrier, to increase the sensitivity and specificity of the hapten molecules in elision of antibodies. Along with it KLH – cyanogenic glycoside conjugates are used as an coating antigen in many procedures [35,45]. Linamarin and amygdalin, are two cyanogenic glycosides for which the periodate cleavage method was used to produce protein-hapten complex. Also, BSA was used as a hapten in this method. The C-C bond between the adjacent -OH group in the molecules of linamarin and amygdalin were cleaved by periodate, forming aldehyde groups which under alkaline condition reacts to the free e-amino groups of lysine of BSA and gives rise to a Schiff base. This Schiff base was further stabilized by sodium borohydride to produce stable conjugates. In this method, the vicinal hydroxyl group is introduced into the amygdalin molecule by using a derivative, as it is necessary for its interaction with BSA [45]. Cyanuric chloride assisted protein-hapten coupling is another method of conjugate preparation. Cyanuric chloride interacts with the hydroxyl group under mild environment and the structure of the target analyte also remains intact in this procedure [49]. Along with it, this method produces a higher hapten:protein ratio, which denotes higher immunogenicity of the complex, which enables more antibody production. A lower hapten:protein ratio denotes higher coating antigen properties [50].

After successful conjugate preparation, characterization, and estimation of the protein-hapten complex, another very important step is to gain confirmation about the quality and quantity of the antibody to be produced by it [35]. After this, the antibodies are produced against the target cyanogenic glycoside and subsequently the ELISA procedure is carried out. For detection of cyanogenic glycosides, indirect competitive ELISA (icELISA) can prove to be very beneficial. A study was performed to quantify the cyanogenic glycosides the commercially available foods, which makes use of the icELISA method [35]. One advantage of using icELISA is that it can determine the cross reactivity with some closely related cyanogenic glycosides in the sample. For example, for the identification of amygdalin, use of icELISA was able to determine the cross reactivity with prunasin, mandelonitrile, linamarin, and dhurrin, which are structurally like amygdalin [35]. The extract kernels of an apricot, a natural cyanogenic plant, were quantified by the icELISA method, and the presence of amygdalin was detected [35]. icELISA is a cost effective, simple, and subtle method, capable of high sample output. However, further research is needed to establish its use in cyanogenic plants.

14.3.2 Pyrazolidine Alkaloids

Pyrazolidine alkaloids (PAs) are basically secondary metabolites of plants. They are naturally distributed among 6,000 plant species including Boraginaceae, Fabaceae, Asteraceae, and some other families [51–53]. PAs are toxic in nature and cause acute hepatotoxicity, mutagenic, carcinogenic, teratogenic, and other cytotoxic/pathological implications [54]. Toxic PAs are categorized as retronecine, heliotridine, otonecine, and platynecine PAs. Amongst them, retronecine, heliotridine, and otonecine PAs possess greater carcinogenic and hepatotoxic properties, while on the other hand, platynecine PAs are comparatively less toxic [55–58]. So far, 660 different PAs have been identified and 50% of them are hepatotoxic and carcinogenic in nature [59]. PA-induced drug-induced liver injury (DILI) is the most common liver-associated pathological implication [60]. PA toxication is gaining attention due to its increased prevalence in recent years.

PAs requires metabolic activation to exert their toxic activities. PAs are enzymatically transformed into their corresponding pyrrolic esters by CYP 450 class of enzymes [61–64]. Pyrrolic esters of PAs are electrophilic in nature, and they react with the nucleophilic functional group situated in the proteins and DNA, which results in protein adducts, mediating PA toxicity [65–68]. The e-amino group derived from the pyrrolic esters plays a very important role in the formation of protein adducts [69].

Some PA-containing plants are gaining interest in the pharmaceutical and nutritional fields. Thus, development of a sensitive, yet specific method is necessary for detection purposes. Generally competitive ELISA is used in the identification of PAs. Polyclonal antibody-assisted immunoassay has already proven to be a sensitive technique [70–72]. However, limitation in the availability of the polyclonal sera is one of the major drawbacks of this process. To overcome this shortcoming, hybridoma-assisted development of monoclonal antibodies is perused and has proven to be very successful. However, as PAs are too small to induce immunological response, PA-bovine thyroglobulin conjugates and pyrrole Nα-acetyl lysine methyl ester-assisted conjugate are used to produce antibodies against PA. In some studies, western blotting has also been employed to ease the antibody and conjugate extraction procedure [69].

14.3.3 Furocoumarins

Furocoumarins are a class of photosensitive DNA interacting compounds, owing to the hydrophobic nature of their tricyclic aromatic core [73]. It forms various types of photoadducts while interacting with the DNA in the presence of light. Amongst these, 4',5'-monoadduct were the most abundant. These photoadducts get cross-linked with the DNA and block the DNA synthesis process, which is very vital for every living being [73]. 8-methoxypsoralen (8-MOP), 5-methoxypsoralen (5-MOP), and 4,5',8-trimethlyprosalen (TMP) are some examples of furocoumarins, which are used in skin diseases like psoriasis and vitiligo [74]. A large assortment of plants belonging to Apiaceae, Rutaceae, Moraceae, and Fabaceae families contain furocoumarins. They are most abundant in plants that have ripe seeds, fruits, and flowering plants [75].

More or less the ELISA technique used for the detection of furocoumarins is same except the antibody front. Specialized antibodies have been developed for sensitive and specific detection of various furocoumarins. In a study, antibody 73E was developed to identify and characterize furocoumarin 6,4,4'-trimethylangelicin (TMA) photoadducts (mainly 4',5'-monoadduct), with non-competitive ELISA [73]. Furocoumarins are also non-immunogenic. In a study involving 8-MOP, conjugates were prepared with the help of BSA. The 8-MOP-BSA conjugate was used to develop an antibody, which was subsequently used in the identification via the ELISA method [74]. A set of monoclonal antibodies (3A8, 5G3, 8G1, 9D8) have also been identified for the detection of 8-MOP-induced altered DNAs [76]. Specific antibodies for identification of DNA-adducts produced by benzo-[a]-pyrene diol epoxide (BPDE-I) and 1-aminopyrene (1-AP), have also been developed. These specific antibodies basically recognize the common predictor that includes the guanine base and the hydrocarbon ring with the hydroxide units. The immunoglobulins cross-react with BPDE-I-dG, a DNA monoadduct, although with a lesser responsiveness than the intact modified DNA. They don't react with acetylaminofluorene (AAF), or 1-AP modified DNA, which both possess guanosine conformational changes at position C-8. Antibodies to 1-AP-modified DNA depict cross-reactivity with 8-nitro-1-aminopyrene- and 6-nitro-1-aminopyrene-modified DNA, as well as BPDE-I-DNA and AAF-DNA [77]. These specific antibodies can detect the presence of food-derived furocoumarins in a highly specific manner.

14.3.4 Glycoalkaloids

Glycoalkaloids (GAs) are nitrogen-containing steroidal glycosides mainly abundant in *Solanum* species (more than 90 different types are found) [78]. Glycoalkaloids can be divided into two major classes, depending on the proportion of alkaloid moiety present in their molecules: solanidanes and spirasolanes. Examples of some widely distributed GAs are α-chaconine, α-solanine, demissine, α-solamrghe, α-solasonine, α-tomatine, and α-solamuine [79].

GAs are mainly found in potatoes and are more concentrated in the tissues that are growing actively, like shoots and berries. Concentration of the GAs is induced by stress conditions, indicating their role as a defense mechanism of the plant. GAs are known to confer resistance to fungi, insects, and potential attack by herbivores [80]. These toxic GAs are very harmful to human health, and they are readily found in commercially cultivated potatoes. α-chaconine, and α-solanine, are the most common GA found in commercially cultivated potatoes [79].

Toxic effects of GAs on humans and animals are well documented in the literature [81,82]. Toxic implications of GAs are demonstrated through gastrointestinal and neurological disturbances [83]. Synergistic toxic effects of various GAs are also found to be a reality. However, the extent of toxicity differs greatly across different GAs. The extent of toxicity depends on both the sugar, as well as the alkaloid moiety and can be affected even by slight changes in the sugar moiety [79,84]. On the other hand, the toxic neurological effects range from mild symptoms like shallow breathing and rapid pulse to serious implications like coma, and is hypothesized to be caused by acetylcholinesterase inhibition [79].

Various detection and analysis techniques like gravimetric, colorimetric, and titrimetric detection have been used in the identification of glycoalkaloids and all of them have a common major drawback, i.e., lack of sensitivity and specificity. Gas chromatography is another commonly used method which becomes obsolete for being expensive, time-consuming, and lacking the ability of individualization of GAs. HPLC is another very good analytical method, but due to the lack of a strong chromophore on the GA, this method had also become obsolete [79].

ELISA has emerged as a very useful technique in the identification of GAs. The basic ELISA procedure for detection of GA is the same. An adjuvant molecule is used along with the antigen to stimulate antibody production. Generally, bacterial cell wall components and oils are used as an adjuvant. Bacterial cell wall components are highly immunogenic, and the oils are used to ensure sustained release of the antigen [85]. As GAs are small molecules, a hapten-carrier protein conjugate is needed to be developed to induce an immunogenic response. Bovine serum albumin (BSA), ovalbumin, keyhole limpet haemocyanin (KLH), and limulus polyphenus haemocyanin (LPH) are some commonly used carrier proteins. None of the carrier proteins furnished above is preferred, but the number of hapten molecules attached to a carrier protein must be high to induce a satisfactory immune response [79].

The choice of linking method is also very important. The functional groups present on the hapten dictate the connection mechanism employed in the first place. Carboxyl, hydroxyl, and amino groups are typical chemical functional elements. When hydroxyl groups are there, they are derivatized to allow for protein binding. The most frequent coupling techniques depend on the interaction of an excited hapten with the amino groups on the protein's lysine regions [86]. Because both aqueous and organic reagents are accessible, different carbodiimides are commonly used as activating agents. The use of a two-step linking process in which an active intermediate is created before linking to the carrier protein is favored. This will prevent the canier protein from cross-linking. Sodium periodate may be utilized to create dialdehydes, which subsequently react with amine residues on the protein in haptens carrying vicinal diols. To make a stable conjugation, the conjugate is reduced using sodium borohydride [79].

Because of the vast variety of feasible Ab specificities existing in serum from an immunized animal, these antibodies are referred to as polyclonal antibodies (PAb). While PAb are easy to make and affordable, their lack of specified specificity and repeatability is a drawback. Monoclonal antibodies (MAb), invented by Kohler and Milstein in 1975 [87], solve this drawback by allowing the creation of antibodies with a predetermined selectivity. After PAb against GAs, production of MAb against the targeted glycoalkaloid is produced, which is further used in the identification process. Generally, the competitive ELISA technique is used in the identification of GAs. BSA-TOM-L and BS A-TOM-H are two conjugates that are used in the detection of GAs via the ELISA method, where BSA-TOM-L acts as a coating conjugate and BSA-TOM-H is used for immunization [79].

Development of monoclonal antibodies against GAs is another critical step in the identification process. A set of MAb was developed against GAs via murine immunization with a solanidine-bovine serum albumin (BSA) conjugate, which binds with two of the most common GAs: α-chaconine and α -solanine [88]. An additional ELISA kit was developed using one of these MAbs, Sol-129, which resulted in enhanced stability and selectivity of the assay [89]. Cleaved sugars associated GA conjugates to produce antibodies, were also [90,91]. As α-chaconine and α -solanine comprise 95% of the GAs present in potato, detection of them can be stated as 'detection of total potato glycoalkaloids'. In another study, an ELISA kit was developed, using a simple solanidine-derived immunogen to elicit antibody production. It was able to detect total potato glycoalkaloids [89]. This new ELISA kit enables quick identification, detection of hostile postharvest upsurges in glycoalkaloid content of potatoes, and detection of the distribution, fate of GAs, and their metabolites at a nanomolecular concentration [89].

Along with the development of desirable antibodies, correct and specific detection of the assay endpoint is equally important. As discussed earlier, endpoints of the ELISA process are determined by color change, catered to an enzymatic reaction between an antibody-conjugated enzyme and substrate. Generally, three types of enzymes, urease, alkaline phosphatase, and horse radish peroxidase, are used in ELISA. Peroxidase, using 2, 2-azinobis (3-ethylbenzthiazoline) 3-sulphonic acid (ABTS) as substrate, was the most effective of the three catalysts, in recognizing positive samples, according to Probit analysis. All potato samples comprising glycoalkaloids at levels larger than the threshold limit were properly identified via visual inspection. Many additional areas of food analysis can benefit from visual evaluation of ELISA data [92].

14.3.5 Lectin

Lectins are carbohydrate-binding proteins or glycoproteins that are of non-immunogenic origin. They recognize and bind covalently with various carbohydrates like "lactose, galactose, mannose, N-acetyl glucosamine, and N-acetyl galactosamine". The carbohydrate-binding properties of lectins often lead to agglutination of eukaryotic cells [93,94]. Additionally, the lectin-carbohydrate interaction is considered as an important part of innate immune response, and this phenomenon is used in pathogen recognition, opsonization, phagocytosis, complement activation, agglutination, and cell adhesion [93,95].

Some dietary lectins are considered as extremely toxic as most of them react with glycoconjugates situated on the epithelial wall of GIT, resulting in severe clinical and sub-clinical effects when large quantities of lectins are consumed [96,97]. Although soybean lectin (SBA) is not normally regarded as hazardous to laboratory models, it may be accountable for development suppression effects when combined with trypsin blockers. Soybean lectin and soybean saponins have been shown to have a combinatorial destabilizing impact on rabbit jejunal epithelium. Both SBA and trypsin blockers have the potential to influence pancreatic functioning, and their cumulative effects might have significant consequences for the utilization of soybean-based diet in human meals [98].

Lectins are abundant throughout plants, animals, and even microorganisms. The edible parts of many crops contain lectins. Soybeans, legumes, roasted peanuts, and wheat germ are some plants that contain lectins [98,99]. In animal lectins, at least one "carbohydrate recognition domain (CRD)" is found, and it interacts with several oligosaccharides found on pathogenic membranes [100]. Animal lectins are divided into 11 basic groups based on their CRD architectures, motif varieties, interaction characteristics, and calcium dependence, such as "C-type, F-type, I-type, L-type, M-type, P-type, R-type, F-box lectins, chitinase-like lectins, ficolins, calnexin, galectins, and intelectins" [101]. C-type lectins (CTLs) are a richly varied group of animal lectins, initially classified as Ca2+-dependent sugar recognition proteins [102]. More than a century has elapsed since the first CTL was discovered, and research into the characteristics, architectures, and functionalities of CTLs has been substantial. Many CTLs include one or more CRDs, which are made up of 115–130 amino acid residues and have a double-loop architecture, disulfide bond locations, and Ca^{2+}-interaction domains [103]. Each CRD has 1–4 Ca^{2+} interaction domains, with Ca^{2+} interaction domain 2 being architecturally preserved and linked to selectivity of the CTL sugar interaction [104]. A preserved sequence, Trp-Asn-Asp (WND) is found at site 2, as well as another substantially preserved Gln-ProAsp (QPD) or Glu-Pro-Asn (EPN) sequence motif. It has been demonstrated that the latter may bind galactose or mannose [104,105].

Before development of immunoassays, lectins were identified and quantified with the help of agglutination assays. As the expression of lectins in the sample, presence of other impurities with agglutination power, and the masking of the agglutination by other compounds, were a reality, the agglutination assays became obsolete. However, agglutination assays were able to satisfactorily detect the presence of hololectins (lectins having several binding sites) in a sample [106–109].

Identification and quantification of lectins with ELISA has gained popularity in recent times. For this purpose, in place of antibodies, carbohydrate moiety (with the target lectin binds specifically) is first immobilized on the surface of the solid phase, and subsequently the ELISA process is performed [110,111]. In a study, glycoproteins (carbohydrate moiety) like fetuin, asialofetuin, mucin, conalbumin, and ovalbumin were first immobilized inside the microtiter plates. These glycoproteins were lectin specific. Then, the sample solution was added to the plates, and subsequently the plates incubated and washed. After washing further binding of lectin to the unbound glycoproteins was performed by adding specific saccharides like – N-acetylglucosamine (GlcNAc) or mannose (Man) to the plates. Then, the addition of an enzyme-linked antibody to the plates was performed, which binds to the glycoprotein-attached lectins. Next, the detection and quantification of lectins were done by adding an alkaline phosphatase substrate (Sigma Fast p-nitrophenyl phosphate), which imparts color to the plates [112]. For detection of lectins in roasted peanuts and wheat germ, the same procedure mentioned above was performed. However, in that case, the wells were coated with asialofetuin and ovalbumin [112]. In another study, during the quantification of soybean derived lectins, SBA, the above-mentioned ELISA process was the same, except, the plate walls in this experiment were coated with asialofetuin (AFET) and fetuin (FET) [98]. Additionally, lectins are also used as a glycoprotein analytical tool, in which the lectins are immobilized on the surface of the solid phase and then the ELISA detection process is carried out [113,114] (Table 14.1).

14.4 STRUCTURAL CHARACTERIZATION AND DETECTION METHODS OF FOOD TOXINS

Structural characterization of various food toxins is eased using the ELISA technique. UV, HPLC-UV, GC-MS, LC-MS/MS, and HPLC are used, along with the ELISA technique, for effective characterization of food toxins. The structural characterization of various food toxins is summarized in Table 14.2.

TABLE 14.1   Identification and Quantification of Food Toxins by Different ELISA Methods

S. no.	Method	Category	Sample	Conjugate proteins	Compound detected	References
1	icELISA	Cyanogenic Glycosides	Fruit seeds and kernels,	Amygdalin-BSA	Amygdalin	[35]
2	ELISA	Cyanogenic Glycosides	Cassava, fruit seeds, and butter bean	Linamarin-BSA and Amygdalin-BSA	Linamarin and amygdalin	[115]
3	Rapid multiplex ELISA	Pyrrolizidine alkaloids	Honey and feed	Horseradish peroxidase	Jacobine, lycopsamine, heliotrine, and senecionine	[116]
4	CI-ELISA	Pyrrolizidine alkaloids	*Senecio genus*	Riddelliine-Succinate	Riddelliine and riddelliine N-oxide	[117]
5	ELISA	Glycoalkaloids	Potato and tomato	Horseradish peroxidase	a-solanine, a- chaconine, B2-chaconine, γ-chaconine, solanidine, demissidine, a- tomatine, tomatidine, solasonine, solamargine, and solasodine	[118,119]
6	Competitive ELISA	Glycoalkaloids	*Solanum* spp	Human serum albumin	Solasodine, solamargine, and khasianine	[120]
7	Multiscan EX ELISA	Furocoumarins	*Ducrosia anethifolia*	–	Pabulenol, oxypeucedanin, imperatorin, isogospherol, heraclenin, and heraclenol	[121]
8	ELISA	Furocoumarins	*Cachrys trifida*	–	Imperatorin, isoimperatorin, and prantschimgin	[122]

TABLE 14.2 Structural Characterization and Detection Methods of Food Toxins

S. no.	Chemical name	Structure	ELISA mode	Detection methods	References
1	Linamarin		ELISA	UV = 200–400 nm	[115]
2	Amygdalin		icELISA & ELISA	HPLC-UV and GC-MS UV= 200–400 nm	[35,115]
3	Jacobine		Rapid multiplex ELISA	LC-MS/MS	[116]
4	Lycopsamine		Rapid multiplex ELISA	LC-MS/MS	[116]
5	Heliotrine		Rapid multiplex ELISA	LC-MS/MS	[116]

6	Senecionine		Rapid multiplex ELISA	LC-MS/MS	[116]
7	Riddelliine		CI-ELISA	UV = 450 nm & NMR	[117]
8	α-solanine		ELISA	HPLC-UV (450 nm) & GC-MS	[118,119]
9	α-chaconine		ELISA	HPLC-UV (450 nm) & GC-MS	[118,119]

(*Continued*)

TABLE 14.2 (Continued)

S. no.	Chemical name	Structure	ELISA mode	Detection methods	References
10	Solanidine		ELISA	HPLC-UV (450 nm) & GC-MS	[118,119]
11	Demissidine		ELISA	HPLC-UV (450 nm) & GC-MS	[118,119]
12	α- tomatine		ELISA	HPLC-UV (450 nm) & GC-MS	[118,119]

13	Tomatidine	ELISA	HPLC-UV (450 nm) & GC-MS	[118,119]
14	Solasonine	ELISA	HPLC-UV (450 nm) & GC-MS	[118,119]
15	Solamargine	ELISA	HPLC-UV (450 nm) & GC-MS	[118–120]

(Continued)

TABLE 14.2   (Continued)

S. no.	Chemical name	Structure	ELISA mode	Detection methods	References
16	Solasodine		ELISA	HPLC-UV (450 nm) & GC-MS	[118–120]
17	Khasianine		Competitive ELISA	HPLC	[120]

18	Pabulenol		Multiscan EX ELISA	Multiscan EX ELISA = 540/630 nm	[121]
19	Oxypeucedanin		Multiscan EX ELISA	Multiscan EX ELISA = 540/630 nm	[121]
20	Imperatorin		Multiscan EX ELISA	Multiscan EX ELISA = 540/630 nm, MS, and NMR	[121,122]
21	Heraclenin		Multiscan EX ELISA	Multiscan EX ELISA= 540/630 nm	[121]

(*Continued*)

TABLE 14.2 (Continued)

S. no.	Chemical name	Structure	ELISA mode	Detection methods	References
22	Heraclenol		Multiscan EX ELISA	Multiscan EX ELISA= 540/630 nm	[121]
23	Isoimperatorin		ELISA	MS & NMR	[122]

14.5 CONCLUSION

ELISA has the potential to replace the ongoing food detection techniques. The main advantages of using an ELISA technique are higher sensitivity, specificity, and lower time consumption and cost. Due to its low cost, it can be a unified member of every agricultural setup. This will enable effective monitoring of the toxin levels in the daily consumables, thus helping in building a healthy society. Although there is some research in the various toxin-specific antibody development aspect, ELISA bears a great potential to be a universal tool in the identification and quantification of food toxins.

REFERENCES

[1] Yalow, R. S. and Berson, S. A. (July 1960). Immunoassay of endogenous plasma insulin in man. *J Clin Invest.*, 39(7), 1157–1175. 10.1172/JCI104130. PMID: 13846364; PMCID: PMC441860.

[2] Engvall, E. and Perlmann, P. (September 1971). Enzyme-linked immunosorbent assay (ELISA). Quantitative assay of immunoglobulin G. *Immunochemistry*, 8(9), 871–874. 10.1016/0019-2791(71)90454-x. PMID: 5135623.

[3] Maeno, H., Wong, P. F., AbuBakar, S., Yang, M., Sam, S. S., Jamil-Abd, J., Shunmugarajoo, A., Mustafa, M., Said, R. M., Mageswaren, E., Azmel, A. and Mat Jelani, A. (November 30, 2021). A 3D microfluidic ELISA for the detection of severe dengue: Sensitivity improvement and vroman effect amelioration by EDC-NHS surface modification. *Micromachines (Basel)*, 12(12), 1503. 10.33 90/mi12121503. PMID: 34945351; PMCID: PMC8715748.

[4] Aydin, S. (October 2015). A short history, principles, and types of ELISA, and our laboratory experience with peptide/protein analyses using ELISA. *Peptides*, 72, 4–15. 10.1016/j.peptides.2015.04.012. Epub 2015 April 20. PMID: 25908411.

[5] Siegle, R. L., Jennings, B. R., Adams, P. L. and King, L. P. (September–October 1980). Development of a model using polypeptide antibodies for scintigraphy of the pancreas. *Invest Radiol.*, 15(5), 457–461. 10.1097/00004424-198009000-00151. PMID: 6783591.

[6] Lattanzio, V. M. T. (July 22, 2020). Introduction to the toxins special issue on improved analytical technologies for the detection of natural toxins and their metabolites in food. *Toxins (Basel)*, 12(8), 467. 10.3390/toxins12080467. PMID: 32707815; PMCID: PMC7472049

[7] Ma, T., Liu, K., Yang, X., Yang, J., Pan, M. and Wang, S. (November 2, 2021). Development of indirect competitive ELISA and visualized multicolor elisa based on gold nanorods growth for the determination of zearalenone. *Foods*, 10(11), 2654. 10.3390/foods10112654. PMID: 34828935; PMCID: PMC8619891.

[8] Uddin, M. J., Bhuiyan, N. H. and Shim, J. S. (January 21, 2021). Fully integrated rapid microfluidic device translated from conventional 96-well ELISA kit. *Sci Rep.*, 11(1), 1986. 10.1038/s41598-021-81433-y. PMID: 33479284; PMCID: PMC7820004.

[9] Hornbeck, P. V. (August 3, 2015). Enzyme-linked immunosorbent assays. *Curr Protoc Immunol.*, 110, 2.1.1–2.1.23. 10.1002/0471142735.im0201s110. PMID: 26237010.

[10] Hornbeck, P., Winston, S. E. and Fuller, S. A. (May 2001). Enzyme-linked immunosorbent assays (ELISA). *Curr Protoc Mol Biol.*, Chapter 11:Unit11.2. 10.1002/0471142727.mb1102s15. PMID: 18265074.

[11] Kohl, T. O. and Ascoli, C. A. (July 5, 2017). Direct competitive enzyme-linked immunosorbent assay (ELISA). *Cold Spring Harb Protoc.*, 2017(7), pdb.prot093740. 10.1101/pdb.prot093740. PMID: 28679705.

[12] Van Weemen, B. K. and Schuurs, A. H. (June 24, 1971). Immunoassay using antigen-enzyme conjugates. *FEBS Lett.*, 15(3), 232–236. 10.1016/0014-5793 (71)80319-8. PMID: 11945853.

[13] Engvall, E. (February 2010). The ELISA, enzyme-linked immunosorbent assay. *Clin Chem.*, 56(2), 319–320. 10.1373/clinchem.2009.127803. Epub 2009 October 22 PMID: 19850633.

[14] Kohl, T. O. and Ascoli, C. A. (July 5, 2017). Indirect competitive enzyme-linked immunosorbent assay (ELISA). *Cold Spring Harb Protoc.*, 2017(7), pdb.prot093757. 10.1101/pdb.prot093757. PMID: 28679706.

[15] Kohl, T. O. and Ascoli, C. A. (June 1, 2017). Immunometric antibody sandwich enzyme-linked immunosorbent assay. *Cold Spring Harb Protoc.*, 2017(6), pdb.prot093716. 10.1101/pdb.prot093716. PMID: 28572187.

[16] Jordan, W. (2005). Competitive enzyme-linked immunosorbent assay. *Methods Mol Biol.*, 295, 215–226. 10.1385/1-59259-873-0:215. PMID: 15596899.

[17] Meulenberg, E. P. (July 1, 1997). Immunochemical detection of environmental and food contaminants: Development, validation and application. *Food Technol. Biotechnol.*, 35(3), 153–163.

[18] Ramesh, A., Thirugnanam, P. E. and Balakrishnamurthy, P. (2007). Hapten synthesis, generation of polyclonal antibodies and development of ELISA for determination of thiamethoxam residues in food and environmental samples. *Indian Journal of Biotech.*, 6, 365–371.

[19] Sathe, M., Ghorpade, R., Merwyn, S., Agarwal, G. S. and Kaushik, M. P. (January 21, 2012). Direct hapten-linked competitive inhibition enzyme-linked immunosorbent assay (CIELISA) for the detection of O-pinacolyl methylphosphonic acid. *Analyst*, 137(2), 406–413. 10.1039/c1an15773f. Epub 2011 November 17. PMID: 22096781.

[20] Quan, Y., Zhang, Y., Wang, S., Lee, N. and Kennedy, I. R. (October 27, 2006). A rapid and sensitive chemiluminescence enzyme-linked immunosorbent assay for the determination of fumonisin B1 in food samples. *Anal Chim Acta.*, 580(1), 1–8. 10.1016/j.aca.2006.07.063. Epub 2006 August 1. PMID: 17723748.

[21] Pastor-Navarro, N., Morais, S., Maquieira, A. and Puchades, R. (July 2, 2007). Synthesis of haptens and development of a sensitive immunoassay for tetracycline residues. *Application to Honey Samples. Anal Chim Acta.*, 594(2), 211–218. 10.1016/j.aca.2007.05.045. Epub 2007 May 29. PMID: 17586117.

[22] Wei, C., Ding, S., You, H., Zhang, Y., Wang, Y., Yang, X. and Yuan, J. (2011). An immunoassay for dibutyl phthalate based on direct hapten linkage to the polystyrene surface of microtiter plates. *PLoS One*, 6(12), e29196. 10.1371/journal.pone.0029196. Epub 2011 December 27. PMID: 22216208; PMCID: PMC3246456.

[23] Kondo, M., Tsuzuki, K., Hamada, H., Yamaguchi Murakami, Y., Uchigashima, M., Saka, M., Watanabe, E., Iwasa, S., Narita, H. and Miyake, S. (February 1, 2012). Development of an enzyme-linked immunosorbent assay (ELISA) for

residue analysis of the fungicide azoxystrobin in agricultural products. *J Agric Food Chem.*, 60(4), 904–911. 10.1021/jf203534n. Epub 2012 January 18 PMID: 22224459.

[24] Sun, Z., Zhang, K., Chen, C., Wu, Y., Tang, Y., Georgiev, M. I., Zhang, X., Lin, M. and Zhou, M. (January 2018). Biosynthesis and regulation of cyanogenic glycoside production in forage plants. *Appl Microbiol Biotechnol.*, 102(1), 9–16. 10.1007/s00253-017-8559-z. Epub 2017 October 12 PMID: 29022076.

[25] Zagrobelny, M., Bak, S., Rasmussen, A. V., Jørgensen, B., Naumann, C. M. and Lindberg Møller, B. (February 2004). Cyanogenic glucosides and plant-insect interactions. *Phytochemistry*, 65(3), 293–306. 10.1016/j.phytochem.2003.10.016. PMID: 14751300.

[26] EFSA Panel on Contaminants in the Food Chain (CONTAM), Schrenk, D., Bignami, M., Bodin, L., Chipman, J. K., Del Mazo, J., Grasl-Kraupp, B., Hogstrand, C., Hoogenboom, L. R., Leblanc, J. C., Nebbia, C. S., Nielsen, E., Ntzani, E., Petersen, A., Sand, S., Vleminckx, C., Wallace, H., Benford, D., Brimer, L., Mancini, F. R., Metzler, M., Viviani, B., Altieri, A., Arcella, D., Steinkellner, H. and Schwerdtle, T. (April 11, 2019). Evaluation of the health risks related to the presence of cyanogenic glycosides in foods other than raw apricot kernels. *EFSA J.*, 17(4), e05662. 10.2903/j.efsa.2019.5662. PMID: 32626287; PMCID: PMC7009189.

[27] Cressey, P. and Reeve, J. (March 2019). Metabolism of cyanogenic glycosides: A review. *Food Chem Toxicol.*, 125, 225–232. 10.1016/j.fct.2019.01.002. Epub 2019 January 4 PMID: 30615957.

[28] Cigolini, D., Ricci, G., Zannoni, M., Codogni, R., De Luca, M., Perfetti, P. and Rocca, G. (September 2011). Hydroxocobalamin treatment of acute cyanide poisoning from apricot kernels. *Emerg Med J.*, 28(9), 804–805. 10.1136/emj.03. 2011.3932rep. PMID: 21856998.

[29] Geller, R. J., Barthold, C., Saiers, J. A. and Hall, A. H. (November 2006). Pediatric cyanide poisoning: Causes, manifestations, management, and unmet needs. *Pediatrics*, 118(5), 2146–2158. 10.1542/peds.2006-1251. PMID: 17079589.

[30] Akyildiz, B. N., Kurtoğlu, S., Kondolot, M. and Tunç, A. (2010). Cyanide poisoning caused by ingestion of apricot seeds. *Ann Trop Paediatr.*, 30(1), 39–43. 10.1179/146532810X12637745451951. PMID: 20196932.

[31] Abraham, K., Buhrke, T. and Lampen, A. (March 2016). Bioavailability of cyanide after consumption of a single meal of foods containing high levels of cyanogenic glycosides: A crossover study in humans. *Arch Toxicol.*, 90(3), 559–574. 10.1007/s00204-015-1479-8. Epub 2015 February 24 PMID: 25708890; PMCID: PMC4754328.

[32] Sanchez-Verlaan, P., Geeraerts, T., Buys, S., Riu-Poulenc, B., Cabot, C., Fourcade, O., Mégarbane, B. and Genestal, M. (January 2011). An unusual cause of severe lactic acidosis: Cyanide poisoning after bitter almond ingestion. *Intensive Care Med.*, 37(1), 168–169. 10.1007/s00134-010-2029-8. Epub 2010 September 16 PMID: 20845029.

[33] Cowan, M., Møller, B. L., Norton, S., Knudsen, C., Crocoll, C., Furtado, A., Henry, R., Blomstedt, C. and Gleadow, R. M. (January 14, 2022). Cyanogenesis in the *sorghum* genus: From genotype to phenotype. *Genes (Basel)*, 13(1), 140. 10.3390/genes13010140. PMID: 35052482; PMCID: PMC8775130.

[34] Appenteng, M. K., Krueger, R., Johnson, M. C., Ingold, H., Bell, R., Thomas, A. L. and Greenlief, C. M. (March 4, 2021). Cyanogenic glycoside analysis in american elderberry. *Molecules*, 26(5), 1384. 10.3390/molecules26051384. PMID: 33806603; PMCID: PMC7961730.

[35] Bolarinwa, I. F., Orfila, C. and Morgan, M. R. (July 9, 2014). Development and application of an enzyme-linked immunosorbent assay (ELISA) for the quantification of amygdalin, a cyanogenic glycoside, in food. *J Agric Food Chem.*, 62(27), 6299–6305. 10.1021/jf501978d. Epub 2014 June 26 PMID: 24905893.

[36] Barceloux, D. G. (June 2009). Cyanogenic foods (cassava, fruit kernels, and cycad seeds). *Dis Mon.*, 55(6), 336–352. 10.1016/j.disamonth.2009.03.010. PMID: 19446677.

[37] Montagnac, J. A., Davis, C. R. and Tanumihardjo, S. A. (July 2009). Nutritional value of cassava for use as a staple food and recent advances for improvement. *Compr Rev Food Sci Food Saf.*, 8(3), 181–194. 10.1111/j.1541-4337.2009.00077.x. PMID: 33467798.

[38] Voldřich, M. and Kyzlink, V. (January 1992). Cyanogenesis in canned stone fruits. *Journal of Food Science*, 57(1), 161–162.

[39] Wasserkrug, K. and Rassi, Z. E. (1997). High performance liquid phase separation of glycosides. I. Reversed phase chromatography of cyanogenic glycosides with UV and pulsed amperometric detection. *J. Liquid Chromat. & Related Technol.*, 20, 335–349.

[40] Sornyotha, S., Kyu, K. L. and Ratanakhanokchai, K. (January 1, 2007). Purification and detection of linamarin from cassava root cortex by high performance liquid chromatography. *Food Chemistry*, 104(4), 1750–1754.

[41] Chassagne, D., Crouzet, J. C., Bayonove, C. L. and Baumes, R. L. (December 18, 1996). Identification and quantification of passion fruit cyanogenic glycosides. *Journal of Agricultural and Food Chemistry*, 44(12), 3817–3820.

[42] Bolarinwa, I. F., Orfila, C. and Morgan, M. R. (2014). Amygdalin content of seeds, kernels and food products commercially-available in the UK. *Food Chem.*, 152, 133–139. 10.1016/j.foodchem.2013.11.002. Epub 2013 November 12 PMID: 24444917.

[43] Jaszczak-Wilke, E., Polkowska, Ż, Koprowski, M., Owsianik, K., Mitchell, A. E. and Bałczewski, P. (April 13, 2021). Amygdalin: Toxicity, anticancer activity and analytical procedures for its determination in plant seeds. *Molecules*, 26(8), 2253. 10.3390/molecules26082253. PMID: 33924691; PMCID: PMC8069783.

[44] Cho, A. Y., Shin, K. J., Chung, J. and Oh, S. (October 2008). A sensitive enzyme immunoassay for amygdalin in food extracts using a recombinant antibody. *J Food Prot.*, 71(10), 2048–2052. 10.4315/0362-028x-71.10.2048. PMID: 18939751.

[45] Bolarinwa, I. F. (July 2015). Synthesis and characterization of hapten-protein conjugates for antibody production against cyanogenic glycosides. *J Food Prot.*, 78(7), 1408–1413. 10.4315/0362-028X.JFP-15-033. PMID: 26197297.

[46] Weber, J., Peng, H. and Rader, C. (March 24, 2017). From rabbit antibody repertoires to rabbit monoclonal antibodies. *Exp Mol Med.*, 49(3), e305. 10.1038/emm.2017.23. PMID: 28336958; PMCID: PMC5382564.

[47] Cho, A. Y., Yi, K. S., Rhim, J. H., Kim, K. I., Park, J. Y., Keum, E. H., Chung, J. and Oh, S. (April 30, 2006). Detection of abnormally high amygdalin content in food by an enzyme immunoassay. *Mol Cells*, 21(2), 308–313. PMID: 16682828.

[48] Jose, D. A., Elstner, M. and Schiller, A. (October 18, 2013). Allosteric indicator displacement enzyme assay for a cyanogenic glycoside. *Chemistry*, 19(43), 14451–14457. 10.1002/chem.201302801. Epub 2013 October 7 PMID: 24123550.

[49] Abuknesha, R. A., Luk, C. Y., Griffith, H. H., Maragkou, A. and Iakovaki, D. (November 30, 2005). Efficient labelling of antibodies with horseradish peroxidase using cyanuric chloride. *J Immunol Methods*, 306(1-2), 211–217. 10.1016/j.jim.2005.09.002. Epub 2005 September 21 PMID: 16223506.

[50] Kemp, H. A. and Morgan, M. R. (November 20, 1986). Studies on the detrimental effects of bivalent binding in a microtitration plate ELISA and possible remedies. *J Immunol Methods*, 94(1-2), 65–72. 10.1016/0022-1759(86)90216-4. PMID: 3782818.

[51] Tamariz, J., Burgueño-Tapia, E., Vázquez, M. A. and Delgado, F. (2018). Pyrrolizidine alkaloids. *Alkaloids Chem Biol.*, 80, 1–314. 10.1016/bs.alkal.2018.03.001. Epub 2018 July 6. PMID: 30001795.

[52] Neuman, M. G., Cohen, L., Opris, M., Nanau, R. M. and Hyunjin, J. (2015). Hepatotoxicity of pyrrolizidine alkaloids. *J Pharm Pharm Sci.*, 18(4), 825–843. 10.18433/j3bg7j. PMID: 26626258.

[53] Schrenk, D. (January 2020). Toxicology of pyrrolizidine alkaloids. *Food Chem Toxicol.*, 135, 110938. 10.1016/j.fct.2019.110938. Epub 2019 November 1 PMID: 31682934.

[54] Zündorf, I., Wiedenfeld, H., Röder, E. and Dingermann, T. (April 1998). Generation and characterization of monoclonal antibodies against the pyrrolizidine alkaloid retrorsine. *Planta Med.*, 64(3), 259–263. 10.1055/s-2006-957421. PMID: 9581524.

[55] Stegelmeier, B. L. (July 2011). Pyrrolizidine alkaloid-containing toxic plants (Senecio, crotalaria, cynoglossum, amsinckia, heliotropium, and echium spp.). *Vet Clin North Am Food Anim Pract.*, 27(2), 419–428. ix. 10.1016/j.cvfa.2011.02.013. PMID: 21575778.

[56] Schrenk, D., Gao, L., Lin, G., Mahony, C., Mulder, P. P. J., Peijnenburg, A., Pfuhler, S., Rietjens, I. M. C. M., Rutz, L., Steinhoff, B. and These, A. (February 2020). Pyrrolizidine alkaloids in food and phytomedicine: Occurrence, exposure, toxicity, mechanisms, and risk assessment – A review. *Food Chem Toxicol.*, 136, 111107. 10.1016/j.fct.2019.111107. Epub 2020 January 3. PMID : 31904473.

[57] Ruan, J., Liao, C., Ye, Y. and Lin, G. (January 21, 2014). Lack of metabolic activation and predominant formation of an excreted metabolite of nontoxic platynecine-type pyrrolizidine alkaloids. *Chem Res Toxicol.*, 27(1), 7–16. 10.1021/tx4004159. Epub 2013 December 13 PMID: 24308637.

[58] Ruan, J., Yang, M., Fu, P., Ye, Y. and Lin, G. (June 16, 2014). Metabolic activation of pyrrolizidine alkaloids: insights into the structural and enzymatic basis. *Chem Res Toxicol.*, 27(6), 1030–1039. 10.1021/tx500071q. Epub 2014 May 22 PMID: 24836403.

[59] Fu, P. P., Xia, Q., Lin, G. and Chou, M. W. (February 2004). Pyrrolizidine alkaloids – Genotoxicity, metabolism enzymes, metabolic activation, and mechanisms. *Drug Metab Rev.*, 36(1), 1–55. 10.1081/dmr-120028426. PMID: 15072438.

[60] Roscioli, T., Ziegler, J. B., Buckley, M. and Wong, M. (February 21, 2007). Hepatic veno-occlusive disease with immunodeficiency. In Adam, M. P.,

Ardinger, H. H., Pagon, R. A., Wallace, S. E., Bean, L. J. H., Gripp, K. W., Mirzaa, G. M. and Amemiya, A. (eds.). *GeneReviews® [Internet]*. Seattle (WA): University of Washington, Seattle, pp. 1993–2022. PMID: 20301448

[61] Yang, X., Li, W., Sun, Y., Guo, X., Huang, W., Peng, Y. and Zheng, J. (February 20, 2017). Comparative study of hepatotoxicity of pyrrolizidine alkaloids retrorsine and monocrotaline. *Chem Res Toxicol.*, 30(2), 532–539. 10.1021/acs.chemrestox.6b00260. Epub 2017 January 31 PMID: 28095673.

[62] Mingatto, F. E., Dorta, D. J., dos Santos, A. B., Carvalho, I., da Silva, C. H., da Silva, V. B., Uyemura, S. A., dos Santos, A. C. and Curti, C. (October 2007). Dehydromonocrotaline inhibits mitochondrial complex I. A potential mechanism accounting for hepatotoxicity of monocrotaline. *Toxicon.*, 50(5), 724–730. 10.1 016/j.toxicon.2007.06.006. Epub 2007 June 26 PMID: 17669457.

[63] Jiang, Y., Fu, P. P. and Lin, G. (2006). Hepatotoxicity of naturally occurring pyrrolizidine alkaloids. *Asian J. Pharmacodyn. Pharmacokinet.*, 6, 187–192.

[64] Mattocks, A. R., Driver, H. E., Barbour, R. H. and Robins, D. J. (April 1986). Metabolism and toxicity of synthetic analogues of macrocyclic diester pyrrolizidine alkaloids. *Chem Biol Interact.*, 58(1), 95–108. 10.1016/s0009-2797(86) 80089-8. PMID: 3708724.

[65] Lin, G., Cui, Y. Y. and Hawes, E. M. (February 1998). Microsomal formation of a pyrrolic alcohol glutathione conjugate of clivorine. Firm evidence for the formation of a pyrrolic metabolite of an otonecine-type pyrrolizidine alkaloid. *Drug Metab Dispos.*, 26(2), 181–184. PMID: 9456306.

[66] Lin, G., Cui, Y. Y. and Hawes, E. M. (December 2000). Characterization of rat liver microsomal metabolites of clivorine, an hepatotoxic otonecine-type pyrrolizidine alkaloid. *Drug Metab Dispos.*, 28(12), 1475–1483. PMID: 11095586.

[67] Lin, G., Wang, J. Y., Li, N., Li, M., Gao, H., Ji, Y., Zhang, F., Wang, H., Zhou, Y., Ye, Y., Xu, H. X. and Zheng, J. (April 2011). Hepatic sinusoidal obstruction syndrome associated with consumption of Gynura segetum. *J Hepatol.*, 54(4), 666–673. 10.1016/j.jhep.2010.07.031. Epub 2010 September 22 PMID: 21146894.

[68] Fashe, M. M., Juvonen, R. O., Petsalo, A., Vepsäläinen, J., Pasanen, M. and Rahnasto-Rilla, M. (April 20, 2015). In silico prediction of the site of oxidation by cytochrome P450 3A4 that leads to the formation of the toxic metabolites of pyrrolizidine alkaloids. *Chem Res Toxicol.*, 28(4), 702–710. 10.1021/tx5004 78q. Epub 2015 February 17 PMID: 25651456.

[69] Cheng, T., Li, W., Yang, X., Wang, H., Zhang, F., Li, N., Lin, G. and Zheng, J. (2020). Antibody-based detection of lysine modification of hepatic protein in mice treated with retrorsine. *J Environ Sci Health C Toxicol Carcinog.*, 38(4), 315–328. 10.1080/26896583.2020.1832411. PMID: 33382013.

[70] Roseman, D. M., Wu, X. and Kurth, M. J. (March–April 1996). Enzyme-linked immunosorbent assay detection of pyrrolizidine alkaloids: Immunogens based on quaternary pyrrolizidinium salts. *Bioconjug Chem.*, 7(2), 187–195. 10.1 021/bc950084e. PMID: 8983340.

[71] Kopp, T., Abdel-Tawab, M. and Mizaikoff, B. (May 13, 2020). Extracting and analyzing pyrrolizidine alkaloids in medicinal plants: A review. *Toxins (Basel)*, 12(5), 320. 10.3390/toxins12050320. PMID: 32413969; PMCID: PMC7290370.

[72] Langer, T., Möstl, E., Chizzola, R. and Gutleb, R. (June 1996). A competitive enzyme immunoassay for the pyrrolizidine alkaloids of the senecionine type. *Planta Med.*, 62(3), 267–271. 10.1055/s-2006-957875. PMID: 8693043.

[73] Miolo, G., Stefanidis, M., Santella, R. M., Dall'Acqua, F. and Gasparro, F. P. (February 1989). 6,4,4'-trimethylangelicin photoadduct formation in DNA: Production and characterization of a specific monoclonal antibody. *J Photochem Photobiol B*, 3(1), 101–112. 10.1016/1011-1344(89)80024-7. PMID: 2498476.

[74] Yin, B. Y., Gasparro, F. P., Bevilacqua, P. M. and Santella, R. M. (December 1991). Quantitation of plasma levels of 8-methoxypsoralen by competitive enzyme-linked immunosorbent assay. *J Invest Dermatol.*, 97(6), 1001–1004. 10.1111/1523-1747.ep12492190. PMID:1748809.

[75] Pathak, M. A., Daniels, F., Jr and Fitzpatrick, T. B. (September 1962). The presenhpy known fistrixuhion of furocoumarins (psoralens) in plants. *J Invest Dermatol.*, 39, 225–239. 10.1038/jid.1962.106. PMID: 13941836.

[76] Santella, R. M., Dharmaraja, N., Gasparro, F. P. and Edelson, R. L. (April 11, 1985). Monoclonal antibodies to DNA modified by 8-methoxypsoralen and ultraviolet A light. *Nucleic Acids Res.*, 13(7), 2533–2544. 10.1093/nar/13.7. 2533. PMID: 4000963; PMCID: PMC341173.

[77] Santella, R. M., Hsieh, L. L. and Perera, F. (1986). Immunologic quantification of carcinogen-DNA adducts. *Basic Life Sci.*, 38, 509–517. 10.1007/978-1-4 615-9462-8_53. PMID: 3090996.

[78] Friedman, M., McDonald, G. M. and Filadelfi-Keszi, M. (January 1, 1997). Potato glycoalkaloids: Chemistry, analysis, safety, and plant physiology. *Critical Reviews in Plant Sciences*, 16(1), 55–132.

[79] Abell, D. C. (1998). Analysis of potato glycoalkaloids by ELISA and matrix-assisted laser desorption/ionization time-of-flight mass spectroscopy. 5208–5208.

[80] Fewell, A. M. and Roddick, J. G. (May 13, 1993). Interactive antifungal activity of the glycoalkaloids α-solanine and α-chaconine. *Phytochemistry*, 33(2), 323–328.

[81] Harvey, M. H., Morris, B. A., McMillan, M. and Marks, V. (September 1985). Measurement of potato steroidal alkaloids in human serum and saliva by radioimmunoassay. *Hum Toxicol.*, 4(5), 503–512. 10.1177/0960327185004 00506. PMID: 4054913.

[82] Morris, S. C. (1984). The toxicity and teratogenicity of solanaceae glycoalkaloids, particularly those of the potato (Solanum tuberosum): A review. *Food Tech Australia*, 36, 118–124.

[83] Nishie, K., Gumbman, U. R. and Keyi, A. C. (1971). Phamiaoology of solanine. *Toxicol. Appl. Pharmacol.*, 19, 81–92.

[84] Keukeng, E.-A. S., de Vrije, T., Fabrie, C. H. J. P., Demel, R. A., Jongen, W. M. F. and de hije, B. (1992). Duai specinciv of sterol-mediated glycoalkaioid lnduced membrane disruption. *Biochim. Biophys. Acta 11*, 10, 127–136.

[85] Coleman, R. M., Lombard, M. F., Sicard, R. E. and Rencricca, N. J. (1989). *Fundamental Immunology*. Dubuque, IA: Wdiam C. Brown Publishers.

[86] Erlanger, B. F. (1980). The preparation of antigenic hapten-carrier conjugates: A survey. *Methods Enqmol.*, 70, 85–104.

[87] Kahler, G. and Milstein, C. (1975). Continuous cultures of fused cells secreting antibody of predeked specificity. *Nature*, 256, 495–497.

[88] Stanker, L. H., Kamps-Holtzapple, C. and Friedman, M. (1994). Development and characterization of monoclonal antibodies that differentiate between potato and tomato glycoalkaloids and aglycons. *J. Agric. Food Chem.*, 42, 2360–2366.

[89] Friedman, M., Bautista, F. F., Stanker, L. H. and Larkin, K. A. (December 21, 1998). Analysis of potato glycoalkaloids by a new ELISA kit. *Journal of Agricultural and Food Chemistry*, 46(12), 5097–5102.

[90] Ward, C. M., Franklin, J. G. and Morgan, M. R. A. (1988). Investigations into the visual assessment of ELISA end points: Application to the determination of glycoalkaloids. *Food Addit. Contam.*, 4, 621–627.

[91] Sporns, P., Abell, D. C., Kwok, A. S. K., Plhak, L. C. and Thompson, C. A. (1996). Immunoassays for toxic glycoalkaloids. In Beier, R. C. and Stanker, L. H. (eds.). *Immunoassays for residue analysis*. ACS Symposium Series 621. Washington, DC: American Chemical Society, pp. 256–272.

[92] Ward, C. M., Franklin, J. G. and Morgan, M. R. (October 1, 1988). Investigations into the visual assessment of ELISA end points: Application to determination of potato total glycoalkaloids. *Food Additives & Contaminants*, 5(4), 621–627.

[93] Soanes, K. H., Figuereido, K., Richards, R. C., Mattatall, N. R. and Ewart, K. V. (November 2004). Sequence and expression of C-type lectin receptors in Atlantic salmon (Salmo salar). *Immunogenetics*, 56(8), 572–584. 10.1007/s00251-004-0719-5. Epub 2004 October 14 PMID: 15490154.

[94] Van Damme, E. J. M., Peumans, W. J., Pusztai, A. and Bardocz, S. (1998). Plant lectins: A special class of plant proteins. In Van Damme, E. J. M., Peumans, W. J., Pusztai, A. and Bardocz, S. (eds.). *Handbook of plant lectins: Properties and biomedical applications*. Chichester, England: Wiley, pp. 4–30.

[95] Vasta, G. R., Nita-Lazar, M., Giomarelli, B., Ahmed, H., Du, S., Cammarata, M., Parrinello, N., Bianchet, M. A. and Amzel, L. M. (December 2011). Structural and functional diversity of the lectin repertoire in teleost fish: Relevance to innate and adaptive immunity. *Dev Comp Immunol.*, 35(12), 1388–1399. 10.1016/j.dci.2011.08.011. Epub 2011 August 30. PMID: 21896283; PMCID: PMC3429948.

[96] Manage, L., Joshi, A. and Sohonie, K. (January 1972). Toxicity to rats and mice of purified phytohaemagglutinins from four Indian legumes. *Toxicon.*, 10(1), 89–91. 10.1016/0041-0101(72)90099-2. PMID: 5015549.

[97] Pusztai, A. (1991). General effects on animal cells. In Phillipson, J. D., Ayres, D. C. and Baxter, H. (eds.). *Plant lectins*. Cambridge, England: University Press, pp. 105–205.

[98] Rizzi, C., Galeoto, L., Zoccatelli, G., Vincenzi, S., Chignola, R. and Peruffo, A. D. (January 1, 2003). Active soybean lectin in foods: Quantitative determination by ELISA using immobilised asialofetuin. *Food Research International*, 36(8), 815–821.

[99] Adamcová, A., Laursen, K. H. and Ballin, N. Z. (November 13, 2021). Lectin activity in commonly consumed plant-based foods: Calling for method harmonization and risk assessment. *Foods*, 10(11), 2796. 10.3390/foods10112796. PMID: 34829077; PMCID: PMC8618113.

[100] Ng, T. B., Fai Cheung, R. C., Wing Ng, C. C., Fang, E. F. and Wong, J. H. (2015). A review of fish lectins. *Curr Protein Pept Sci.*, 16(4), 337–351. 10.2174/1389203716041504291608 50. PMID: 25929869.

[101] Wang, X. W. and Wang, J. X. (January–February 2013). Diversity and multiple functions of lectins in shrimp immunity. *Dev Comp Immunol.*, 39(1–2), 27–38. 10.1016/j.dci.2012.04.009. Epub 2012 May 2 PMID: 22561073.

[102] Drickamer, K. and Taylor, M. E. (1993). Biology of animal lectins. *Annu Rev Cell Biol.*, 9, 237–264. 10.1146/annurev.cb.09.110193.001321. PMID: 8280461.

[103] Drickamer, K. (July 15, 1988). Two distinct classes of carbohydrate-recognition domains in animal lectins. *J Biol Chem.*, 263(20), 9557–9560. PMID: 3290208.

[104] Zelensky, A. N. and Gready, J. E. (December 2005). The C-type lectin-like domain superfamily. *FEBS J.*, 272(24), 6179–6217. 10.1111/j.1742-4658.2 005.05031.x. PMID: 16336259.

[105] Weis, W. I., Kahn, R., Fourme, R., Drickamer, K. and Hendrickson, W. A. (December 13, 1991). Structure of the calcium-dependent lectin domain from a rat mannose-binding protein determined by MAD phasing. *Science*, 254(5038), 1608–1615. 10.1126/science.1721241. PMID: 1721241.

[106] Nachbar, M. S. and Oppenheim, J. D. (November 1980). Lectins in the United States diet: A survey of lectins in commonly consumed foods and a review of the literature. *Am J Clin Nutr.*, 33(11), 2338–2345. 10.1093/ajcn/33.11.2338. PMID: 7001881.

[107] Whitmore, F. A. (February 1, 1992). A hemagglutinating substance in chitin. *Biotechniques*, 12(2), 202–207.

[108] Carratu, G., Carafa, A. M., Corsaro, M. M. and Giannattasio, M. (1995). Pollen hemagglutinating activity is not related to lectin. *Sex. Plant Reprod.*, 8, 91–94.

[109] Peumans, W. J. and Van Damme, E. J. M. (1996). Prevalence, biological activity and genetic manipulation of lectins in foods. *Trends Food Sci. Technol.*, 7, 132–138.

[110] Shao, M. C. (1992). The use of streptavidin-biotinylglycans as a tool for characterization of oligosaccharide-binding specificity of lectin. *Anal. Biochem.*, 205, 77–82.

[111] Satoh, A., Fukui, E., Yoshino, S., Shinoda, M., Kojima, K. and Matsumoto, I. (November 15, 1999). Comparison of methods of immobilization to enzyme-linked immunosorbent assay plates for the detection of sugar chains. *Anal Biochem.*, 275(2), 231–235. 10.1006/abio.1999.4329. PMID: 10552909.

[112] Vincenzi, S., Zoccatelli, G., Perbellini, F., Rizzi, C., Chignola, R., Curioni, A. and Peruffo, A. D. (October 23, 2002). Quantitative determination of dietary lectin activities by enzyme-linked immunosorbent assay using specific glycoproteins immobilized on microtiter plates. *J Agric Food Chem.*, 50(22), 6266–6270. 10.1021/jf011585z. PMID: 12381101.

[113] Duk, M., Lisowska, E., Wu, J. H. and Wu, A. M. (September 1994). The biotin/avidin-mediated microtiter plate lectin assay with the use of chemically modified glycoprotein ligand. *Anal Biochem.*, 221(2), 266–272. 10.1006/abio.1994.1410. PMID: 7810865.

[114] Dawson, R. M., Paddle, B. M. and Alderton, M. R. (September-October 1999). Characterization of the Asialofetuin microtitre plate-binding assay for evaluating inhibitors of ricin lectin activity. *J Appl Toxicol.*, 19(5), 307–312. 10.1002/(sici) 1099-1263(199909/10)19:5<307::aid-jat581>3.0.co;2-p. PMID: 10513675.

[115] Bolarinwa, I. F. (2015). Synthesis and characterization of hapten-protein conjugates for antibody production against cyanogenic glycosides. *Journal of Food Protection*, 78(7), 1408–1413. 10.4315/0362-028X.JFP-15-033

[116] Oplatowska, M., Elliott, C. T., Huet, A. C., McCarthy, M., Mulder, P. P., von Holst, C., Delahaut, P., Van Egmond, H. P. and Campbell, K. (2014). Development and validation of a rapid multiplex ELISA for pyrrolizidine alkaloids and their N-oxides in honey and feed. *Analytical and Bioanalytical Chemistry*, 406(3), 757–770. 10.1007/s00216-013-7488-7

[117] Lee, S. T., Schoch, T. K., Stegelmeier, B. L., Gardner, D. R., Than, K. A. and Molyneux, R. J. (2001). Development of enzyme-linked immunosorbent assays for the hepatotoxic alkaloids riddelliine and riddelliine N-oxide. *Journal of Agricultural and Food Chemistry*, 49(8), 4144–4151. 10.1021/jf010042m

[118] Friedman, M., Bautista, F. F., Stanker, L. H. and Larkin, K. A. (1998). Analysis of potato glycoalkaloids by a new ELISA kit. *Journal of Agricultural and Food Chemistry*, 46(12), 5097–5102.

[119] Friedman, M. (2004). Analysis of biologically active compounds in potatoes (Solanum tuberosum), tomatoes (Lycopersicon esculentum), and jimson weed (Datura stramonium) seeds. *Journal of Chromatography. A*, 1054(1–2), 143–155. 10.1016/j.chroma.2004.04.049

[120] Putalun, W., Tanaka, H., Yahara, S., Lhieochaiphan, S. and Shoyama, Y. (2000). Survey of solasodine-type glycoalkaloids by western blotting and ELISA using an anti-solamargine monoclonal antibody. *Biological & Pharmaceutical Bulletin*, 23(1), 72–75. 10.1248/bpb.23.72

[121] Mottaghipisheh, J., Nové, M., Spengler, G., Kúsz, N., Hohmann, J. and Csupor, D. (2018). Antiproliferative and cytotoxic activities of furocoumarins of ducrosia anethifolia. *Pharmaceutical Biology*, 56(1), 658–664. 10.1080/13 880209.2018.1548625

[122] Abad, M. J., De las Heras, B., Silvan, A. M., Pascual, R., Bermejo, P., Rodriguez, B. and Villar, A. M. (August 2001). Effects of furocoumarins from cachrys trifida on some macrophage functions. *Journal of Pharmacy and Pharmacology*, 53(8), 1163–1168.

CHAPTER 15

Detection and Quantification of Food Toxins of Plant Origin Using Biosensors

Saher Islam[1], Devarajan Thangadurai[2], and Ravichandra Hospet[3]

[1]Department of Biotechnology, Lahore College for Women University, Lahore, Pakistan
[2]Department of Botany, Karnatak University, Dharwad, Karnataka, India
[3]Department of Food Protectants and Infestation Control, CSIR-Central Food
Technological Research Institute, Mysore, Karnataka, India

CONTENTS

15.1	Introduction	245
15.2	Food Toxins of Plant Origin	246
15.3	Electrochemical Biosensors for Food Analysis	248
	15.3.1 Amperometric Sensors	248
	15.3.2 Impedimetric Sensors	248
	15.3.3 Potentiometric Sensors	249
15.4	Quartz Crystal Microbalance Biosensors for Food Analysis	249
15.5	Optical Biosensors for Food Analysis	249
15.6	Natural Biosensors for Food Analysis	249
	15.6.1 Enzyme-Based Sensors	249
	15.6.2 Antibody-Based Sensors	250
	15.6.3 Nucleic Acid–Based Sensors	250
	15.6.4 Whole-Cell-Based Sensors	250
15.7	Emerging Biosensors for Food Analysis	251
	15.7.1 Nano-Biosensors	251
	15.7.2 Cell-Free Biosensors	251
	15.7.3 Microfluidics-Based Biosensors	252
	15.7.4 Magnetic-Bead-Based Biosensors	252
15.8	Conclusion	253
References		253

15.1 INTRODUCTION

Several agricultural products including fresh vegetables and fruits (Sanzani et al. 2016; Gonçalves et al. 2018), nuts (Kluczkovski 2019); fluids such as grape juice, wine (Welke 2019), beer (Pascari et al. 2018), and milk products (Viegas et al. 2019); cereals like rice,

DOI: 10.1201/9781003222194-18

245

wheat, and maize (Varzakas 2015); feed (Kebede et al. 2020); cocoa and coffee (Huertas-Pérez et al. 2016; Bessaire et al. 2019); and herbs and spices (Gambacorta et al. 2019) can get naturally occurring food toxins at any stage of the food chain. Natural toxins are produced in plants as defense mechanisms against insects, microorganisms, predators, or climate stress. These toxins can pose adverse health threats to humans and other livestock species. Cyanogenic glycosides as phytotoxins are present in around 2,000 plant varieties including sorghum, cassava, bamboo roots, almonds, and stone fruits (Yulvianti and Zidorn 2021). Furocoumarins are stress toxins present in celery roots, parsnips, citrus plants, and medicinal plants and may cause gastral problems in humans. Many beans contain lectins, such as kidney beans, that can cause diarrhea, vomiting, and stomachache (Dolan et al. 2010). Mycotoxins as natural occurring toxins are produced by molds that grow on foodstuffs like cereals, spices, and dried fruits (Fletcher and Netzel 2020). Pyrrolizidine alkaloids are toxic compounds produced by at least 600 plants including families Asteraceae, Fabaceae, and Boraginaceae. Pyrrolizidine alkaloids cause adversative health effects such as DNA damage and cancer (Hama et al. 2021). Sensitive analysis for naturally occurring food toxins requires appropriate methods for detection because many toxins express toxicity even at low levels. Biosensors are a tremendously useful approach for the identification of toxins in foods of plant origin (Tothill 2011; Neethirajan et al. 2018; Oliveira et al. 2019).

Biosensors can contribute to the food industry by reducing the traces of food toxins as they provide significant advantages such as easy, inexpensive, and fast sample analysis, accuracy, reproducibility, and on-site analysis of samples (Pirinçci et al. 2018). Biosensors provide valuable tools for the agro-food diagnosis as they are portable, convenient, and do not require special skills to use (Thakur and Ragavan 2013). They are often used for the nutrient analysis, detecting the antinutrients and natural toxins, for detecting genetically modified species, and monitoring the food processing quality (Ahmed et al. 2016). With different immunogenic and enzymatic reactions, biosensors may also detect pesticides, proteins, antibiotics, fatty acids, and vitamin B complexes present in food (Kim and Cho 2011).

Recently, several advanced applications of biosensors for the analysis of food toxins have been reported (Perumal and Hashim 2014; Hammond et al. 2016; Xiang et al. 2018; Hossain et al. 2019; Nan et al. 2019). The types of biosensors that are primarily used for detection of food toxins are optical, piezoelectric, and electrochemical (Tothill 2011; Oliveira et al. 2019; Schulz et al. 2019) (Figure 15.1). Recognition elements for biosensors include enzymes, antibodies, peptides, nucleic acids, cells, and aptamers (Oliveira et al. 2019). Metallic nanoparticles, nanofibers, and carbon nanotubes have been examined to tweak the sensitivity and accuracy of biosensors as these compounds are sufficiently biocompatible and characterized by certain physicochemical features such as surface-to-volume ratio (Doria et al. 2012; Malekzad et al. 2017). This chapter discusses the innovative biosensors used for the assessment of naturally occurring toxins in food.

15.2 FOOD TOXINS OF PLANT ORIGIN

Food toxins of plant origin are naturally present in plants such as fruits and vegetables. Usually, these toxicants are secondary metabolites produced by plants to protect against bacteria, insects, fungi, and other plant predators, and are further responsible for particular characteristics such as colors and flavors. Toxins may present in food plants due to natural selection and breeding methods that influence these protective mechanisms

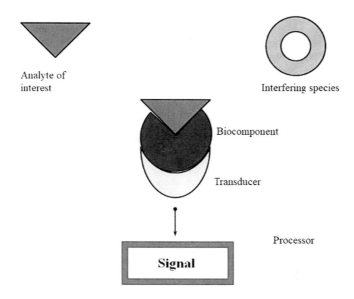

FIGURE 15.1 A schematic representation of biosensors [adapted from Girigoswami et al. (2021), ©Creative Commons Attribution International 4.0].

(Hajšlová et al. 2018). Natural toxins in food plants include lectins in green beans and red and white kidney beans; cyanogenic glycosides in bitter bamboo shoots, apricot seeds, flaxseeds, and cassava; glycoalkaloids in potatoes; 4-methoxypyridoxine in ginkgo seeds; colchicine in fresh lily flowers; and muscarine in some wild mushrooms (Ogori 2019).

Alkaloids are the organic compounds with nitrogen in a heterocyclic ring, derived from amino acids and most of which exhibit strong physiological activity. Glycosides are substances that comprise a non-sugar moiety. Cyanogenic glycosides are well known to release prussic acid. Tannins can precipitate proteins. Several protein toxins produced by plants enter eukaryotic cells and inhibit protein synthesis enzymatically. Proteins may also be considered as inherent plant toxicants; mainly lectins bind specifically to carbohydrates and are present in many food raw materials, particularly in legumes. Phytohemaglutinin (PHA) is a lectin observed in red kidney beans, known to agglutinate mammalian red blood cells (Noah et al. 1980). The toxic proteins include abrin (rosary pea), ricin (castor plant), and white acacia (Essers et al. 1998; Ogori 2019).

The common potato produces various glycoalkaloids, mainly α-chaconine and α-solanine, from cholesterol. Generally observed in the peel, its concentration depends on multiple factors such as cultivar, sprouting, and storage conditions. Many reports were highlighted human poisoning from intake of potatoes rich in glycoalkaloids. The common symptoms of substandard glycoalkaloids poisoning include diarrhea, acute gastrointestinal illness, abdominal pain, and vomiting (Schilter et al. 2014; Ogori 2019).

Oxalic acid and oxalates are another class of toxins usually observed in the juice or sap of crystals. These toxins can irritate the skin, mouth, tongue, and throat, resulting in throat swelling, breathing difficulties, burning pain, and stomach upset. These substances may be present in trichomes or raphides. They can provoke mechanical irritation. Oxalate in the blood binds calcium to form the insoluble calcium oxalate. With advances in food science and technology, food quality has continuously improved over the last 20 years. Significant developments have been observed on several sensory, safety, and nutritional attributes. In this connection, scientific efforts are progressively focused on designing and developing functional food products of sustainable health benefits. With respect to chemical composition, plant-based foods can be a consortium of chemical compounds, classified as intrinsic factors with inherent constituents of plants and extrinsic factors of both natural and industrial origin. Intrinsic factors comprise a 'wide array of chemical compounds with several health impacts. Anti-nutrients are compounds that suppress the uptake of nutrients, further reducing nutritional importance of food plants. A few examples include blocking of protein digestion by protease inhibitors and phytate inhibiting absorption of iron. Cyanogenic glucosides and glycoalkaloids in potatoes are known as inherent toxicants (Schilter et al. 2014).

Cassava, a tropical shrub, is a major staple food in Asia and Africa. Cassava roots produce two cyanogenic glycosides, such as linamarin and lotaustralin. These are biotransformed into thiocyanate during detoxification in mammalians. During food processing, the toxicological concern arises when glucosides come into contact with linamarase.

Pyrrolizidine alkaloids (PAs) are secondary plant metabolites commonly found in many flowering plants. Chemically, they are heterocyclic compounds of wide structural activity and comprise four necine bases: platynecine, retronecine, otonecine, and heliotridine (Edgar et al. 2011). The toxicity of individual compounds directly correlates with their chemical structure. After ingestion, PAs are rapidly absorbed from the gastrointestinal tract and activated to toxic metabolites in the liver through various cytochrome P450 enzymes (Edgar et al. 2011; Schilter et al. 2014).

15.3 ELECTROCHEMICAL BIOSENSORS FOR FOOD ANALYSIS

The literature on biosensors for analysis of food toxins of plant origin is very scarce. That is why a number of methods are cited for other toxic substances.

15.3.1 Amperometric Sensors

Amperometric sensors require three (counter, reference, and working) or two (reference and working) electrodes systems. These biosensors calculate the flow of current that is produced by an electroactive species. A mediator input is able to improve the efficacy of amperometric sensors by increasing the electron transfer rate (Grieshaber et al. 2008; Hammond et al. 2016). Inert metals, including gold or platinum or otherwise graphene and carbon nanotubes, can be utilized as working electrodes. Currently, disposable imprinted electrodes are in use because they tend to be cheaper and designed at a larger scale (Ricci et al. 2007; Ferreira et al. 2013). These sensors have shown successful results for ochratoxin A in red wine (Tang et al. 2018) and zearalenone in corn products (Liu et al. 2016; He and Yan 2019b).

15.3.2 Impedimetric Sensors

The electrochemical impedance spectroscopy approach has been established to analyze the toxins present in food. This approach monitors the variations observed in an interface between a redox and electrode probe (Ram et al. 2016). An impedimetric sensor comprises three electrodes named reference electrode, counter electrode, and working electrode. These sensors are being effectively used for the detection of ochratoxin A, patulin, aflatoxin B1, and M1 (Karczmarczyk et al. 2017a; Yagati et al. 2018; Khan et al. 2019; Nan et al. 2019).

15.3.3 Potentiometric Sensors

Potentiometric sensors employ ion-selective electrode systems including three (counter, reference, and working) or two (reference and working) electrodes. Evidence of the recognition process is recorded by alterations in a circuit potential signal between reference and working electrodes (Perumal and Hashim 2014). Voltammetry methods including cyclic, square wave, and differential pulse have been utilized for determination of food toxins (Xu et al. 2016). Recently, potentiometric sensors are successfully being assessed for zearalenone in maize (He and Yan 2019a), aflatoxin B1 in corn (Zhang et al. 2016), patulin in juice (He and Dong 2018), and ochratoxin A in red wine and grape juice (Xiang et al. 2018).

15.4 QUARTZ CRYSTAL MICROBALANCE BIOSENSORS FOR FOOD ANALYSIS

Quartz crystal microbalance-based sensors have been positively experimented for pathogen monitoring and food toxin analysis. The transducer contains a gold-coated crystal quartz that sends an electrical signal and ultimately alters the resonance frequency. Moreover, the quartz surface has a sensory layer that causes particular vibrations and mass change (Montagut et al. 2011). These biosensors have also been investigated for aflatoxin B1 in pistachio, peanut, wheat, and rice (Tang et al. 2018; Gu et al. 2019) and ochratoxin A in red wine (Karczmarczyk et al. 2017b).

15.5 OPTICAL BIOSENSORS FOR FOOD ANALYSIS

Optical biosensors show attributes such as high-level sensitivity, specificity, cost effectiveness, and most importantly real-time detection. Optical biosensors follow the fluorescence resonance energy transfer and surface plasmon resonance methods. Surface plasmon resonance is an innovative and simple analytical approach that provides immediate results with great sensitivity. This approach involves real-time quantitative and qualitative evaluation of multiplexed toxins and label-free identification (Hill 2014; Damborský et al. 2016). Surface plasmon resonance-based sensors have been investigated for deoxynivalenol, T-2 toxin, and zearalenone in wheat (Hossain et al. 2019), aflatoxin M1 in milk products (Karczmarczyk et al. 2017a), and ochratoxin A in coffee (Rehmat et al. 2019).

15.6 NATURAL BIOSENSORS FOR FOOD ANALYSIS

15.6.1 Enzyme-Based Sensors

The particular attributes including ability to exactly recognize the substrate and accordingly catalyze the reaction makes enzymatic bodies a suitable tool for analytical devices. Concentration of analyte can easily be analyzed with the catalytic transformation rate of analyte (Velusamy et al. 2010). Enzymes can also be immobilized with a transducer surface for enhancing the accuracy of detection (Mehrotra 2016). Enzyme-based sensors have been widely used as detectors of naturally occurring particles in biomedical, pharmaceutical, industrial, environmental, and food processing analyses (Leca-Bouvier and Blum 2010). Abdullah et al. (2006) studied the tyrosinase-based sensor to quantify phenolic compounds.

For the design of a biosensor sensitive to steroidal glycoalkaloids, pH-sensitive field effect transistors as transducers and immobilized butyrylcholinesterase as a biorecognition element have been used (Korpan et al. 2002). The total potato glycoalcaloids can be measured by this biosensor in the concentration range 0.5–100 µM with detection limits of 0.5 µM for α-chaconine and of 2.0 µM for α-solanine and solanidine, respectively. The responses of the developed biosensors were reproducible with a relative standard deviation of about 1.5% and 5% for intra- and inter-sensor responses (both cases, n = 10, for an alkaloid concentration of 5 µM), respectively. Moreover, due to the reversibility of the enzyme inhibition, the same sensor chip with immobilized butyrylcholinesterase can be used several times (for at least 100 measurements) after a simple washing by a buffer solution and can be stored at 4°C for at least 3 months without any significant loss of the enzymatic activity.

15.6.2 Antibody-Based Sensors

Antibodies, either monoclonal or polyclonal, are a prime choice as a recognition part of biosensors because of their affinity and target specificity. Monoclonal antibodies are more common as immune biosensors because they have homogeneous binding characteristics and molecular structures and can also be engineered in bulk quantities (Saerens et al. 2008).

15.6.3 Nucleic Acid–Based Sensors

Nucleic acid biosensors employ immobilization of single-strand oligonucleotides onto the surface of a transducer for the detection of a target sequence. This DNA hybridization reaction then gets converted into a signal. Application of engineered nucleic acids has afforded a new trend for the recognition of nucleic acids in more sensitive manner. DNA dendrimers can be immobilized on a transducer and are comprised of several single-strand sequences that can bind to complementary targets to attain higher sensitivity (Wang 2002).

15.6.4 Whole-Cell-Based Sensors

Whole cells, mostly obtained from engineered bacterial strains, have been used as biorecognition constituents of biosensors due to the capability of microbes to recognize and respond to environmental stimuli (Justino et al. 2015). Products like ammonia and protons released during the metabolic reactions can be detected as a digital signal by a transducer (Su et al. 2011). Whole-cell-based sensors are used to recognize changes in cells' vicinity by examining electrical signals, thus declaring this biosensor a potential tool for pathogen detection in a variety of food samples (Adley 2014) (Figure 15.2).

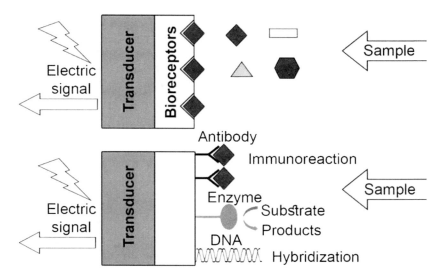

FIGURE 15.2 The schematic diagram of affinity-based biosensors showing the specificity of a bioreceptor and the interaction between antigen-antibody, enzyme-substrate, and DNA hybridization, which is to be detected by the biosensor [adapted from Girigoswami et al. (2021), ©Creative Commons Attribution International 4.0].

15.7 EMERGING BIOSENSORS FOR FOOD ANALYSIS

15.7.1 Nano-Biosensors

Using nanoparticles in biosensors is an approach to intensify the signals because of their remarkable biocompatibility, conductivity, and loading strength of signal compounds, which ultimately leads to improved performance of biosensors (Jafari et al. 2015; Karimi Pur et al. 2016). Nanoparticles can be implemented as a bio-recognition element of surface plasmon resonance-based sensing elements and as an immobilization agent of biosensors to intensify the signal alterations (Kwon et al. 2012). Covalent linkage of biomaterials to nanoparticles has the advantages of reproducibility, stability, and lowering the non-specific physisorption (Holzinger et al. 2014). Various nanomaterials including carbon nanomaterials, metal nanoparticles, semiconductors, and nanowires can also be used for amplification of electrochemical signals (Norouzi et al. 2010). Wee et al. (2019) designed voltammetric tyrosinase biosensor to detect the phenolic compounds by using carbon nanotubes. Using these nanomaterials, sensitivity of the biosensor was greatly enhanced. Bhardwaj et al. (2020) developed surface plasmon resonance-based biosensors using gold nanoparticles to detect aflatoxin B1 in foods with an amplified signal.

15.7.2 Cell-Free Biosensors

Developments in technologies related to cell-free synthesis have led to use transcription factors, aptamers, aminoacyl tRNA synthetases, and toehold switches as recognition elements. Characteristics of cell-free biosensors (*in vitro*) lead to higher operational facility and flexibility compared to microbial whole-cell-based biosensors for target analytes (Lee and Kim 2019). Moreover, cell-free biosensors have higher specificity, sensitivity, and the least error rate in comparison to a whole-cell-based system (Pellinen et al. 2004; Karig 2017). Song et al. (2017) used muscarinic acetylcholine M2 bioreceptor based on resonant waveguide grating system for the detection of hazardous natural products such as hyoscyamine, scopolamine, sanguinarine, and chelerythrine. In the drug profiling system, a resonant waveguide grating biosensor is used to explore the intracellular adjustments induced by protein relocalization by measuring the optical signals.

15.7.3 Microfluidics-Based Biosensors

Application of biosensors in microfluidic scales provides rapid operation, higher sensitivity, reliability, and controllability. It uses less sample volume and minimizes the cross-contamination and waste production (Selmi et al. 2017; Nikoleli et al. 2018). Microfluidics confine preprocessed biomolecules in a particular region and thus allow detection from the least sample size (Rivet et al. 2011). Olcer et al. (2014) developed the biosensor in an integrated microfluidic scale comprising of 10 µl channel capacity to detect the mycotoxins in a wheat grain by using antibodies as a sensing agent. Labroo and Cui (2014) developed a graphene-enzyme-based biosensor with a microfluidic system for the simultaneous detection of various metabolites. In this biosensor, each metabolite is predictable by an electrochemical signal released by oxidase enzyme activity on a metabolite sample. This sensor identified the glucose, lactate, cholesterol, and xanthine in a few minutes for all of the testing metabolites. Fournel et al. (2012) used the surface acoustic wave-based sensor for analysis of phycotoxin (okadaic acid) in a real-time manner. Integrating the microfluidic paper in this biosensor enhanced the mass convection and flow rate on the bio-functional site, which resulted in an increase in the response quality with higher sensitivity.

15.7.4 Magnetic-Bead-Based Biosensors

Magnetic beads can also be used in biosensing elements to increase the concentration of analytes with gradient magnetic field from external source. Magnetic beads attached with particular sites can be relocated within a sensor to reach the target point (Wagner et al. 2016). With the use of magnetic nanoparticles, mass alterations can be enhanced and nonspecific binding can be reduced with analyte surface, which makes the detection more sensitive (Kwon et al. 2012). Xin et al. (2014) analyzed the mast cells by using immobilized nano-sized magnetic beads for the detection of botulinum neurotoxin type B. This sensor detected 100 pM botulinum neurotoxin type B in a couple of minutes. Lin et al. (2015) developed the field-effected enzymatic biosensor to quantify urea, hydrogen ions, glucose, and particular proteins in a sample solution. Calcium alginate was used to immobilize the enzyme in magnetic powder. Then, magnetic beads functioned as an enzyme carrier by settling enzymes on the surface of the biosensor using a magnetic field from an external source. This kind of biosensor can measure the target by directly interacting with the surfaces, thus making quantitative measurement of analyte more feasible.

15.8 CONCLUSION

Most of the non-nutritional components that humans ingest are natural constituents of food plants. Presently, detection and evaluation of food toxins of plant origin have become a concern in assuring food safety and hygiene. In the view of public health, focusing on detection systems of specific inherent plant-based food toxicants may be more crucial. Advancements in the application of a wide range of biosensors for the comprehensive analyses of food toxins of plant origin were well discussed. The reported biosensors were successfully applied to the detection of various food toxins. As a proof of concept, the disposable sensor with high sensitivity and selectivity exhibited great application potential for the risk control of plant toxins in food. Biosensors can be used as an advanced analytical tool for screening a wide range of food toxicants of plant origin. Biosensors have a great potential in the determination of the concentration of food toxins, indirectly monitor their interaction with biological species, or find their intrinsic function in a biological process. Hence, biosensors can detect specific food toxins among a mixture of samples. Moreover, the overall performance of a biosensor highly depends on the materials used in its process design and selection of specific detection methods. Standardization of these key parameters can further optimize the performance of biosensors. The emerging advancements in microfluidics, magnetic beads, and nanotechnology may improve the efficacy of biosensors, leading to faster, cheaper, and more accurate detection of food toxins of plant origin.

REFERENCES

Abdullah, J., Ahmad, M., Karuppiah, N., Yook, L. and Sidek, H. (2006). Immobilization of tyrosinase in chitosan film for an optical detection of phenol. *Sensor Actuator B Chem.*, 114, 604–609. 10.1016/j.snb.2005.06.019

Adley, C. C. (2014). Past, present and future of sensors in food production. *Foods*, 3(3), 491–510. 10.3390/foods3030491

Ahmed, M. U., Hossain, M. M., Safavieh, M., Wong, Y. L., Abd Rahman, I., Zourob, M. and Tamiya, E. (2016). Toward the development of smart and low cost point-of-care biosensors based on screen printed electrodes. *Crit Rev Biotechnol.*, 36(3), 495–505. 10.3109/07388551.2014.992387

Bessaire, T., Perrin, I., Tarres, A., Bebius, A., Reding, F. and Theurillat, V. (2019). Mycotoxins in green coffee: Occurrence and risk assessment. *Food Control*, 96, 59–67.

Bhardwaj, H., Sumana, G. and Marquette, C. A. (2020). A label-free ultrasensitive microfluidic surface plasmon resonance biosensor for aflatoxin B1 detection using nanoparticles integrated gold chip. *Food Chem.*, 307, 125530. 10.1016/j.foodchem.2019.125530

Damborský, P., Švitel, J., Katrlík, J. and Vitel, J. (2016). Optical biosensors. *Essays Biochem.*, 60, 91–100.

Dolan, L. C., Matulka, R. A. and Burdock, G. A. (2010). Naturally occurring food toxins. *Toxins*, 2(9), 2289–2332. 10.3390/toxins2092289

Doria, G., Conde, J., Veigas, B., et al. (2012). Noble metal nanoparticles for biosensing applications. *Sensors*, 12, 1657–1687.

Edgar, J. A., Colegate, S. M., Boppré, M. and Molyneux, R. J. (2011). Pyrrolizidine alkaloids in food: A spectrum of potential health consequences. *Food Additives and Contaminants: Part A*, 28(3), 308–324.

Essers, A. A., Alink, G. M., Speijers, G. J., et al. (1998). Food plant toxicants and safety: Risk assessment and regulation of inherent toxicants in plant foods. *Environ Toxicol Pharmacol.*, 5(3), 155–172.

Ferreira, A. A. P., Venturini, C., Castilho, M. D. S., et al. (2013). Amperometric biosensor for diagnosis of disease. In Rinken, T. (ed.). *State of the art in biosensors*. Rijeka, Croatia: InTech, pp. 253–289.

Fletcher, M. T. and Netzel, G. (2020). Food safety and natural toxins. *Toxins (Basel)*, 12(4), 236. 10.3390/toxins12040236

Fournel, F., Baco, E., Mamani-Matsuda, M., et al. (2012). Love wave biosensor for real-time detection of okadaic acid as DSP phycotoxin. *Sensor Actuator B Chem.*, 170, 122–128. 10.1016/j.snb.2011.02.056

Gambacorta, L., El Darra, N., Fakhoury, R., Logrieco, A. and Solfrizzo, M. (2019). Incidence and levels of *Alternaria* mycotoxins in spices and herbs produced worldwide and commercialized in Lebanon. *Food Control*, 106, 106724.

Girigoswami, A., Ghosh, M. M., Pallavi, P., Ramesh, S. and Girigoswami, K. (2021). Nanotechnology in detection of food toxins – Focus on the dairy products. *Biointerface Research in Applied Chemistry*, 11(6), 14155–14172.

Gonçalves, B. L., Coppa, C. C., De Neeff, D. V., Corassin, C. H. and De Oliveira, C. A. F. (2018). Mycotoxins in fruits and fruit-based products: Occurrence and methods for decontamination. *Toxin Rev.*, 38, 263–272.

Grieshaber, D., MacKenzie, R., Vörös, J. and Reimhult, E. (2008). Electrochemical biosensors – Sensor principles and architectures. *Sensors*, 8, 1400–1458.

Gu, Y., Wang, Y., Wu, X., et al. (2019). Quartz crystal microbalance sensor based on covalent organic framework composite and molecularly imprinted polymer of poly (o-aminothiophenol) with gold nanoparticles for the determination of aflatoxin B1. *Sens. Actuators B Chem.*, 291, 293–297.

Hajšlová, J., Schulzová, V., Botek, P. and Lojza, J. (2018). Natural toxins in food crops and their changes during processing. *Czech J Food Sci.*, 22, 29–34.

Hama, J. R. and Strobel, B. W. (2021). Occurrence of pyrrolizidine alkaloids in ragwort plants, soils and surface waters at the field scale in grassland. *The Science of the Total Environment*, 755(Pt 1), 142822.

Hammond, J. L., Formisano, N., Carrara, S. and Tkac, J. (2016). Electrochemical biosensors and nanobiosensors. *Essays Biochem.*, 60, 69–80.

He, B. and Dong, X. (2018). Aptamer based voltammetric patulin assay based on the use of ZnO nanorods. *Microchim. Acta*, 185, 462.

He, B. and Yan, X. (2019a). A "signal-on" voltammetric aptasensor fabricated by hcPt@AuNFs/PEI-rGO and Fe_3O_4NRs/rGO for the detection of zearalenone. *Sens. Actuators B Chem.*, 290, 477–483.

He, B. and Yan, X. (2019b). An amperometric zearalenone aptasensor based on signal amplification by using a composite prepared from porous platinum nanotubes, gold nanoparticles and thionine-labelled graphene oxide. *Microchim. Acta*, 186, 383.

Hill, R. T. (2014). Plasmonic biosensors. *Wiley Interdiscip. Rev. Nanomed. Nanobiotechnol.*, 7, 152–168.

Holzinger, M., Goff, A. L. and Cosnier, S. (2014). Nanomaterials for biosensing applications: a review. *Front Chem.*, 2, 1–10. 10.3389/fchem.2014.00063

Hossain, Z., Busman, M. and Maragos, C. M. (2019). Immunoassay utilizing imaging surface plasmon resonance for the detection of cyclopiazonic acid (CPA) in maize and cheese. *Anal. Bioanal. Chem.*, 411, 3543–3552.

Huertas-Pérez, J. F., Arroyo-Manzanares, N., García-Campaña, A. and Gámiz-Gracia, L. (2016). Solid phase extraction as sample treatment for the determination of ochratoxin A in foods: A review. *Crit. Rev. Food Sci. Nutr.*, 57, 3405–3420.

Jafari, S., Faridbod, F., Norouzi, P., et al. (2015). Detection of *Aeromonas hydrophila* DNA oligonucleotide sequence using a biosensor design based on ceria nanoparticles decorated reduced graphene oxide and fast Fourier transform square wave voltammetry. *Anal Chim Acta*, 895, 80–88. 10.1016/j.aca.2015.05.055

Justino, C. I. L., Freitas, A. C., Pereira, R., Duarte, A. C. and Santos, T. A. P. R. (2015). Recent developments in recognition elements for chemical sensors and biosensors. *Trends Anal. Chem.*, 68, 2.

Karczmarczyk, A., Baeumner, A. J. and Feller, K.-H. (2017a). Rapid and sensitive inhibition-based assay for the electrochemical detection of Ochratoxin A and Aflatoxin M1 in red wine and milk. *Electrochim. Acta*, 243, 82–89.

Karczmarczyk, A., Haupt, K. and Feller, K. -H. (2017b). Development of a QCM-D biosensor for Ochratoxin A detection in red wine. *Talanta*, 166, 193–197.

Karig, D. K. (2017). Cell-free synthetic biology for environmental sensing and remediation. *Curr Opin Biotechnol.*, 45, 69–75. 10.1016/j.copbio.2017.01.010

Karimi Pur, M. R., Hosseini, M., Faridbod, F., Dezfuli, A. S. and Ganjali, M. R. (2016). A novel solid-state electrochemiluminescence sensor for detection of cytochrome C based on ceria nanoparticles decorated with reduced graphene oxide nanocomposite. *Anal Bioanal Chem.*, 408, 7193–7202. 10.1007/s00216-016-9856-6

Kebede, H., Liu, X., Jin, J. and Xing, F. (2020). Current status of major mycotoxins contamination in food and feed in Africa. *Food Control*, 110, 106975.

Khan, R., Ben Aissa, S., Sherazi, T. A., Catanante, G., Hayat, A. and Marty, J. L. (2019). Development of an impedimetric aptasensor for label free detection of patulin in apple juice. *Molecules*, 24, 1017.

Kim, N. S. and Cho, Y. J. (2011). *Emerging technologies for food quality and food safety evaluation*. Boca Raton, FL: CRC Press, p. 257.

Kluczkovski, A. M. (2019). Fungal and mycotoxin problems in the nut industry. *Curr. Opin. Food Sci.*, 29, 56–63.

Korpan, Y. I., Volotovsky, V. V., Martelet, C., Jaffrezic-Renault, N., Nazarenko, E. A., El'skaya, A. V. and Soldatkin, A. P. (2002). A novel enzyme biosensor for steroidal glycoalkaloids detection based on pH-sensitive field effect transistors. *Bioelectrochemistry*, 55(1–2), 9–11.

Kwon, M. J., Lee, J., Wark, A. W. and Lee, H. J. (2012). Nanoparticle-enhanced surface plasmon resonance detection of proteins at attomolar concentrations: Comparing different nanoparticle shapes and sizes. *Anal Chem.*, 84, 1702–1707. 10.1021/ac202957h

Labroo, P. and Cui, Y. (2014). Graphene nano-ink biosensor arrays on a microfluidic paper for multiplexed detection of metabolites. *Anal Chim Acta*, 813, 90–96. 10.1016/j.aca.2014.01.024

Leca-Bouvier, B. D. and Blum, L. J. (2010). Enzyme for biosensing applications. In Zourob, M. (ed.). *Recognition receptors in biosensors*. New York: Springer, pp. 177–220.

Lee, K. H. and Kim, D. M. (2019). In vitro use of cellular synthetic machinery for biosensing applications. *Front Pharmacol.*, 10, 1–9. 10.3389/fphar.2019.01166

Lin, Y. H., Chu, C. P., Lin, C. F., Liao, H. H., Tsai, H. H. and Juang, Y. Z. (2015). Extended-gate field-effect transistor packed in micro channel for glucose, urea and protein biomarker detection. *Biomed Microdevices*, 17, 1–9. 10.1007/s10544-015-0020-4

Liu, N., Nie, D., Tan, Y., et al. (2016). An ultrasensitive amperometric immunosensor for zearalenones based on oriented antibody immobilization on a glassy carbon electrode modified with MWCNTs and AuPt nanoparticles. *Microchim. Acta*, 184, 147–153.

Malekzad, H., Zangabad, P. S., Mirshekari, H., Karimi, M. and Hamblin, M. R. (2017). Noble metal nanoparticles in biosensors: Recent studies and applications. *Nanotechnol. Rev.*, 6, 301–329.

Mehrotra, P. (2016). Biosensors and their applications - a review. *J Oral Biol Craniofacial Res.*, 6, 153–159.

Montagut, Y., Garcia, J., Jiménez, Y., March, C., Montoya, A. and Arnau, A. (2011). QCM technology in biosensors. In Serra, P. A. (ed.). *Biosensors – Emerging materials and applications*. Rijeka, Croatia: IntechOpen, pp. 153–178.

Nan, M., Bi, Y., Xue, H., Xue, S., Long, H., Pu, L. and Fu, G. (2019). Rapid determination of Ochratoxin A in grape and its commodities based on a label-free impedimetric aptasensor constructed by layer-by-layer self-assembly. *Toxins*, 11, 71.

Neethirajan, S., Ragavan, K. and Weng, X. (2018). Agro-defense: Biosensors for food from healthy crops and animals. *Trends Food Sci. Technol.*, 73, 25–44.

Nikoleli, G. P., Siontorou, C. G., Nikolelis, D. P., Bratakou, S., Karapetis, S. and Tzamtzis, N. (2018). Biosensors based on microfluidic devices lab-on-a-chip and microfluidic technology. In Nikolelis, D. P. and Nikoleli, G. P. (eds.). *Nanotechnology and biosensors: Advanced nanomaterials*. Amsterdam: Elsevier, pp. 375–394. 10.1016/B978-0-12-813855-7.00013-1

Noah, N. D., Bender, A. E., Reaidi, G. B. and Gilbert, R. J. (1980). Food poisoning from raw red kidney beans. *British Medical Journal*, 281, 236–237.

Norouzi, P., Faridbod, F., Rashedi, H. and Ganjali, M. R. (2010). Flow injection glutamate biosensor based on carbon nanotubes and Pt-nanoparticles using FFT continuous cyclic voltammetry. *Int J Electrochem Sci.*, 5, 1713–1725.

Ogori, A. F. (2019). Plant toxins. *Am J Biomed Sci Res.*, 4(3), AJBSR.MS.ID.000793. 10.34297/AJBSR.2019.04.000793

Olcer, Z., Esen, E., Muhammad, T., Ersoy, A., Budak, S. and Uludag, Y. (2014). Fast and sensitive detection of mycotoxins in wheat using microfluidics based real-time electrochemical profiling. *Biosens Bioelectron.*, 62, 163–169. 10.1016/j.bios.2014.06.025

Oliveira, I. S., Junior, A. G. D. S., De Andrade, C. A. S. and Oliveira, M. D. L. (2019). Biosensors for early detection of fungi spoilage and toxigenic and mycotoxins in food. *Curr. Opin. Food Sci.*, 29, 64–79.

Pascari, X., Ramos, A. J., Marín, S. and Sanchis, V. (2018). Mycotoxins and beer. Impact of beer production process on mycotoxin contamination: A review. *Food Res. Int.*, 103, 121–129.

Pellinen, T., Huovinen, T. and Karp, M. (2004). A cell-free biosensor for the detection of transcriptional inducers using firefly luciferase as a reporter. *Anal Biochem.*, 330, 52–57. 10.1016/j.ab.2004.03.064

Perumal, V. and Hashim, U. (2014). Advances in biosensors: Principle, architecture and applications. *J. Appl. Biomed.*, 12, 1–15.

Pirinçci, S., Ertekin, Ö., Laguna, D. E., Özen, F. S., Ozturk, Z. Z. and Öztürk, S. (2018). Label-free QCM immunosensor for the detection of Ochratoxin A. *Sensors*, 18(4), 1161.

Ram, Y., Yoetz-Kopelman, T., Dror, Y., Freeman, A. and Shacham-Diamand, Y. (2016). Impact of molecular surface charge on biosensing by electrochemical impedance spectroscopy. *Electrochim. Acta*, 200, 161–167.

Rehmat, Z., Mohammed, W., Sadiq, M. B., Somarapalli, M. and Anal, A. K. (2019). Ochratoxin A detection in coffee by competitive inhibition assay using chitosan-based surface plasmon resonance compact system. *Colloids Surf. B Biointerfaces*, 174, 569–574.

Ricci, F., Volpe, G., Micheli, L. and Palleschi, G. (2007). A review on novel developments and applications of immunosensors in food analysis. *Anal. Chim. Acta*, 605, 111–129.

Rivet, C., Lee, H., Hirsch, A., Hamilton, S. and Lu, H. (2011). Microfluidics for medical diagnostics and biosensors. *Chem Eng Sci.*, 66, 1490–1507. 10.1016/j.ces.2010. 08.015

Saerens, D., Huang, L., Bonroy, K. and Muyldermans, S. (2008). Antibody fragments as probe in biosensor development. *Sensors (Basel)*, 8(8), 4669–4686. 10.3390/ s8084669

Sanzani, S. M., Reverberi, M. and Geisen, R. (2016). Mycotoxins in harvested fruits and vegetables: Insights in producing fungi, biological role, conducive conditions, and tools to manage postharvest contamination. *Postharvest Boil. Technol.*, 122, 95–105.

Schilter, B., Constable, A. and Perrin, I. (2014). Naturally occurring toxicants of plant origin. In Motarjemi, Y. and Lelieveld, H. (eds.). *Food safety management: A practical guide for the food industry*. Amsterdam, The Netherlands: Elsevier, pp. 45–57.

Schulz, K., Pöhlmann, C., Dietrich, R., Märtlbauer, E. and Elßner, T. (2019). Electrochemical biochip assays based on anti-idiotypic antibodies for rapid and automated on-site detection of low molecular weight toxins. *Front. Chem.*, 7, 31.

Selmi, M., Gazzah, M. H. and Belmabrouk, H. (2017). Optimization of microfluidic biosensor efficiency by means of fluid flow engineering. *Sci Rep.*, 7, 1–11. 10.1038/ s41598-017-06204-0

Song, H. P., Wang, H., Zhao, X., et al. (2017). Label-free pharmacological profiling based on dynamic mass redistribution for characterization and authentication of hazardous natural products. *J. Hazard Mater.*, 333, 265–274. 10.1016/j.jhazmat. 2017.03.025

Su, L., Jia, W., Hou, C. and Lei, Y. (2011). Microbial biosensors: A review. *Biosens Bioelectron.*, 26(5), 1788–1799. 10.1016/j.bios.2010.09.005

Tang, J., Huang, Y., Cheng, Y., Huang, L., Zhuang, J. and Tang, D. (2018). Two-dimensional MoS_2 as a nano-binder for ssDNA: Ultrasensitive aptamer based amperometric detection of Ochratoxin A. *Microchim. Acta*, 185, 162.

Thakur, M. S. and Ragavan, K. V. (2013). Biosensors in food processing. *J Food Sci Technol.*, 50(4), 625–641. 10.1007/s13197-012-0783-z

Tothill, I. E. (2011). Biosensors and nanomaterials and their application for mycotoxin determination. *World Mycotoxin J.*, 4, 361–374.

Varzakas, T. (2015). Quality and safety aspects of cereals (wheat) and their products. *Crit. Rev. Food Sci. Nutr.*, 56, 2495–2510.

Velusamy, V., Arshak, K., Korostynska, O., Oliwa, K. and Adley, C. (2010). An overview of foodborne pathogen detection in the perspective of biosensors. *Biotechnol Adv.*, 28(2), 232–254. 10.1016/j.biotechadv.2009.12.004

Viegas, S., Assunção, R., Twaruzek, M., Kosicki, R., Grajewski, J. and Viegas, C. (2019). Mycotoxins feed contamination in a dairy farm – potential implications for milk contamination and workers' exposure in a One Health approach. *J. Sci. Food Agric.*, 100, 1118–1123.

Wagner, T., Vornholt, W., Frederik, C., et al. (2016). Physics in medicine light-addressable potentiometric sensor (LAPS) combined with magnetic beads for pharmaceutical screening. *Phys Med.*, 1, 2–7. 10.1016/j.phmed.2016.03.001

Wang, J. (2002). Electrochemical nucleic acid biosensors. *Analytica Chimica Acta*, 469(1), 63–71. 10.1016/S0003-2670(01)01399-X

Wee, Y., Park, S., Kwon, Y. H., Ju, Y., Yeon, K. M. and Kim, J. (2019). Tyrosinase-immobilized CNT based biosensor for highly-sensitive detection of phenolic compounds. *Biosens Bioelectron.*, 132, 279–285. 10.1016/j.bios.2019.03.008

Welke, J. E. (2019). Fungal and mycotoxin problems in grape juice and wine industries. *Curr. Opin. Food Sci.*, 29, 7–13.

Xiang, Y., Camarada, M. B., Wen, Y., et al. (2018). Simple voltammetric analyses of ochratoxin A in food samples using highly-stable and anti-fouling black phosphorene nanosensor. *Electrochim. Acta*, 282, 490–498.

Xin, W., Yao, W., Gao, X., et al. (2014). Development of aequorin-based mast cell nanosensor for rapid identification of botulinum neurotoxin type B. *J Biomed Nanotechnol.*, 10, 3318–3328. 10.1166/jbn.2014.2026

Xu, L., Zhang, Z., Zhang, Q. and Li, P. (2016). Mycotoxin determination in foods using advanced sensors based on antibodies or aptamers. *Toxins*, 8, 239.

Yagati, A. K., Chavan, S. G., Baek, C., Lee, M.-H. and Min, J. (2018). Label-free impedance sensing of Aflatoxin B1 with polyaniline nanofibers/Au nanoparticle electrode array. *Sensors*, 18, 1320.

Yulvianti, M. and Zidorn, C. (2021). Chemical diversity of plant cyanogenic glycosides: An overview of reported natural products. *Molecules*, 26, 719.

Zhang, X., Li, C. -R., Wang, W. -C., et al. (2016). A novel electrochemical immunosensor for highly sensitive detection of aflatoxin B1 in corn using single-walled carbon nanotubes/chitosan. *Food Chem.*, 192, 197–202.

Index

Note: *Italicized* page numbers refer to figures, **bold** page numbers refer to tables

abrin, 72, **163**, 247
Abrus precatorius, 72, **163**
absinthe, **153**
acetylcholinesterase, 59–60
3-acetyldeoxynivalenol, 196
15-acetyldeoxynivalenol, 196
Achillea setacea, 204
ackee fruits (*Blighia sapida*), 14
Aconitum napellus, **163**
acrylamide, VIS/NIR analysis of, 100
acrylic adhesives, **186**
adhesins, 106
adsorption chromatography, 148
adulterants, NMR detection of, 135–136
affinity-based biosensors, *251*
aflatoxins, 196
Alexa austrate, 59
Alexa leiopetala, 59, **61**
alkenylbenzene, 180
allergens, 156
allo-psoralen, *69*
Alternaria toxins, 166–167
2-amino-2-thiazoline-4-carboxylic acid
        (ATCA), 38–39
amperometric sensors, 248
amygdalin, 16, 27, 28, 103, 204.
            See also cyanogenic glycosides
    enzymatic hydrolysis of, 34, *35*
    sources of, 32, *33*
    structure of, *29*, *103*, **120**, **208**, **228**
α-amylase inhibitors, 9
andafocoumarins, 123–124
Angelica acutiloba, 73
Angelica archangelica, 71
Angelicae dahurica, 123
angelicins, 67, 69, 72, 105, *106*, 123, *124*.
            See also furanocoumarins
angular furanocoumarins, 67, 105.
            See also furanocoumarins
antibody-based biosensors, 250.
            See also biosensors
anti-thiamine compounds, 9–10
Apiaceae, 8, 67, 105
Apocynaceae, 121, 180

apoptosis, 85
apples (*Malus domestica*), 32
apricot fruits (*Prunus armeniaca*), 32–33
archangelicin, 105, *106*, *124*
Argemone mexicana, **163**
Arnebia euchroma, 197
Artemisia absinthium (grand wormwood), 5
artificial colors, 135
asialofetuin, 226
Aspergillus, 166, 197
Aspergillus flavus, 197
Asteraceae, 10, 54, 121, 180
atmospheric pressure photo-ionization
        (APPI), 165
atmospheric-pressure chemical ionization
        (APCI), 165
authentication, 133–134

Bacillus, 9–10
bamboo shoots, 31
beans (*Phaseolus* spp.), 9
bee pollen, 104, 122
benzaldehyde, *35*
benzoates, 135
α-benzopyrone, *69*
bergamot, 70
bergamottin, *69*, 70
bergapten, *69*, 74
bergaptol, *69*
bicoloridine, 182
biomolecules, 151
biosensors, 245–253
    affinity-based, *251*
    cell-free, 251–252
    electrochemical, 248–249
        amperometric sensors, 248
        impedimetric sensors, 248–249
        potentiometric sensors, 248–249, *249*
    magnetic-bead-based, 252
    microfluidics-based, 252
    nano-biosensors, 251
    natural, 249–252
        antibody-based biosensors, 250
        enzyme-based biosensors, 249–250
        nucleic acid-based biosensors, 250

259

260   Index

whole-cell-based biosensors, 250
 optical, 249
 overview, 245–246
 quartz crystal microbalance biosensors, 249
 recognition elements for, 246
 schematic representation of, *247*
 types of, 246
Blighia sapida (ackee fruits), 14
BMAA (β-methylamino-L-alanine), 17
Boraginaceae, 10, 54, 104, 121, 180
Boraginoideae, 55
bovine serum albumin (BSA), 221, 224
Brassica, 7
Brassicaceae, 8
brinjal. *See* eggplant (*Solanum melongena*)
Brosimum gaudichaudii, 75
BSA (bovine serum albumin), 221, 224
byakangelicin, *69*
byakangelicol, *69*

Calotropis gigantea, **163**
Canavalia spp., 74
Candida, 10
canola oil, 8
capillary electrochromatography (CEC), 203
capillary electrophoresis (CE), 201–202
 in analysis of plant toxins, 204–212, **205–206**
  cyanogenic glycosides, 204
  pyrrolizidine alkaloids, 207
 capillary electrochromatography, 203
 capillary gel electrophoresis, 203
 capillary isoelectric focusing, 203–204
 capillary isotachophoresis, 204
 capillary zone electrophoresis, 202–203
 instrumentation, *202*
 micellar electrokinetic capillary
  chromatography, 203
 overview, 201
 structural characterization and detection of food
  toxins, **208–211**
 types of, *202*, 202–204
capillary gel electrophoresis (CGE), 203
capillary isoelectric focusing (CIEF), 203–204
capillary isotachophoresis (CITP), 204
capillary zone electrophoresis (CZE), 202–203
Capsicum annum (peppers), 207
carbohydrate recognition domain (CRD), 225
carbohydrates, 151
Carica papaya, 204
cassava (*Manihot esculenta*), 6, 16, 27, 28,
  30–31, 34, 39, 40, 41–43, **102**, **119**,
  120, **153**, **163**, 196, 220, **227**, 246,
  247, 248
Cassia alata, 197
Castanospermum australe, **61**
castor oil plant (*Ricinus communis*), 9

cedar leaf oil, 6
cedarwood oil, 6
celery, **153**
cell-free biosensors, 251–252
Cerbera odollam, **163**
chaconine, 16, 81–82. *See also* glycoalkaloids
 IR spectroscopy, 100
 LC-MS analysis of, 169
 pharmacological actions of, 87–88
 structure of, *101*, *118*, **210**, **229**
chemical structures, 117
Chinese star anise (*Illicium verum*), 15–16
cis-13-docosenoic acid, 8
Citrus paradisi (grapefruits), 71
Citrus plants, 70–71
Claviceps, 167
Clostridium, 9–10
Cnidium monnieri, 71, 74
Codex Alimentarius Commission, 4, 5
Codex Committee on Contaminants in Foods
  (CCCF), 4
Colocasia esculenta, **163**
coltsfoot (*Tussilago farfara*), 121
comfrey (*Symphytum officinale*), 121
competitive ELISA (antigen/antibody
  screening), 219, *220*
Compositae, 26, 104, 121
concanavalin A, 74
condelphine, 182
Conium maculatum, **163**
Convallaria majalis, **163**
copper sulphate, 135
coumarins, 67
cretinism, 42–43
Crotalaria, 54, 123
Crotalaria assamica, **62**
Crotalarieae, 54, 55
Cryptantha crassipes, 123
C-type lectins (CTLs), 226
cucurbitacins, 11–12, **12**
cyanide, 6
 antidotes, 40
 detoxification, 38–39
 diseases caused by, 40, 42–44
  cyanide poisoning, 81–82
  growth retardation, 43
  iodine deficiency diseases, 42–43
  konzo, 43
  tropical ataxic neuropathy (TAN), 43
 human health effect of, 39
 poisoning, 43–44
 toxicity, 35–38
 treatment of cyanide poisoning, 40
cyanide bomb, 28
cyanocobalamin (vitamin B12), 38

cyanogenesis, 28, 34, 102
cyanogenic glycosides, 25–45
    analysis of
        capillary electrophoresis, 204
        ELISA, 219–222
        IR spectroscopy, 102–104
        LC-MS, 167–168
        mass spectrometry, 119–121
    basic framework of, 102
    biosynthesis of, 33–40, *34*
    chemical structures of, *29, 103, 119, 120*
    cyanide detoxification in human, 38–39
    cyanogenesis, 28
    diseases caused by cyanide, 42–44
        cyanide poisoning, 43–44
        growth retardation, 43
        iodine deficiency diseases, 42–43
        konzo, 43
        tropical ataxic neuropathy (TAN), 43
    distribution, 26–27
    enzymatic hydrolysis of, 34–35
    in food plants, 28–33
        apples, 32
        apricot fruits, 32–33
        bamboo shoots, 31
        cassava, 30–31
        lima beans, 32
        linseed, 32
        sorghum, 31–32
    human health effect of, 39, **40**
    mechanism of how it affects the body, 37–38
    as natural toxins, 26
    occurrence in human food, **102, 119**
    overview, 16
    preventing the effects of, 40–42
        cooking, 42
        drying, 42
        by processing, 40–41
        by soaking and fermentation, 41–42
    quantification of, 44
    as secondary metabolites, 27
    synthesis/production of, 27–28
    toxicity, 35–37
    treatment of cyanide poisoning, 40
cyanogens, 28
cyanohydrins, 28, 34
cycasin, 17
cyclopseudohypericin, 7
Cynoglossum, 55
Cynoglossum officinale, **62**

Datura, 180
Datura stromonium, **163**
dehydrotomatine, *101*
delpheline, 182
Delphinium, 182
demissidine, **211, 230**
deoxynivalenol, 196–197

detectors, HPLC, 149–150
    electrochemical detector, 150
    evaporation light scattering detector, 150
    fluorescent detector, 149
    mass detector, 150
    photodiode array detector (PDA) detector, 149
    refractive index (RI) detector, 149
    ultraviolet-visible light detector, 149
dhurrin, *27, 29, 35. See also* cyanogenic
        glycosides
    enzymatic hydrolysis of, *36*
    sources of, 32, 33
    structure of, *103, 120*
dhurrinase, 35
diacetoxyscirpenol, 196
Dieffenbachia sp., **163**
6,'7'-dihydrobergamottin, *69*
dihydrofuranocoumar, 105
dihydrosafrole, 15
Dipteryx odorata, 106
direct ELISA, 218, *220*
disaccharide intolerance, **145**
docosenoic acid, 8
Dorstenia, 124

Echium, 55
Echium confusum, 60, **61**
Echium plantagineum, 122, 170
ecstasy (MDMA, *N*-methyl-3,4-
        methylenedioxymethamphetamine), 14
eggplant (*Solanum melongena*), **82**, 82, 85, 117,
        154, 207
α-elaterin, 11
electrochemical biosensors, 248–249.
        See also biosensors
    amperometric sensors, 248
    impedimetric sensors, 248–249
    potentiometric sensors, 249
electrochemical detector, 150
electrospray ionization (ESI), 164–165
environmental contaminants, 17
enzyme-based biosensors, 249–250.
        See also biosensors
enzyme-linked immunosorbent assay (ELISA),
        215–235
    in analysis of plant toxins, 219–226
        cyanogenic glycosides, 219–222
        furocoumarins, 222–223
        glycoalkaloids, 223–225
        lectins, 225–226
        pyrrolizidine alkaloids, 222
    competitive (antigen/antibody screening),
        219, *220*
    defined, 216–217
    direct, 218, *220*
    indirect, 218–219, *220*

overview, 215–216
sandwich (antibody screening), 219, *220*
structural characterization and detection of food toxins, 226, **227–234**
types of, 217–219, *218, 220*
working principle of, 217
ergot alkaloids, 167
erucic acid, 8
essential oil, **186**
Eupatorieae, 55
European Council's Directive on food flavourings 88/388/EEC, 15
European Food Safety Authority (EFSA), 4, 8
evaporation light scattering detector, 150
extraction methods, 146–147
gel permeation chromatography, 147
liquid extraction, 146
liquid/liquid extraction, 147
solid-phase extraction, 147
sorbent-based microextraction, **155**
Soxhlet extraction, 147
steam distillation, 146
supercritical fluid, 146

Fabaceae, 8, 26, 54, 121, 180
Fabaceae, 10
favism, **145**
Federal Food and Drug Act, 162
fetuin, 226
Flabellaria paniculata, 75
flatoxins, 166
flavonoids, 7, 152
Flavourings in Food Regulations 1992 (United Kingdom), 15
flaxseed, 32
fluorescent detector, 149
food additives, 152
Food and Agriculture Organization (FAO), 4–5
Food and Drug Administration (FDA), 4
food composition, 133–134
Fritillaria, 117
β-fructofuranosidase, **90**
Fucaceae, 67
fumonisins, 166
furanocoumarins, 67–75
analysis of
ELISA, 222–223
GC-MS, 188
HPLC, 154
HPTLC, 196
IR spectroscopy, 105–106
mass spectrometry, 123–124
angular, 67, 105
basic structure, 105

chemical structures of, 68, *69, 124*
chemistry of, 72–73
linear, 67, 105
overview, 8, 67–70
pharmacological actions of, 73–75
as photosensitizing chemicals, 68
sources of, 70–72
toxicological reports of, 75
furans, **186**
fusarenone-x, 196
Fusarium, 166

Garcinia, 134
gas chromatography-mass spectrometry (GC-MS), 179–189
in analysis of food toxins, **185–186**
in analysis of plant toxins, 188
fast, 180–181, **181**
GC-TOF-MS, 181–182
low-pressure, 182–184
methoxyamination, 182
overview, 180
supersonic molecular beam, 184–188
techniques, **183**
gel permeation chromatography, 147
gempylotoxin, 17
General Food Law, 4
genistein, **153**
gluten intolerance, **145**
glycoalkaloids, 81–90
analysis of
ELISA, 223–225
HPLC, 154
HPTLC, 197
IR spectroscopy, 99–101
LC-MS, 168–169
mass spectrometry, 117–119
biosynthesis of, *84–85*
chemistry of, 83
overview, 16, 81–82
pharmacological actions of, 85–86, **87**
sources of, 82–83
toxicological activities of, 86–89, **90**
glycyrrhizic acid, **153**
goiter, 7
goitre, 42–43
goitrin, 7
goitrogens, 7–8
grand wormwood (*Artemisia absinthium*), 5
grapefruits (*Citrus paradisi*), 71
grapevine (*Vitis vinifera* L.), 168
grayatoxins, 17
growth retardation, 43
Gutierrezia sarothrae, **163**
Gynura segetum, 207

haptens, 221

heliotridine, *56, 57, 104*. *See also* pyrrolizidine alkaloids

heliotrine, **228**

Heliotropioideae, 55

Heliotropium, 55

Heliotropium crossifolium, 105

Heliotropium digynum, **61**

Heliotropium europaeum, **62**, 170

Heliotropium indicum, **61**

hemagglutinins, 106

heraclenin, **227, 233**

heraclenol, **234**

Heracleum candicans, 154, 196

heterocyclic aromatic amines (HAAs), 17

high-performance liquid chromatography (HPLC), 143–157
  analysis of food compounds, 151–156
    biomolecules, 151
    carbohydrates, 151
    flavonoids, 152
    food additives, 152
    lipids, 151
    phenolic compounds, 152
    pigments, 152
    proteins, 151
    vitamins, 151–152
  analysis of plant toxins and alkaloids, 150
  analysis of unwanted compounds/food toxins, 153–156
    allergens, 156
    mycotoxins, 154–156
    pesticides, 156
  detectors, 149–150
    electrochemical detector, 150
    evaporation light scattering detector, 150
    fluorescent detector, 149
    mass detector, 150
    photodiode array detector (PDA) detector, 149
    refractive index (RI) detector, 149
    ultraviolet-visible light detector, 149
  extraction methods, 146–147
    gel permeation chromatography, 147
    liquid extraction, 146
    liquid/liquid extraction, 147
    solid-phase extraction, 147
    Soxhlet extraction, 147
    steam distillation, 146
    supercritical fluid, 146
  isolation mechanisms, 148
  methods, 149–150
  normal phase chromatography, 149
  overview, 144–145
  principles of, 148
  reversed-phase chromatography, 149
  sample preparation, 145–147
    liquids, 147
    solids, 146–147
  working principle of, 148

high-performance thin-layer chromatography (HPTLC), 195–197
  in analysis of food toxins, 196–197
  overview, 195
  principles of, 195–196

Hippomane mancinella, **163**

honey, 122

hydrocyanic acid, 6

hydrogen cyanide (HCN), 6, 28, 30–31, *36, 36–37*

hydroxocobalamin (vitamin B12a), 38, 40

7-hydroxycoumarin, 105

hydroxynitrile, 34

hypericin, 6–7

Hypericum perforatum, 6–7

hypoglycin, 14

Illicium verum (Chinese star anise), 15–16

impedimetric sensors, 248–249

imperatorin, *69*, 73–74, **233**. *See also* furanocoumarins

indirect competitive ELISA (icELISA), 221–222

indirect ELISA, 218–219, *220*

infrared spectroscopy, 97–107
  in food toxin analysis, 99–107
    cyanogenic glycosides, 102–104
    furanocoumarins, 105–106
    glycoalkaloids, 99–101
    lectins, 106–107
  need for, 99
  overview, 98–99
  pros and cons of, **99**
  pyrrolizidine alkaloids, 104–105

insecticides, 15

interferon, 74–75

iodine deficiency diseases, 42–43

ion-exchange chromatography, 148

isohypericin, 7

iso-imperatorin, *69*, **234**

iso-oxypeucedanin, *69*

isoretronecanole, *58*

isosafrole, 15

iso-schaftoside, **208**

jacobine, **228**

Jamaica vomiting sickness, 14

Japanese star anise, 15–16

Joint FAO/WHO Expert Committee on Food Additives (JECFA), 4–5, *5*

α-ketoglutarate, 39

α-ketoglutarate cyanohydrin (α-KGCN), 39

keyhole limpet hemocyanin (KLH), 221, 224

khasianine, **232**
konzo, 43

lactones, **186**
Lathyrus vulgaris, **163**
lectins, 67–75
    analysis of
        ELISA, 225–226
        IR spectroscopy, 106–107
    chemistry of, 72–73
    overview, 9
    pharmacological actions of, 73–75
    sources of, 70–72
    toxicological reports of, 75
Leguminoceae, 104
Leguminosae, 26, 67, 121
licorice, **153**
lima beans (*Phaseolus lunatus* L.), 32
limulus polyphenus haemocyanin (LPH), 224
Linaceae, 26
linamarase, *30, 30*, 35
linamarin, 16, 27, 28, 30, 102.
        See also cyanogenic glycosides
    enzymatic hydrolysis of, 34–35, *36*
    sources of, 32, *33*
    structure of, *29, 30, 103, 120*, **228**
    toxic effects of, **153**
linear furanocoumarins, 67, 105.
        See also furanocoumarins
linseed (*Linum usitatissimum*), 32
linustatin, 102
    sources of, 32
    structure of, *103, 120*
Liparis nervosa, 54, 59, **61**
lipids, 151
liquid chromatography-mass spectrometry (LC-MS), 161–173
    in analysis of plant toxins, 167–172
        cyanogenic glycosides, 167–168
        glycoalkaloids, 168–169
        pyrrolizidine alkaloids, 169–171
        steroidal glycoalkaloids, 169
    atmospheric pressure photo-ionization, 165
    atmospheric-pressure chemical ionization, 165
    electrospray ionization, 164–165
    instrumentation, 164–165
    ionization techniques, 164
    LC conditions, 172–173
    mass analyzers, 165
    overview, 162–164
    preparation techniques, 165–166
    in quality analysis of food compounds, 166–167
    quantitation process, 166
liquid extraction, 146
liquid/liquid extraction, 147
lotaustralin, 16, 27, *29, 30*, 33, 34–35, 102, *103, 120*

low-pressure gas (LP) chromatography-mass spectrometry (GC-MS), 182–184
    applications of, 184
    instrumentation, 183–184
    significances of, 184
    working principle of, 183–184
lycopene, **90**
lycopsamine, **228**

magnetic-bead-based biosensors, 252
malachite green, 135
Malus domestica (apples), 32
mandelonitrile, *35*
Manihot esculenta (cassava), 6, 27, 30–31, 41–42, **153**, *163*, 248
Mareya micrantha, 11
mass analyzers, 165
mass detector, 150
mass spectrometry (MS), 115–124
    chemical ionization (CIMSI), 120
    electron impact (EI), 120
    field desorption (FDMS), 120
    in food toxin analysis, 117–124
        cyanogenic glycosides, 119–121
        furanocoumarins, 123–124
        glycoalkaloids, 117–119
        pyrrolizidine alkaloids, 121–123
    overview, 116
Matricaria chamomilla, 136
MDMA (*N*-methyl-3,4-methylenedioxymethamphetamine), 14
melamine, 135
Mentha (mint), 5
metabolomics, 133
methoxalen, *69*
8-methoxypsoralen, **153**
methoxysafrole, 15
micellar electrokinetic capillary chromatography (MECK), 203
microfluidics-based, biosensors, 252
Momordica, 11
momordicosides, 11
monoclonal antibodies (MAb), 224
monocrotaline, 122
Moraceae, 8, 67
mustard oil, 8
mycotoxins, 154–156, 196–197
Myristica fragrans, 15
myristicin, 15
Mytilus trossulus, 74–75

nano-biosensors, 251
natural biosensors, 249–252
natural toxins

from animals of non-seafood sources, 17
contaminants formed during processing, 17
defined, 4–5
environmental contaminants, 17
health effects of, 4
health risks, minimizing, 5
of plant origin, 5–7
risk assessment, 5
in seafood, 17
necic acid, *57*
neobyakangelicol, *69*
neolinustatin, 32, 102, *103, 120*
neosolaniol, 196
Nerium indicum, **163**
Nicotiana tabacum, **163**
nivalenol, 196
nodakenetin, 105, *106, 124*
normal phase chromatography, 149
notopterol, *69*
nuclear magnetic resonance (NMR)
   spectroscopy, 131–136
   in analysis of food composition and
      authentication, 133–134
   in detection of food toxins and adulterants,
      135–136
   fundamentals of, 132–133
   metabolomics and, 133
nucleic acid-based biosensors, 250.
   See also biosensors

ochratoxins, 166, 197
oil of thuja, 6
Oospora, 10
optical biosensors, 249
Orchidaceae, 104
osteocalcin, 74
osthole, *69*
otonecine, *56, 57, 58, 104.*
   See also pyrrolizidine alkaloids
ovalbumin, 224
oxalates, 10–11, 247
oxalic acids, 11, 247
oxypeucedanin, *69*, **233**
oxypeucedanin hydrate, *69*

pabulenol, **233**
pangelin, *69*
Parthenium hysterophorus, **163**
partition chromatography, 148
Passiflora edulis, 204
Passiflora quandrangularis, **163**
patulin, 166
Penicillium, 166
peppers (*Capsicum annum*), 207
Persea sp., **163**

pesticides, 135, 156, **185–186**
petasites (*Petasites japonicus*), 121
Petasites japonicus (petasites), 121
Petroselinum crispum, 136
Phaseolus coccineus (runner beans), 9
Phaseolus lunatus L. (lima beans), 32
Phaseolus spp. (beans), 9
Phaseolus vulgaris (Zihua snap bean seeds), 74,
   106, **163**
phellotorin, *69*
phenolic compounds, 152
Philodendron spp., **163**
photodiode array detector (PDA) detector, 149
phototoxin, 68–70
phytates, 12–14
phytic acid, 12–14
phytohemagglutinins, 74, 106
phyto-photodermatitis, 68
pigments, 152
plant toxins, 5–17
   α-amylase inhibitors, 9
   anti-thiamine compounds, 9–10
   chaconine, 16
   cucurbitacins, 11–12, **12**
   cyanogenic glycosides, 16, 25–45
   cycasin, 17
   erucic acid, 8
   furanocoumarins, 8, 67–75
   glycoalkaloids, 16, 81–90
   goitrogens, 7–8
   hypericin, 6–7
   hypoglycin, 14
   Japanese star anise, 15–16
   lectins, 9
   myristicin, 15
   oxalates, 10–11
   phytates, 12–14
   phytic acid, 12–14
   plant species and toxicants, **163**
   prussic acid, 6
   pyrrolizidine alkaloids, 10, 53–62
   safrole, 14–15
   sanguinarine, 17
   solanines, 16
   thujone, 5–6
platynecine, *56, 57, 104. See also* pyrrolizidine
   alkaloids
polychlorinated biphenyls (PCBs), **185**
polyclonal antibodies (PAb), 224
polygalacturonase, **90**
potatoes (*Solanum tuberosum*), 82, 85, 117,
   153, 207
potentiometric sensors, 249
proteins, 9, 30, 43, 60, 73, 74, 104, 106, 144,
   148, 151, 156, 203, 216, 222, 224,
   225, **227**, 246, 247, 252

protohypericin, 7
protopseudohypericin, 7
prunasin, 16, 27, *29*, 33, *35*, *103*, *120*, **208**
Pruni domesticae, 103–104
Prunus armeniaca (apricot fruits), 32–33
Prunus serotina (wild cherry bark), 16
prussic acid, 6
pseudohypericin, 7
Psoralea corylifolia, 71
psoralen, 67, *69*, 72, *105*, *106*, 123, *124*.
      See also furanocoumarins
putrescine, *58*
pyrethroids, **185**
pyrimidine, 10
pyrrolizidine alkaloids, 53–62
    acetylcholinesterase inhibitor activity, 59–60
    analysis of
        capillary electrophoresis, 207
        ELISA, 222
        HPTLC, 197
        IR spectroscopy, 104–105
        LC-MS, 169–171
        mass spectrometry, 121–123
    anticancer activity of, 59
    anti-HIV activity of, 59
    anti-inflammatory activity of, 57–58
    antimicrobial activity of, 58–59
    antiulcer activity of, 59
    biosynthesis of, *58*
    chemical structures of, *56*, *57*, *104*, *122*
    chemistry of, 55–56
    hepatotoxic effects of, 62
    limits in food products and spices, **55**
    mutagenic effects of, 60
    overview, 10, 53–54
    pharmacological actions of, 56–60
    plants containing, **61**
    sources of, 54–55
    toxicological effects of, 60–62, **62**
    types of, *56*, *57*

quartz crystal microbalance biosensors, 249.
      See also biosensors

Ranunculaceae, 121
rape seed, 8
refractive index (RI) detector, 149
refractory sprue, **145**
retronecine, *56*, *57*, *58*, *104*.
      See also pyrrolizidine alkaloids
retrorsine, 122
reversed-phase chromatography, 149
rhodanese, 38
ricin, 9, 72
Ricinus communis, 72, **163**
Ricinus communis (castor oil plant), 9

riddelliine, **229**
Rosaceae, 7, 8, 26, 32
runner beans (*Phaseolus coccineu*s), 9
Runt-associated transcription thing 2
      (RUNX2), 74
Rutaceae, 8, 67, 71, 72, 105, 223

safrole, 14–15
Salmonella typhimurium, 60
sambunigrin, *103*, *120*, **209**
sandalwood oil, **186**
sandwich ELISA (antibody screening), 219, *220*
sanguinarine, 17
Sassafras albidum, 14
sassafras oil, 15
schaftoside, **208**
Scopolia, 180
Scrophulariaceae, 121
seafood toxins, 17
selenium, 17
Senecio, 122, 123
Senecio brasiliensis, 59, **61**
Senecio jacobaea, **61**
Senecio longilobus, **62**
Senecioneae, 55
senecionine, 122, 197, 207, **209**, **229**
seneciphylline, 122, 207, **210**
senkirkine, 122, 207, **209**
size exclusion chromatography, 148
solamargine, *101*, *118*, 154, **231**.
      See also glycoalkaloids
    pharmacological actions of, 87
Solanaceae, 16, 81, 117
solanidine, 117, **210**, **230**.
      See also glycoalkaloids
solanines, 16, 81–82. *See also* glycoalkaloids
    LC-MS analysis of, 169
    pharmacological actions of, 87–88
    structure of, *101*, *118*, **211**, **229**
    toxic effects of, 90, **153**
    VIS/NIR analysis of, 100
Solanum, 16
Solanum lycocarpum, 154
Solanum lycopersicum (tomatoes), **82**, *82*, 85,
      117, 207
Solanum melongena (eggplant), 82, **82**, 85, 117,
      154, 207
Solanum tuberosum (potatoes), 82, 85, 117,
      153, 207
solasodine, **211**, **232**. *See also* glycoalkaloids
    pharmacological actions of, 87
solasonine, *101*, *118*, 154, **211**, **231**.
      See also glycoalkaloids

Index   267

Solenanthus lanatus, 60, **61**
solid-phase extraction, 147
sorbates, 135
sorbent-based microextraction, **155**
sorbic acid, 135
sorghum, 31–32
Soxhlet extraction, 147
soybean, **153**
spirosolane, *84–85. See also* glycoalkaloids
St. John's wort, 6–7
steam distillation, 146
steroidal glycoalkaloids (SGAs), 169, 188
Strychnos nux-vomica, **163**
Sudan red, 135
supercritical fluid, 146
supersonic molecular beam gas chromatography-mass spectrometry (SMB GC-MS), 184–188
Symphytum, 55
Symphytum officinale (comfrey), 121
synthetic colors, 135

taxiphyllin, 27, 28, *29*, 31, *103*, *120*
terpene, 6
tetramine, 17
Thevetia peruviana, **163**
thiaminase disease, 10
thiaminase I/II, 9–10
thiazol, 10
thin-layer chromatography (TLC), 195–196
thiocyanates, 7
  excretion, 39
Thuja, 6
Thuja occidentalis, 6
thujone, 5–6, **153**
Ticocarpus crinitus, 74
tomatidine. *See also* glycoalkaloids
  pharmacological actions of, 88
  structure of, **211**, **231**
tomatine, 16, 81. *See also* glycoalkaloids

  pharmacological actions of, 87–88
  structure of, *101*, *118*, **210**, **230**
  toxic effects of, **90**
tomatoes (*Solanum lycopersicum*), **82**, 82, 85, 117, 207
toxic squash syndrome, 12
tremetol, 17
trichothecenes, 166
triglochinin, *29*
trimethylamine oxide, 17
trimethylangelicin (TMA), 223
tropical ataxic neuropathy (TAN), 43
tumor necrosis factor (TNF), 74
Tussilago farfara, **62**, 197, 207
Tussilago farfara (coltsfoot), 121

ultraviolet-visible light detector, 149
umbelliferone, 105, *106*, 123, *124*
Urtica dioica, **163**

vanilla extracts, 106
Veratrum, 16, 117
vicenin, **209**
vitamin B12a (hydroxocobalamin), 38, 40
vitamins, 151–152
Vitis vinifera L. (grapevine), 168

wheat germ agglutinin (WGA), 107
whole-cell-based biosensors, 250.
  See also biosensors
wild cherry bark (*Prunus serotina*), 16
Wnt/ff-catenin pathway, 74
World Health Organization (WHO), 4–5
wormwood oil, **153**

Xanthium strumarium, **163**

zearalenone, 166, 196
Zihua snap bean seeds (*Phaseolus vulgaris*), 74, 107